Computational Aeroacoustics

ICASE/NASA LaRC Series

Stability of Time Dependent and Spatially Varying Flow
D.L. Dwoyer and M.Y. Hussaini (eds.)

Studies of Vortex Dominated Flows
M.Y. Hussaini and M.D. Salas (eds.)

Finite Elements: Theory and Application
D.L. Dwoyer, M.Y. Hussaini and R.G. Voigt (eds.)

Instability and Transition, Volumes I and II
M.Y. Hussaini and R.G. Voigt (eds.)

Natural Laminar Flow and Laminar Flow Control
R.W. Barnwell and M.Y. Hussaini (eds.)

Major Research Topics in Combustion
M.Y. Hussaini, A. Kumar and R.G. Voigt (eds.)

Instability, Transition, and Turbulence
M.Y. Hussaini, A. Kumar and C.L. Streett (eds.)

Algorithmic Trends in Computational Fluid Dynamics
M.Y. Hussaini, A. Kumar and M.D. Salas (eds.)

Computational Aeroacoustics
Jay C. Hardin and M.Y. Hussaini (eds.)

Jay C. Hardin M.Y. Hussaini
Editors

Computational Aeroacoustics

With 179 Illustrations

Springer-Verlag
New York Berlin Heidelberg London Paris
Tokyo Hong Kong Barcelona Budapest

Jay C. Hardin
NASA Langley Research Center
Mail Stop 462
Hampton, VA 23681
USA

M.Y. Hussaini
ICASE
NASA Langley Research Center
Mail Stop 132-C
Hampton, VA 23681
USA

Library of Congress Cataloging-in-Publication Data
Computational aeroacoustics / [edited by] Jay C. Hardin
and M.Y. Hussaini
p. cm. --(ICASE/NASA LaRC series)
Presentations at the Workshop on Computational Aeroacoustics
sponsored by ICASE and the Acoustics Division of NASA LaRC
on April 6-9, 1992.
Includes index.

ISBN-13: 978-1-4613-8344-4 e-ISBN-13: 978-1-4613-8342-0
DOI: 10.1007/978-1-4613-8342-0

1. Aerodynamic noise--Mathematics--Congresses. I. Hardin, Jay C.
II. Hussaini, M. Yousuff. III. Series.
TL574.N6C66 1993
629.132'3--dc20 93-17486

Printed on acid-free paper.

Production managed by Jim Harbison; manufacturing supervised by Vincent Scelta.
Photocomposed pages prepared from the editors' LaTeX files.

9 8 7 6 5 4 3 2 1

PREFACE

Computational aeroacoustics is rapidly emerging as an essential element in the study of aerodynamic sound. As with all emerging technologies, it is paramount that we assess the various opportunities and establish achievable goals for this new technology. Essential to this process is the identification and prioritization of fundamental aeroacoustics problems which are amenable to direct numerical simulation. Questions, ranging from the role numerical methods play in the classical theoretical approaches to aeroacoustics, to the correct specification of well-posed numerical problems, need to be answered.

These issues provided the impetus for the Workshop on Computational Aeroacoustics sponsored by ICASE and the Acoustics Division of NASA LaRC on April 6-9, 1992. The participants of the Workshop were leading aeroacousticians, computational fluid dynamicists and applied mathematicians. The Workshop started with the opening remarks by M. Y. Hussaini and the welcome address by Kristin Hessenius who introduced the keynote speaker, Sir James Lighthill. The keynote address set the stage for the Workshop. It was both an authoritative and up-to-date discussion of the state-of-the-art in aeroacoustics. The presentations at the Workshop were divided into five sessions – i) Classical Theoretical Approaches (William Zorumski, Chairman), ii) Mathematical Aspects of Acoustics (Rodolfo Rosales, Chairman), iii) Validation Methodology (Allan Pierce, Chairman), iv) Direct Numerical Simulation (Michael Myers, Chairman), and v) Unsteady Compressible Flow Computational Methods (Douglas Dwoyer, Chairman).

There was also a panel discussion on the last day. The panel was chaired by Sir James Lighthill and consisted of the chairmen of the various sessions. The goals of the panel discussion were to delimit the state-of-the-art in computational aeroacoustics and ascertain the most fruitful avenues for immediate research effort. A report of these final discussions by Sir James Lighthill is included in this volume. In addition, each of the session chairman has included some of their own views in their papers in this volume.

One significant discussion of the panel concerned whether computational aeroacoustics represented a totally new approach to the field or merely heralded a "second golden age" of aeroacoustic research. On the basis of the work presented, it was concluded that the exact acoustic analogy approach, perhaps utilizing numerical methods for source computation, would continue to have a significant role to play for certain problems and that direct numerical simulations provided additional opportunities that might properly initiate a "second golden age."

JH, MYH

CONTENTS

MATHEMATICAL ASPECTS OF ACOUSTICS

VALIDATION METHODOLOGY

DIRECT SIMULATION

UNSTEADY COMPRESSIBLE FLOW COMPUTATIONAL METHODS

CONTRIBUTED PAPERS

REPORT ON FINAL PANEL DISCUSSION

WELCOME ADDRESS

As the Director of the Aeronautical Research Division at NASA Headquarters, I represent research interests in roughly 10 disciplines of science and engineering that are of critical importance to aeronautics. Aeroacoustics is one such discipline. The overarching research goals within these critical disciplines are to gain scientific understanding in order to pioneer new and revolutionary concepts in aeronautics and to improve the theoretical, experimental, and computational tools for the design and analysis of advanced aerospace systems. Fully one-third of NASA resources in aeronautics are committed to these overarching goals in the aeronautical critical disciplines. A discipline is considered "critical," in the NASA vernacular, if advancements in the state-of-the-art will lead to significant improvements in performance capability, economics, safety, or in environmental capability. I consider aeroacoustics to have this potential, with impact for vehicles across the entire speed regime.

Consider, for example, a commercial high-speed civil transport (see Figure 1). NASA currently is conducting a program to develop technology to minimize the environmental impacts of such a vehicle. Source noise and community noise are elements of this program. The NASA goal is to provide proven concepts for reducing sideline noise to 20 dB below Concorde levels. We believe from previous experience that the technology is in hand to reach the Federal Aviation Regulation, FAR 36 Stage 2, but our ultimate goal is to reach the more stringent Stage 3 levels that exist now for new subsonic aircraft. This is a challenge for our prediction techniques and validation procedures requiring them to confidently produce an extra 8 to 10 dB below existing concepts. This goal will be accomplished not only by reducing source noise, but by analyzing the vehicle as a complete system. We must examine engine/airframe integration and account for improvements in high-lift aerodynamics where reduced thrust on take-off could provide us with a reduction of a few dB. But these improvements will only be achieved if we perform the systems integration very carefully and very cleverly. Furthermore, the complexity of the geometries that need to be considered and the high temperatures and pressures of the flight regime will be challenges to our current predictive techniques.

In the subsonic transport arena, we have done well in the sense that our existing technology meets FAR 36 Stage 3 regulation levels. And so our challenge is to gain even further improvement. New noise reduction concepts for subsonic transports will be an incredible competitive advantage for the industry. The cartoon in Figure 2 displays the benefit of an average 10 dB reduction in noise over a B-757 operating with a 1000 arrivals and departures per day. The footprint of unacceptable noise, the day/night level 65 dB contour, is shown at the bottom of the figure. With a 10 dB reduction, on the average, the footprint of unacceptable noise is reduced from 2.5 square miles to roughly 0.4 of a square mile. Clearly, this magnitude of improvement will permit us to increase the capacity of our nearly saturated air traffic control system by eliminating or at least reducing airport curfews at some of our busier airports. Predictive techniques enabling this sort of noise reduction will be challenged by the complexities of the geometries involved and the difficulties in predicting the low amplitude fan tones.

For fighter aircraft and for hypersonic aircraft, such as the National Aerospace Plane, our principal concern is the vibrational structural loads on the vehicle and on weapons that may be in the vicinity of the aircraft. Figure 3 depicts the typical, structural sonic fatigue life curve for a hypersonic vehicle that is undergoing fluctuating pressure loads. Note the log scale of time to failure and the steepness of the curve. In fact, if we could achieve a "mere" 5 dB reduction, we could add an order of magnitude or more to the fatigue life of a particular vehicle, depending on the material, of course.

The challenge posed by practical problems, such as the ones that I have just described, suggests that there is high leverage in working to improve our computational predictive techniques, especially since computational simulation is playing an ever greater role in the concurrent engineering environment of our aerospace industry. Significant advances have been made in the last decade in computer speed and memory and in numerical algorithms. With those advances has come increasing confidence in the use of computational fluid dynamics, "CFD," to predict the mean flow field. These advances need to be exploited to pioneer novel noise reduction concepts. As you know, it is not a simple extension of computational fluid dynamics techniques into the acoustic arena. Acoustic pressure fluctuations can be 60 dB below those of interest to fluid dynamicists, with frequencies in the 100 to 1000 Hz range. So it is fitting that a group that a group of multidisciplinary scientists: acousticians, fluid dynamicists, physicists, and mathematicians, has assembled for this four-day workshop to better define the emerging field of computational aeroacoustics and to plan, and hopefully greatly accelerate, its progress. As a NASA Headquarters manager, I look forward to the results of your deliberations. I know that they will greatly affect the content of the NASA program in the future, and I have no doubt that they also will influence the field of acoustics, considering the credentials of those of you here assembled.

Kristin A. Hessenius
Director, Aeronautical Research Division
National Aeronautics and Space Administration
Washington, DC

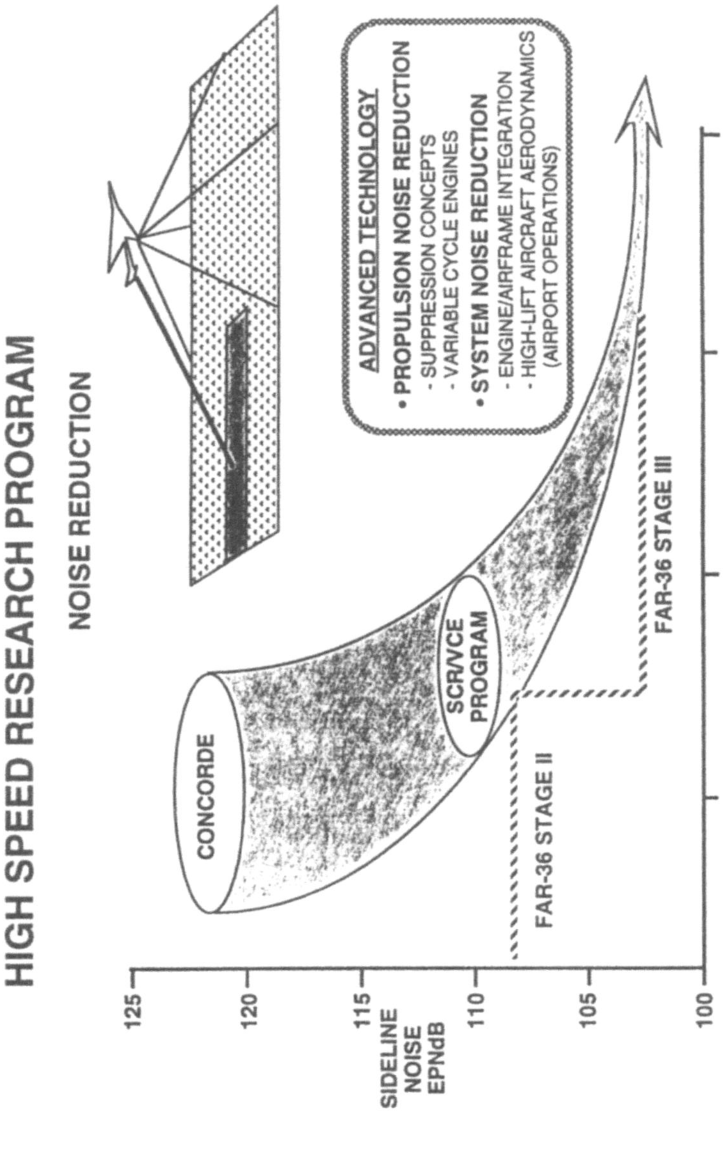

HIGH SPEED RESEARCH PROGRAM
NOISE REDUCTION
SIDELINE NOISE EPNdB
125
120
115
110
105
100
1970
1980
1990
2000
YEAR
CONCORDE
SCR/VCE PROGRAM
FAR-36 STAGE II
FAR-36 STAGE III
ADVANCED TECHNOLOGY
• PROPULSION NOISE REDUCTION
- SUPPRESSION CONCEPTS
- VARIABLE CYCLE ENGINES
• SYSTEM NOISE REDUCTION
- ENGINE/AIRFRAME INTEGRATION
- HIGH-LIFT AIRCRAFT AERODYNAMICS
(AIRPORT OPERATIONS)

Figure 1

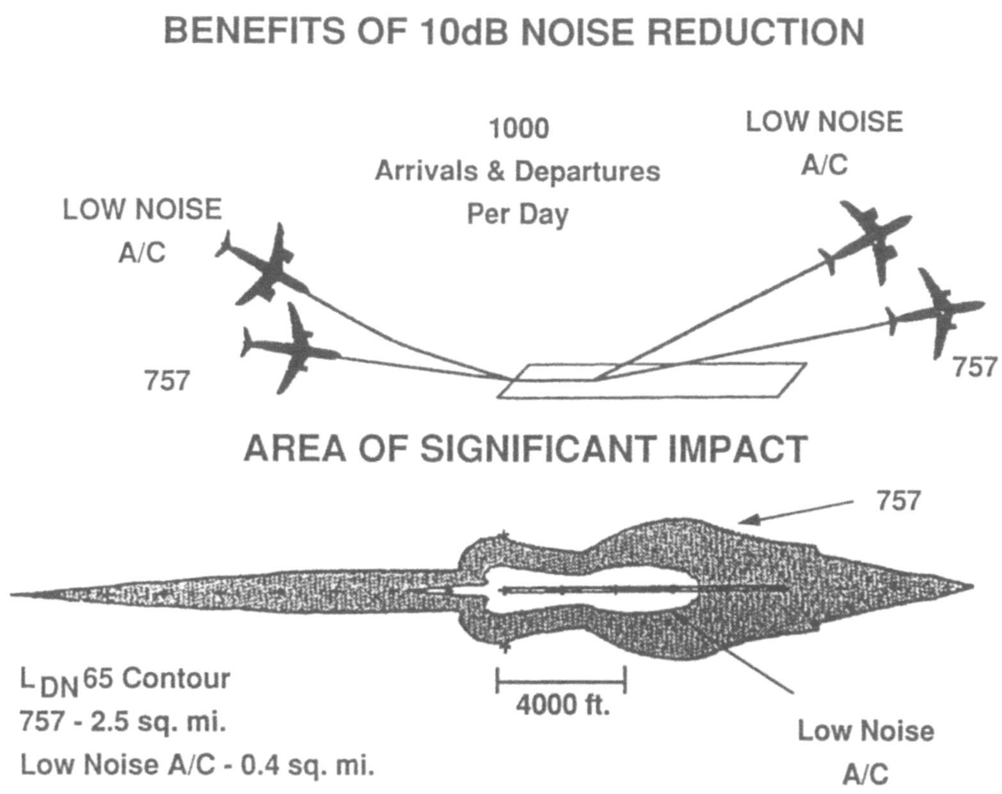
BENEFITS OF 10dB NOISE REDUCTION
1000
Arrivals & Departures
Per Day
LOW NOISE
A/C
757
LOW NOISE
A/C
757
AREA OF SIGNIFICANT IMPACT
757
LDN65 Contour
757 - 2.5 sq. mi.
Low Noise A/C - 0.4 sq. mi.
4000 ft.
Low Noise
A/C

Figure 2

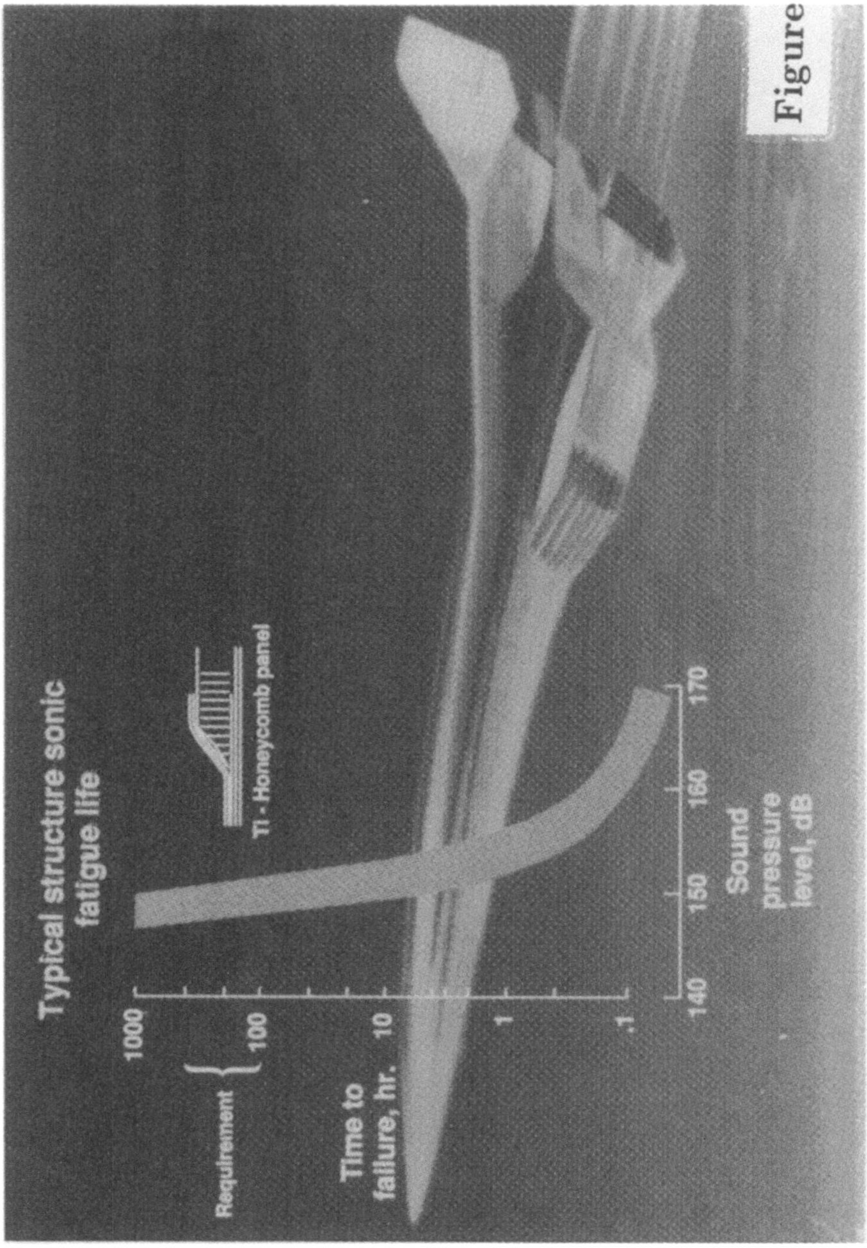

Typical structure sonic fatigue life
Time to failure, hr.
Requirement
1000
100
10
1
.1
Ti - Honeycomb panel
140
150
160
170
Sound pressure level, dB

Figure 3

CONTRIBUTORS

Ramesh Agarwal
McDonnell Douglas Research Laboratories
P.O. Box 516
St. Louis, MO 63166

John T. Batina
NASA Langley Research Center
Mail Stop 173
Hampton, VA 23681-0001

Alvin Bayliss
Department of Engineering Sciences
Technological Institute
Northwestern University
Evanston, IL 60208

Charles H. Berman
AeroChem Research Laboratories, Inc.
P.O. Box 12
Princeton, NJ 08542

Kenneth S. Brentner
NASA Langley Research Center
Mail Stop 461
Hampton, VA 23681-0001

Genevieve Comte-Bellot
Centre Acoustique
Ecole Centrale de Lyon
B P 163
69131 Ecully
Cedex
FRANCE

Sylvie Courtier-Arnoux
Département "Acoustique et Mécanique Vibratoire"
EDF/Direction des Etudes et Recherches
1, Av. du Gl de Gaulle
F-92141 Clamart Cedex
FRANCE

David G. Crighton
Department of Applied Mathematics & Theoretical Physics
University of Cambridge
Cambridge CB3 9EW
UNITED KINGDOM

F. Farassat
NASA Langley Research Center
Mail Stop 460
Hampton, VA 23681-0001

James F. Geer
Watson School of Engineering and Applied Science
State University of New York
Binghamton, NY 13092

M. E. Goldstein
NASA Lewis Research Center
Mail Stop 3-17
Cleveland, OH 44135

Jay C. Hardin
NASA Langley Research Center
Mail Stop 462
Hampton, VA 23681-0001

Kristin Hessenius
Code RJR
NASA Headquarters
Washington, DC 20546

M. Y. Hussaini
ICASE
NASA Langley Research Center
Mail Stop 132C
Hampton, VA 23681-0001

T. Kambe
Department of Physics
University of Tokyo
Hongo, Bunkyo-ku
Tokyo 113
JAPAN

Abbas Khavaran
Sverdrup Technology, Inc.
NASA Lewis Research Center
2001 Aerospace Parkway
Brook Park, OH 44142

Heinz-Otto Kreiss
Department of Mathematics
University of California
Los Angeles, CA 90024

Sanjiva K. Lele
Department of Mechanical
Engineering
Stanford University
Stanford, CA 94305-3030

Sir James Lighthill
Department of Mathematics
University College London
Gower Street
London WCIE 6BT
UNITED KINGDOM

Geoffrey M. Lilley
Department of Aeronautics
and Astronautics
University of Southampton
Southampton
UNITED KINGDOM

Philip J. Morris
Department of Aerospace
Engineering
The Pennsylvania State
University
153D Hammond Building
University Park, PA 16802

M. K. Myers
The George Washington
University
JIAFS
Mail Stop 269
NASA Langley Research Center
Hampton, VA 23681-0001

Allan D. Pierce
The Pennsylvania State
University
Graduate Program in Acoustics
157 Hammond Building
University Park, PA 16802

Man Mohan Rai
NASA Ames Research Center
Mail Stop 258-1
Moffett Field, CA 94035

Sutanu Sarkar
ICASE
NASA Langley Research Center
Mail Stop 132C
Hampton, VA 23681-0001

John Seiner
NASA Langley Research Center
Mail Stop 165
Hampton, VA 23681-0001

Gautam SenGupta
CFD Laboratory
The Boeing Company
Mail Stop 7H-91
P.O. Box 3707
Seattle, WA 98124-2207

Christopher K. W. Tam
Department of Mathematics
Florida State University
Tallahassee, FL 32306-3027

Willie R. Watson
NASA Langley Research Center
Mail Stop 460
Hampton, VA 23681-0001

William E. Zorumski
NASA Langley Research Center
Mail Stop 461
Hampton, VA 23681-0001

"SPEAKERS"

1. James Geer
2. Fereidoun Farassat
3. Sir James Lighthill
4. Ramesh Agarwal
5. Charles Berman
6. Guatam SenGupta
7. Hafiz M. Atassi
8. Sylvie Courtier-Arnoux
9. Rudolfo Rosales
10. Geoffrey M. Lilley
11. Genevieve Comte-Bellot
12. Phillip Morris
13. T. Kambe
14. Kenneth S. Brentner
15. Willie Watson
16. Andrew Majda
17. Jay Hardin
18. James D. Baeder
19. Sanjiva Lele
20. S. I. Hariharan

KEYNOTE ADDRESS

A GENERAL INTRODUCTION TO AEROACOUSTICS AND ATMOSPHERIC SOUND

James Lighthill

Department of Mathematics
University College London
Gower Street, London WCIE 6BT
UNITED KINGDOM

1. Broad Overview

This general introductory paper is devoted to Interactions of Sound with Air, including transmission through the atmosphere and

both generation of sound by } airflows
and propagation of sound in }

(e.g., manmade flows – around aircraft or air machinery – or natural winds) as affected by the air's boundaries and atmospheric composition; with (conversely) generation of airflows by sound (acoustic streaming).

From linear acoustics I utilize the properties of the wave equation, including

(i) the short-wavelength ray-acoustics approximation (Lighthill, 1978b, hereinafter denoted WF, pp. 67-) and

(ii) the theory of multipole sources (WF, pp. 31-) – with the long-wavelength "compact-source" approximation (source region of size ℓ with $\omega\ell/c$ small, where ω = radian frequency, radiates like a concentrated source);

while from nonlinear acoustics I use (WF, pp. 150-) the physics of waveform shearing and shock formation.

Techniques special to aeroacoustics and atmospheric sound are centered on the momentum equation for air. Its difference from a wave-equation approximation include

A. Linear effects, of gravity acting on air stratified as meteorologists observe; effects which allow independent propagation (WF, pp. 292-) of "internal" gravity waves and of sound, except at wavelengths of many kilometers when the atmosphere becomes a waveguide (WF, pp. 425-) for global propagation of interactive acoustic-gravity waves;

and (still more importantly) include

B. Nonlinear effects, of the momentum flux $\rho u_i u_j$; i.e. the flux – rate of transport across unit area – of any ρu_i momentum component by any u_j velocity component. This term, neglected in linear acoustics, acts like a stress (i.e. force per unit area – since rate of change of momentum is force). In particular,

(i) an airflow's momentum flux $\rho u_i u_j$ generates sound like a distribution of (time-varying) imposed stresses; thus not only do forces between the airflow and its boundary radiate sound as distributed dipoles, ($\rightarrow \mathbf{F} \simeq - + \mathbf{F}$) but also such stresses (acting on fluid elements with equal and opposite dipole-like forces) radiate (WF, pp. 63-) as distributed quadrupoles; ($\rightarrow\leftarrow \;\simeq - + + \; -$) (Lighthill, 1952 and Lighthill, 1962)

(ii) the mean momentum flux $< \rho u_i u_j >$ in any sound waves propagating through a sheared flow (with shear $\partial V_i/\partial x_j$) is a stress on that flow (Lighthill, 1972 and WF, pp. 329-), and the consequent energy exchange (from sound to flow when positive, vice versa when negative) is

$$< \rho u_i u_j > \partial V_i/\partial x_j; \tag{1}$$

(iii) even without any pre-existing flow, energy-flux attenuation in a sound wave allows streaming to be generated (WF, pp. 337-) by unbalanced stresses due to a corresponding attenuation in acoustic momentum flux – essentially, then, as acoustic energy flux is dissipated into heat, any associated acoustic momentum flux is transformed into a mean motion (Lighthill, 1978a and WF).

And another (less crucial) momentum-equation/wave-equation difference is

C. Nonlinear deviation of pressure excess $p - p_0$ from a constant multiple, $c_0^2(\rho - \rho_0)$, of density excess.

(i) For sound generation by airflows, this adds an isentropic term to the quadrupole strength per unit volume (WF, pp. 60-)

$$T_{ij} = \rho u_i u_j + \left[(p - p_0) - c_0^2(\rho - \rho_0)\right] \delta_{ij}, \qquad (2)$$

the last term being considered important mainly for flows at above-ambient temperatures (Lighthill, 1952 and Lighthill, 1963);

(ii) for propagation of sound with energy density E through flows of air with adiabatic index γ, the mean deviation is about $\left(\frac{1}{2}\right)(\gamma - 1)E$, and the total radiation stress (Bretherton and Garrett, 1968)

$$< \rho u_i u_j > + \frac{1}{2}(\gamma - 1)E\delta_{ij} \qquad (3)$$

adds an isotropic pressure excess to the mean momentum flux (although the energy exchange (1) is unchanged in typical cases with $\partial V_i/\partial x_i$ essentially zero).

(And we note that the very special case of sound waves interacting on themselves [in other words, nonlinear acoustics] may be interpreted (WF, p. 148) as a combined operation of the "self-convection effect" B and the (smaller) "sound-speed deviation" C.)

2. Compact Source Regions

2.1. Sound generation by low-Mach-number airflows

The main nondimensional parameters governing airflows of characteristic speed U and length-scale L are Mach number $M = U/c$ (where, in aeroacoustics, c is taken as the sound speed in the atmosphere into which sound radiates) and Reynolds number $R = UL/\nu$ where ν = kinematic viscosity. Low-Mach-number airflows are compact sources of sound, with frequencies

narrow-banded at moderate R	when flow instabilities lead to	regular flow oscillations;
broad-banded at high R		extremely irregular turbulence.

Since, in either case, a typical frequency ω scales as U/L (Strouhal scaling), the compactness condition $\omega L/c$ small is satisfied if $M = U/c$ is small (Lighthill, 1962).

A solid body which, because of flow instability, is subjected to a fluctuating aerodynamic force F scaling as $\rho U^2 L^2$ (at frequencies scaling as U/L), radiates as an acoustic dipole of strength F, with mean radiated power $< \dot{F}^2 > /12\pi\rho c^3$.

This acoustic power scales as $\rho U^6 L^2/c^3$ (a sixth-power dependence on flow speed). Therefore acoustic efficiency, defined as the ratio of acoustic power to a rate of delivery (scaling as $\rho U^3 L^2$) of energy to the flow, scales as $(U/c)^3 = M^3$.

(Exceptions to compactness include bodies of high aspect-ratio; thus, a long wire in a wind [where the scale L determining frequency is its diameter] radiates as a lengthwise distribution of dipoles.)

Away from any solid body a compact flow (oscillating or turbulent, with frequencies scaling as U/L) leads to quadrupole radiation (see B(i) above) with total quadrupole strength scaling as $\rho U^2 L^3$. Acoustic power then scales as $\rho U^8 L^2/c^5$: an eighth-power dependence (Lighthill, 1952 and Lighthill, 1962) on flow speed. In this case acoustic efficiency (see above) scales as $(U/c)^5 = M^5$.

Such quadrupole radiation, though often important, may become negligible near a solid body when dipole radiation due to fluctuating body force (with its sixth-power dependence) is also present (Curle, 1955 and Ffowcs Williams and Hawkings, 1969).

Near not necessarily compact bodies a more refined calculation – using Green's functions not for free space but for internally bounded space – leads in general to the same conclusion: that quadrupole radiation with its eighth-power dependence is negligible alongside the sixth-power dependence of dipole radiation due to fluctuating body forces; but important exceptions to this rule include sharp-edged bodies, where features of the relevant Green's function imply a fifth-power dependence on flow speed of acoustic radiation from turbulence (Ffowcs Williams and Hall, 1970; Crighton and Leppington, 1970; Crighton and Leppington, 1971; and Crighton, 1981).

2.2. Sound generation by turbulence at not so low Mach number

The chaotic character of turbulent flow fields implies that velocity fluctuations at points P and Q, although they are well correlated when P and Q are very close, become almost uncorrelated when P and Q are not close to one another.

Reminder: statisticians define correlation coefficient C for the velocities $\mathbf{u}_P$ and $\mathbf{u}_Q$ as $C = < \mathbf{v}_P \cdot \mathbf{v}_Q > / < \mathbf{v}_P^2 >^{1/2} < \mathbf{v}_Q >^{1/2}$ in terms of the deviations, $\mathbf{v}_P = \mathbf{u}_P - < \mathbf{u}_P >$ and $\mathbf{v}_Q = \mathbf{u}_Q - < \mathbf{u}_Q >$, from their means. When two uncorrelated quantities are combined, their mean square deviations are added up:

$$< \mathbf{v}_P + \mathbf{v}_Q >^2 = < \mathbf{v}_P^2 > + 2 < \mathbf{v}_P \cdot \mathbf{v}_Q > + < \mathbf{v}_Q^2 >$$

$$= < \mathbf{v}_P^2 > + < \mathbf{v}_Q^2 > \text{ if } C = 0.$$

Theories of turbulence define a correlation length ℓ, with

$$\mathbf{u}_P \text{ and } \mathbf{u}_Q \left\{ \begin{array}{c} \text{well correlated} \\ \text{uncorrelated} \end{array} \right\} - - C \text{ close to } \left\{ \begin{array}{c} 1 \\ 0 \end{array} \right\} - -$$

$$\text{when } PQ \text{ is substantially} \left\{ \begin{array}{c} < \ell \\ > \ell \end{array} \right\}.$$

Roughly speaking, different regions of size ℓ ("eddies") generate sound independently, and the mean square radiated noise is the sum of the mean square outputs from all the regions (Lighthill, 1954). Typical frequencies in the turbulence are of order $\omega = v/\ell$, where v is a typical root mean square velocity deviation $< \mathbf{v}^2 >^{1/2}$, so that for each region the compactness condition $\omega\ell/c$ small is satisfied if v/c is small. Compactness, then, requires only that a r.m.s. velocity deviation v (rather than a characteristic mean velocity U) be small compared with c – which is less of a restriction on $M = U/c$ and can be satisfied at "not so low" Mach number (Lighthill, 1963).

3. Doppler Effect

How is the radiation from such "eddies" modified by the fact that they are being convected at "not so low" Mach number? The expression Doppler effect, covering all aspects of how the movement of sources of sound alters their radiation patterns, comprises (i) frequency changes (WF), (ii) volume changes (Lighthill, 1952 and Lighthill, 1962), (iii) compactness changes (Lighthill, 1963 and Ffowcs Williams, 1963).

3.1. Frequency changes

When a source of sound at frequency ω approaches an observer at velocity w, then in a single period $T = 2\pi/\omega$ sound emitted at the beginning travels a distance cT while at the end of the period sound is being emitted from a source that is closer by a distance wT. The wavelength λ (distance between crests) is reduced to

$$\lambda = cT - wT = 2\pi(c - w)/\omega \tag{4}$$

(Figure 1) and the frequency heard by the observer (2π divided by the time λ/c between arrival of crests) is increased to the Doppler-shifted value (WF)

$$\omega_r = \frac{\omega}{1 - (w/c)} : \text{the relative frequency} \tag{5}$$

that results from relative motion between source and observer. For an observer located on a line making an angle θ with a source's direction of motion at speed V, the source's velocity of approach towards the observer is $w = V\cos\theta$ (Figure 1) and the relative frequency becomes

$$\omega_r = \frac{\omega}{1 - (V/c)\cos\theta} : \left\{ \begin{matrix} \text{augmented} \\ \text{diminished} \end{matrix} \right\} \text{when } \theta \text{ is an} \left\{ \begin{matrix} \text{acute} \\ \text{obtuse} \end{matrix} \right\} \text{angle.} \tag{6}$$

Such Doppler shifts in frequency are familiar everyday experiences.

3.2. Volume changes

When an observer is approached at velocity w by a source whose dimension (in the direction of the observer) is ℓ, sounds arriving simultaneously (Figure 2) from the source's $\left\{ \begin{matrix} \text{far} \\ \text{near} \end{matrix} \right\}$ sides have been emitted $\left\{ \begin{matrix} \text{earlier} \\ \text{later} \end{matrix} \right\}$ by a time τ (say).

In the time t for sound from the far side to reach the observer, after travelling a distance ct, the relative distance of the near side in the direction of the observer was increased from ℓ to $\ell + \omega\tau$ before it emitted sound which then travelled a distance $c(t - \tau)$. Both sounds arrive simultaneously if

$$ct = \ell + w\tau + c(t - \tau), \text{ giving } \tau = \frac{\ell}{c - w}$$

$$\text{and } \ell + w\tau = \frac{\ell}{1 - (w/c)} = \ell\omega_r/\omega. \tag{7}$$

The source's effective volume during emission is increased, then, by the Doppler factor ω_r/ω (since dimension in the direction of the observer is so increased whilst other dimensions are unaltered: Lighthill, 1952 and Lighthill, 1962).

If turbulent "eddies" are effectively being convected, relative to the air into which they are radiating, at velocity V, then Equation (6) gives, for radiation at angle θ, the Doppler factor ω_r/ω which modifies both the frequencies at which they radiate and the effective volume occupied by a radiating eddy.

But Equation (2) specifies the quadrupole strength T_{ij} per unit volume for such an eddy. Without convection the pattern of acoustic intensity around a compact eddy of volume ℓ^3 and quadrupole strength $\ell^3 T_{ij}$ would be

$$< (\ell^3 \ddot{T}_{ij} x_i x_j r^{-2})^2 > /16\pi^2 r^2 \rho_0 c^5; \tag{8}$$

and, since different eddies of volume ℓ^3 radiate independently, we can simply add up mean squares in the corresponding expressions for their far-field intensities. This gives

$$\ell^3 < (\ddot{T}_{ij} x_i x_j r^{-2})^2 > /16\pi^2 r^2 \rho_0 c^5 \tag{9}$$

as the intensity pattern radiated by unit volume of turbulence. The Doppler effect modifies this, when the compactness condition is satisfied, by five factors ω_r/ω (one for the change in source volume ℓ^3 and four for the frequency change as it affects the mean square of a multiple of the second time-derivative of T_{ij}) and this intensity modification by a factor

$$[1 - (V/c)\cos\theta]^{-5} \tag{10}$$

brings about an important preference for forward emission (Ffowcs Williams, 1963).

3.3. Compactness changes

As (V/c) increases, however, the Doppler effect tends to degrade the compactness of aeroacoustic sources in relation to forward emission. Not only does $\omega\ell/c$ increase in proportion to Mach number, but an even greater value is taken by $\omega_r\ell/c$, the ratio which must be small if convected sources are to be compact. A restriction on the extent (10) of intensity enhancement for forward emission as V/c increases is placed by these tendencies (Lighthill, 1963; Ffowcs Williams, 1963; and Dowling et al., 1978).

They can develop, indeed, to a point where the compact-source approximation may appropriately be replaced by its opposite extreme: the ray-acoustics approximation (WF). Thus, for supersonic source convection $(V/c > 1)$, the relative frequency (6) becomes infinite in

$$\text{the Mach direction } \theta = \cos^{-1}(c/V), \tag{11}$$

and radiation from the source proceeds (Figure 3) along rays emitted at this angle (Ffowcs Williams and Maidanik, 1965).

Explanatory note: the source's velocity of approach w towards an observer positioned at an angle (11) to its direction of motion is the sound speed c; thus, not only is the generated wavelength (4) reduced indefinitely (the ray-acoustics limit) but, essentially, different parts of a signal are observed simultaneously: the condition of stationary phase satisfied on rays (WF). Sounds emitted (WF, p. 196) by a source approaching an observer at a speed w exceeding c are heard by him in reverse order ("pap pep pip pop pup" becomes "pup pop pip pep pap"!) but when $w = c$ all the sound (vowels and consonants!) are heard together as one single "boom."

Further note: the influences placing a limit on the signal propagated along rays may include the duration δ of well-correlated emission from turbulent "eddies"; and, also, may include nonlinear effects (see §4.2. Supersonic booms).

3.4. Uniformly valid Doppler-effect approximations

Just as a correlation length ℓ for turbulence was specified in §2.2, so a correlation duration δ can be characterized by the requirement that moving eddies have $\left\{ \begin{array}{c} \text{well correlated} \\ \text{uncorrelated} \end{array} \right\}$ velocities at times differing by substantially $\left\{ \begin{array}{c} < \delta \\ > \delta \end{array} \right\}$. Combined use of correlation length ℓ and duration δ affords an approximation to the radiation pattern from convected "eddies" that has some value at all Mach numbers, spanning the areas of applicability of the compact-source and ray-acoustics approximations.

Figure 4 uses space-time diagrams where the space-coordinate (abscissa) is distance in the direction of the observer. Diagram (a) for unconvected "eddies" approximates the region of good correlation as an ellipse with axes ℓ (in the space direction) and δ (in the time direction). Diagram (b) shows such a region for convected "eddies" whose velocity of approach towards the observer is w; thus, it is Diagram (a) sheared by distance w per unit time.

Signals from far points F and near points N, in either case, reach the observer simultaneously – as do signals from other points on the line FN – if this line slopes by distance c (the sound speed) per unit time.

Compact-source case (i) with w/c small: the space component of FN in Diagram (b) is $\ell[1-(w/c)]^{-1}$, just as in Equation (7) for normal Doppler effect (neglecting finite δ).

Ray-acoustics case (ii) with $w/c = 1$: the space component of FN is $c\delta$.

Intermediate case (iii) with w/c "moderately" < 1: the space component of FN is ℓ multiplied by an enhancement factor

$$[(1-w/c)^2 + (\ell/c\delta)^2]^{-1/2} \tag{12}$$

which represents the effective augmentation of source volume due to convection (Ffowcs Williams, 1963).

This enhancement factor (12) is applied not only to the volume term ℓ^3 in the quadrupole field (9) but also twice to each of the pair of twice-differentiated terms inside the mean square; essentially, because time-differentiations in quadrupole fields arise (WF) from differences in the time of emission by different parts of the quadrupole source region (and the time component of FN in Diagram (b) is simply the space component divided by c). As before, then, five separate factors (12) enhance the intensity field; and, with w replaced by $V\cos\theta$, expression (10) for the overall intensity modification factor is replaced by

$$\{[1-(V/c)\cos\theta]^2 + (\ell/c\delta)^2\}^{-5/2} \tag{13}$$

This modification factor (13) affords us an improved description of the influence of Doppler effect not only on the preference for forward emission but also on the overall acoustic power output from convected turbulence (Lighthill, 1963 and Ffowcs Williams, 1963). For example, Diagram (c) gives (plain line) a log-log plot of the average (spherical mean) of (13) as a function of V/c on the reasonable assumption that $\ell = 0.6V\delta$. As V/c increases this average modification factor rises a little at first, but falls drastically like $5(V/c)^{-5}$ for V/c significantly > 1.

Now low-Mach-number turbulence away from solid boundaries (§2.1) should radiate sound with an acoustic efficiency scaling as $(U/c)^5$ where U is a characteristic velocity in the flow. With V taken as that characteristic velocity (although in a jet a typical velocity V of eddy convection would be between 0.5 and 0.6 times the jet exit speed), the modification of (say) an acoustic efficiency of $10^{-3}(V/c)^5$ for low Mach number by the average modification factor would cause acoustic efficiency to follow the broken-line curve in Diagram (c), tending asymptotically to a constant value, 0.005, (aeroacoustic saturation) at high Mach number. Such a tendency is often observed for sound radiation from "properly expanded" supersonic jets (see below).

4. Introduction to Aircraft Noise

4.1. Aero-engine and airframe noise

How are aeroacoustic principles applied to practical problems – such as those of studying aircraft noise with a view to its reduction (Crighton, 1975; Goldstein, 1976; and Goldstein, 1984)?

In any analysis of the generation of sound by airflows, we may need first of all to ask whether the geometry of the problem has features that tend to promote resonance. For example, a long wire in a wind (§2.1) generates most sound when vortex-shedding frequencies $\left\{\begin{array}{c}\text{are fairly close to}\\ \text{and so can "lock on" to}\end{array}\right\}$ the wire's lowest natural frequency of vibration; giving good correlation of sideforces, and so also of dipole strengths, all along the wire.

Again, a jet emerging from a thin slit may interact with a downstream edge (parallel to the slit) in a resonant way (Curle, 1953 and Powell, 1953b); with very small directional disturbances at the jet orifice being amplified by flow instability as they move downstream to the edge, where they produce angle-of-attack variations. Dipole fields associated with the resulting sideforces can at particular frequencies renew the directional disturbances at the orifice with the right phase to produce a resonant oscillation. Some musical wind instruments utilize such jet-edge resonances, reinforced by coincidence with standing-wave resonances in an adjacent pipe.

But in the absence of such resonances (leading to enhanced acoustic generation at fairly well defined frequencies) airflows tend to generate acoustic "noise" whose reaction on the flow instability phenomena themselves is negligible.

Resonances analogous to the above which need to be avoided in aircraft design include, for example,

(a) panel flutter, generated at a characteristic frequency as an unstable vibration of a structural panel in the presence of an adjacent airflow (Dowell, 1975);

(b) screeching of supersonic jets from nozzles which, instead of being "properly expanded" so that an essentially parallel jet emerges, produce a jet in an initially non-parallel form followed by shock waves in the well known recurrent "diamond" shock-cell pattern; the first of these, replacing the edge in the above description, can through a similar feedback of disturbances to the jet orifice generate a powerful resonant oscillation (Powell, 1953a; Powell, 1953b; and Howe and Ffowcs Williams, 1978).

And undesirable resonances may also be associated with aeroengine combustion processes (Candel and Poinsot, 1988). But we turn now to the aircraft noise of a broad-banded character that remains even when resonances have been avoided.

Then aero-engine jet noise proper (Lighthill, 1963) (that is, the part unrelated to any interaction of jet turbulence with solid boundaries) tends to follow a broad trend similar to that in Figure 4; where, however, because the eddy convection velocity V is between 0.5 and 0.6 times the jet exit speed U, the acoustic efficiency makes a transition between a value of around $10^{-4}M^5$ in order-of-magnitude terms for subsonic values of $M = U/c$ and an asymptotically constant value of 10^{-2} or a little less for M exceeding about 2.

The above tendency for $M < 1$ implies that noise emission from jet engines may be greatly diminished if a given engine power can be achieved with a substantially lower jet exit speed, requiring of course a correspondingly larger jet diameter, L. Furthermore, with acoustic power output scaling as $\rho U^8 L^2/c^5$ (§2.1) and jet thrust as $\rho U^2 L^2$, noise emission for given thrust can be greatly reduced if U can be decreased and L increased by comparable factors.

Trends (along these lines) in aero-engine design towards large turbofan engines with higher and higher bypass ratios, generating very wide jets at relatively modest mean Mach numbers, have massively contributed to jet noise suppression (whilst also winning advantages of reduced fuel consumption). On the other hand, such successes in suppressing jet noise proper (originally, the main component of noise from jet aircraft) led to needs for a dedicated focusing of attention upon parallel reductions of other aircraft-noise sources (Crighton, 1972):

(a) those associated with the interaction of jet turbulence with solid boundaries – where sharp-edged boundaries (§2.1) pose a particular threat;

(b) fan noise emerging from the front of the engine and turbine noise from the rear;

(c) airframe noise including acoustic radiation from boundary-layer turbulence and from interaction of that turbulence with aerodynamic surfaces for control purposes or lift enhancement.

Some key areas of modern research on aero-engine and airframe noise are:

for jet noise, techniques for relating acoustic output to vorticity distributions (Powell, 1964 and Mohring, 1978), and to any coherent structures (Ribner, 1964 and Ffowcs Williams and Kempton, 1978), in jet turbulence; and for taking into account (cf §5.4) propagation through the sheared flow in a wide jet (Phillips, 1960 and Mani, 1976);

for noise from fans and propellers, mathematically sophisticated ways of reliably estimating the extent of cancellation of dipole radiation from different parts of a rotating-blade system (alongside a good independent estimate of quadrupole radiation: Parry and Crighton, 1991);

for airframe noise, a recognition (Powell, 1960 and Crighton, 1984) that massive cancellations act to minimize noise radiation from boundary-layer turbulence on a flat surface of uniform compliance –

and, therefore, that avoidance of sharp nonuniformities in airframe skin compliance may promote noise reduction.

4.2. Supersonic booms

In addition to aero-engine and airframe noise, any aircraft flying at a supersonic speed V emits a concentrated "boom"-like noise along rays (Figure 3) in the Mach direction (11). I sketch the theory of supersonic booms with the atmosphere approximated as isothermal (so that the undisturbed sound speed takes a constant value c even though the undisturbed density ρ varies with altitude): a case permitting quite a simple extension of the nonlinear analysis of waveform shearing and shock formation (WF and Whitham, 1956). Then the rays continue as straight lines at the Mach angle for reasons summarized in the explanatory note below expression (11). (Actually, the slight refraction of rays by temperature stratification in the atmosphere, when taken into account in a generalized version of the theory, produces only somewhat minor modifications of the results.)

As such straight rays stretch out from a straight flight path along cones with semi-angle (11), any narrow tube of rays has its cross-sectional area A increasing in proportion to distance r along the tube (WF and Lighthill, 1956). On linear theory (WF), acoustic energy flux $u^2\rho cA$ is propagated unchanged along such a ray tube (so that $u(\rho r)^{1/2}$ is unchanged) where u is air velocity along it. On nonlinear theory, $u(\rho r)^{1/2}$ is propagated unchanged but at a signal speed altered to

$$c + \frac{\gamma+1}{2}u \tag{14}$$

by self-convection and excess-wavespeed effects.

This property can be described (WF) by an equation

$$\left[\left\{\frac{1}{c} - \frac{\gamma+1}{2}\frac{u}{c^2}\right\}\frac{\partial}{\partial t} + \frac{\partial}{\partial r}\right] u(\rho r)^{1/2} = 0 \tag{15}$$

(WF, pp. 187-) where the quantity in braces is the altered value of the reciprocal of the signal speed (14). Now a simple transformation of variables

$$x_1 = r - ct,\ t_1 = \int_0^r (\rho r)^{-1/2} dr,\ u_1 = \frac{\gamma+1}{2}\frac{u}{c}(\rho r)^{1/2} \tag{16}$$

is found to convert Equation (15) into the familiar form

$$\frac{\partial u_1}{\partial t_1} + u_1 \frac{\partial u_1}{\partial x_1} = 0 \tag{17}$$

which describes the waveform shearing at a uniform rate that is associated with shock formation and propagation in nonlinear plane-wave acoustics (Figure 5).

From amongst this equation's physically relevant solutions – namely, those with area-conserving discontinuities (representing shocks) – the famous N-wave solution is the one produced by an initial signal (such as an aircraft's passage through the air) that is first compressive and then expansive. The rules (WF) governing N-wave solutions of Equation (17) are that the discontinuity Δu_1 at each shock falls off like $t_1^{-1/2}$ while the space (change Δx_1 in x_1) between shocks increases like $t_1^{1/2}$. These rules for the transformed variables (16) have the following consequences for the true physical variables: at a large distance r from the flight path the velocity change Δu at each shock and the time interval Δt between the two shocks vary as

$$\Delta u \approx \left[(\rho r) \int_0^r (\rho r)^{-1/2} dr\right]^{-1/2} \quad \text{and} \quad \Delta t \approx \left[\int_0^r (\rho r)^{-1/2} dr\right]^{1/2} \tag{18}$$

On horizontal rays (at the level where the aircraft is flying), ρ is independent of r and the Equations (18) take the greatly simplified form

$$\Delta u \approx r^{-3/4} \quad \text{and} \quad \Delta t \approx r^{1/4} \tag{19}$$

appropriate to conical N-waves in a homogeneous atmosphere (which supersonically convected eddies may also generate). Actually, the rules (19) apply also to the propagation of cylindrical blast waves generated by an exploding wire; since, here also, ray tube areas increase in proportion to r.

On downward pointing rays in an isothermal atmosphere ρ increases exponentially in such a way that the time interval Δt between shocks approaches the constant value obtained in (18) by making the integral's upper limit infinite (WF and Lighthill, 1956). On the other hand the shock strength (proportional to the velocity change Δu) includes the factor $(\rho r)^{-1/2}$ where the large increase in ρ from the flight path to the ground (as well as in r) enormously attenuates the supersonic boom. Below Concorde cruising at Mach 2, for example, an observer on the ground hears two clear shocks with an interval of around 0.5s between them, and yet with strengths $\Delta p/p$ only about 0.001.

5. Propagation of Sound through Steady Mean Flows

5.1. Adaptations of ray acoustics

Useful information on sound propagation (including sound of aeroacoustic origin) through steady mean flows (Blokhintsev, 1956 and Lighthill, 1972) can be obtained by adaptations of the ray-acoustics approximation. I sketch these here before, first, applying them (in §5.3 below) to propagation through sheared stratified winds and, secondly, giving indications of how effects of such parallel mean flows are modified at wavelengths too large for the applicability of ray acoustics.

Sound propagation through a steady airflow represents an autonomous mechanical system: one governed by laws that do not change with time. Then small disturbances can be Fourier-analyzed in the knowledge that propagation of signals with different frequencies ω must proceed without exchange of energy between them.

General Theory of Such Systems (WF, pp. 317-)

Such disturbances of frequency ω involve pressure changes in the form $P \cos\alpha$ where P varies with position and the phase α is a function of position and time satisfying

$$\frac{\partial\alpha}{\partial t} = \omega : \text{ the frequency; and } -\frac{\partial\alpha}{\partial x_i} = k_i : \text{ the wavenumber,} \quad (20)$$

a vector with its direction normal to crests and its magnitude 2π divided by a local wavelength.

In ray theory for any wave system (WF and Lighthill, 1972), we assume that the wavelength is small enough (compared with distances over which the medium – and its motion, if any – change significantly) for a well-defined relationship

$$\omega = \Omega(k_i, x_i) \quad (21)$$

to link frequency with wavenumber at each position. Equations (20) and (21) require that

$$-\frac{\partial k_j}{\partial t}=\frac{\partial^2\alpha}{\partial x_j\partial t}=\frac{\partial\omega}{\partial x_j}=\frac{\partial\Omega}{\partial k_i}\left(-\frac{\partial^2\alpha}{\partial x_i\partial x_j}\right)+\frac{\partial\Omega}{\partial x_j}=\frac{\partial\Omega}{\partial k_i}\frac{\partial k_j}{\partial x_i}+\frac{\partial\Omega}{\partial x_j}, \tag{22}$$

yielding the basic law (in Hamiltonian form) for any wave system:

$$\text{on rays satisfying } \frac{dx_i}{dt}=\frac{\partial\Omega}{\partial k_i} \text{ wavenumbers vary as } \frac{dk_j}{dt}=-\frac{\partial\Omega}{\partial x_j}; \tag{23}$$

equations easy to solve numerically for given initial position and wavenumber. However, the variations (23) of wavenumber ("refraction") produce no change of frequency along rays:

$$\frac{d\omega}{dt}=\frac{\partial\Omega}{\partial k_i}\frac{dk_i}{dt}+\frac{\partial\Omega}{\partial x_i}\frac{dx_i}{dt}=\frac{\partial\Omega}{\partial k_i}\left(-\frac{\partial\Omega}{\partial x_i}\right)+\frac{\partial\Omega}{\partial x_i}\frac{\partial\Omega}{\partial k_i}=0, \tag{24}$$

so that rays are paths of propagation of the excess energy, at each frequency, associated with the waves' presence.

For sound waves we write k as the magnitude of the wavenumber vector, expecting that at any point the value of the relative frequency in a frame of reference moving at the local steady flow velocity u_{fi} will be $c_f k$ (the local sound speed times k); this implies (WF and Lighthill, 1972) that

$$\omega_r=\frac{\partial\alpha}{\partial t}+u_{fi}\frac{\partial\alpha}{\partial x_i}=\omega-u_{fi}k_i, \text{ giving } \omega=\omega_r+u_{fi}k_i=c_f k+u_{fi}k_i \tag{25}$$

as the acoustic form of the relationship (21).

[Note: this rule (25) for relative frequency agrees with the Doppler rule (6), since the velocity of a source of frequency ω relative to stationary fluid into which it radiates is minus the velocity of the fluid relative to a frame in which the acoustic frequency is ω.]

Use of this form (25) of the relationship (21) in the basic law (23) tells us that

$$\frac{dk_j}{dt}=-k\frac{\partial c_f}{\partial x_j}-k_i\frac{\partial u_{fi}}{\partial x_j} \text{ on rays with } \frac{dx_i}{dt}=c_f\frac{k_i}{k}+u_{fi}; \tag{26}$$

where the last terms in these equations represent adaptations of ray acoustics associated with the mean flow. For example, the velocity of propagation along rays is the vector sum of the mean flow velocity

u_{fi} with a wave velocity of magnitude c_f and direction normal to crests.

5.2. Energy exchange between sound waves and mean flow

The excess energy (say, E per unit volume) associated with the presence of sound waves is propagated along such rays; in particular, if attenuation of sound energy is negligible, then

$$\text{flux of excess energy along a ray tube} = \text{constant.} \tag{27}$$

Note: this excess energy density E is by no means identical with the sound waves' energy density

$$E_r = < \frac{1}{2}\rho_f u_{si}u_{si} > + < \frac{1}{2}c_f^2\rho_f^{-1}\rho_s^2 > = c_f^2\rho_f^{-1} < \rho_s^2 > \tag{28}$$

(where the subscript s identifies changes due to the sound waves and the equality of the kinetic and potential energies makes E_r simply twice the latter) in a frame of reference moving at the local flow velocity (compare the definition (25) of ω_r). The kinetic-energy part of the excess energy density E is

$$< \frac{1}{2}(\rho_f + \rho_s)(u_{fi} + u_{si})^2 > - \frac{1}{2}\rho_f u_{fi}u_{fi}, \tag{29}$$

which includes an extra term

$$< \rho_s u_{fi}u_{si} > = < \rho_s u_{fi}\frac{c_f}{\rho_f}\rho_s\frac{k_i}{k} > = E_r\frac{u_{fi}k_i}{c_f k} = E_r\left(\frac{\omega}{\omega_r} - 1\right); \tag{30}$$

and E is the sum of expressions (28) and (30), giving

$$E = E_r\frac{\omega}{\omega_r}; \text{ or, equivalently, either } E_r = E\frac{\omega_r}{\omega} \text{ or } \frac{E_r}{\omega_r} = \frac{E}{\omega}. \tag{31}$$

The quantity E/ω, called action density in Hamiltonian mechanics, is identical in both frames of reference, and Equations (24) and (27) tell us that its flux along a ray is constant (WF and Lighthill, 1972).

But Equation (31) shows too that energy is exchanged between (i) the acoustic motions relative to the mean flow and (ii) the mean flow itself. For example, where sound waves of frequency ω enter a region of opposing flow (or leave a region where the mean flow is along their direction of propagation) the ratio ω_r/ω increases and so therefore does E_r/E: the sound waves gain energy at the expense of the mean flow.

The rate of exchange of energy takes the value (1) written down in §1. This is readily seen from the laws governing motion in an accelerating frame of reference, which feels

$$\text{an inertial force} = -(\text{mass}) \times (\text{acceleration of frame}). \qquad (32)$$

If we use at each point of space a local frame of reference moving with velocity u_{fi} then fluid in that frame has velocity u_{si} but is subject to an additional force (32); where, per unit volume, mass is ρ_f and the frame's acceleration is

$$u_{sj}\frac{\partial u_{fi}}{\partial x_j} \text{ giving force } -\rho_f u_{sj}\frac{\partial u_{fi}}{\partial x_j} \text{ doing work } -\rho_f < u_{si}u_{sj} > \frac{\partial u_{fi}}{\partial x_j} \qquad (33)$$

per unit time on the local relative motions. This rate of energy exchange (33) proves to be consistent with the fact that it is the flux, not of E_r but of action E_r/ω_r, that is conserved along ray tubes.

Energy can be extracted from a mean flow, then, not only by turbulence but also by sound waves; and, in both cases, the rate of extraction takes the same form (33) in terms of perturbation velocities u_{si}. It represents the effect (§1) of the

$$\text{mean momentum flux } \rho_f < u_{si}u_{sj} > \qquad (34)$$

or Reynolds stress (Reynolds, 1895) with which either the sound waves or the turbulent motions act upon the mean flow. For sound waves, by Equation (28) for E_r and by the substitution

$$u_{si} = \frac{c_f}{\rho_f}\rho_s\frac{k_i}{k}, \text{ mean momentum flux } = E_r\frac{k_i}{k_j}k^2; \qquad (35)$$

so that the Reynolds stress is a uniaxial stress in the direction of the wavenumber vector having magnitude E_r.

[Note: strictly speaking, the complete

$$\text{radiation stress } E_r\left(\frac{k_ik_j}{k^2} + \frac{\gamma-1}{2}\delta_{ij}\right) \text{ for sound waves} \qquad (36)$$

includes not only the momentum flux (35) but also the waves' mean pressure excess

$$\frac{1}{2}\left(\frac{\partial^2 p}{\partial \rho^2}\right)_{\rho=\rho_f} < \rho_s^2 > = \frac{\gamma-1}{2}\frac{c_f^2}{\rho_f} < \rho_s^2 > = \frac{\gamma-1}{2}E_r \qquad (37)$$

acting equally in all directions (Bretherton and Garrett, 1968); however (§1) this isotropic component produces no energy exchange with solenoidal mean flows.]

5.3. Propagation through sheared stratified winds

The extremely general ray-acoustics treatment outlined above for sound propagation through fluids in motion has far-reaching applications (in environmental and, also, in engineering acoustics) which, however, are illustrated below only by cases of propagation through parallel flows, with stratification of velocity as well as of temperature (WF and Lighthill, 1972). The x_1-direction is taken as that of the mean flow velocity $V(x_3)$ which, together with the sound speed $c(x_3)$, depends only on the coordinate x_3. Thus, V replaces u_{f1} in the general theory while c replaces c_f (and, for atmospheric propagation, x_3 is altitude). [Note: the analysis sketched here is readily extended to cases of winds veering with altitude, where u_{f2} as well as u_{f1} is nonzero.]

Either the basic law (23) or its ray-acoustics form (26) provides, in general, "refraction" information in the form of three equations for change of wavenumber; while the single, far simpler, Equation (24) is a consequence of, but is by no means equivalent to, those three. By contrast, in the particular case when u_{fi} and c_f are independent of x_1 and x_2, Equations (24) and, additionally, (26) in the cases $j = 1$ and 2 give three simple results,

$$\omega = \text{ constant, } k_1 = \text{ constant and } k_2 = \text{ constant along rays,} \tag{38}$$

that may be shown fully equivalent to the basic law.

If now we write the wavenumber (a vector normal to crests) as

$$(k_1, k_2, k_3) = (\kappa \cos\psi, \kappa \sin\psi, \kappa \cot\theta), \tag{39}$$

so that κ is its constant horizontal resultant, ψ its constant azimuthal angle to the wind direction, and θ its variable angle to the vertical (Figure 6), and use Equation (25) in the form

$$\omega = c(x_3)k + V(x_3)k_1 = c(x_3)\kappa\text{cosec}\theta + V(x_3)\kappa\cos\psi, \tag{40}$$

we obtain the important

$$\text{extension } \sin\theta = \frac{c(x_3)}{\omega\kappa^{-1} - V(x_3)\cos\psi} \text{ to Snell's Law} \tag{41}$$

from the classical case ($V = 0$) when the denominator is a constant. This extended law (41) tells us how θ varies with x_3 along any ray – whose path we can then trace, using Equations (26) in the form

$$\frac{dx_1}{dt} = c(x_3)\cos\psi\sin\theta + V(x_3), \frac{dx_2}{dt} = c(x_3)\sin\psi\sin\theta, \qquad (42)$$

$$\frac{dx_3}{dt} = c(x_3)\cos\theta,$$

by simply integrating dx_1/dx_3 and dx_2/dx_3 with respect to x_3.

It follows that a ray tube covers the same horizontal area at each altitude, so that conservation of the flux of wave action E_r/ω_r along it implies that the vertical component

$$(E_r/\omega_r)(dx_3/dt) = E_r\kappa^{-1}\sin\theta\cos\theta \qquad (43)$$

of wave action flux is constant along rays; from which, with Equation (28), sound amplitudes are readily derived.

Wind shear is able to reproduce all the main types of ray bending (WF) associated with temperature stratification, and often to an enhanced extent. Roughly, the downward curvature of near-horizontal rays in $(\text{km})^{-1}$ comes to

$$3\left\{V'(x_3)\cos\psi + c'(x_3)\right\} \qquad (44)$$

where the velocity gradients are in s^{-1} and the factor $3(\text{km})^{-1}s$ outside the braces is an approximate reciprocal of the sound speed (WF and Lighthill, 1972).

Cases when (44) is negative: curvature is upward; its magnitude with zero wind is at most 0.018 $(\text{km})^{-1}$ (because temperature lapse rate in stable atmospheres cannot exceed 10°C per km, giving $c' = -0.006s^{-1}$) but with strong wind shear can take much bigger values for upwind propagation ($\psi = \pi$). In either case Figure 7(a) shows how the lowest ray emitted by a source "lifts off" from the ground, leaving below it a zone of silence (on ray theory – actually, a zone where amplitudes decrease exponentially with distance below that ray).

Cases when (44) is positive: curvature is downward, as found with zero wind in temperature-inversion conditions (e.g., over a calm cold lake) and even more with strong wind shear for downwind propagation ($\psi = 0$). Figure 7(b) shows how this leads to signal enhancement through multiple-path communication.

In summary, then, the very familiar augmentation of sound levels downwind, and diminution upwind, of a source are effects of the wind's shear (increase with altitude).

5.4. Wider aspects of parallel-flow acoustics

The propagation of sound through parallel flows at wavelengths too great for the applicability of ray acoustics can be analyzed by a second-order ordinary differential equation. Thus, a typical Fourier component of the sound pressure field takes the form

$$p_s(x_3)e^{i(\omega t - k_1 x_1 - k_2 x_2)} \text{ with } \rho \frac{d}{dx_3}\left[\frac{1}{\rho(\omega - Vk_1)^2}\frac{dp_s}{dx_3}\right] + \left[\frac{1}{c^2} - \frac{k_1^2 + k_2^2}{(\omega - Vk_1)^2}\right] p_s = 0. \tag{45}$$

Equation (45) can be used to improve on ray acoustics

(a) near caustics (envelopes of rays) where it allows a uniformly valid representation of amplitude in terms of the famous Airy function, giving "beats" between superimposed waves on one side of the caustic, and exponential decay on the other (WF);

(b) at larger wavelengths by abandoning ray theory altogether in favor of extensive numerical solutions of Equation (45); and

(c) to obtain waveguide modes for sound propagation in a two-dimensional duct (between parallel planes) (Pridmore-Brown, 1958; Mungur and Gladwell, 1969; and Shankar, 1971).

On the other hand, in the case of a three-dimensional duct carrying parallel flow $V(x_2, x_3)$ in the x_1-direction, Equation (45) is converted into a partial differential equation (the first term being supplemented by another with d/dx_2 replacing d/dx_3, while k_2 is deleted) which is used

(d) to obtain waveguide modes in such ducts;

(e) in calculations of propagation of sound through the wide jets – modelled as parallel flows – typical (§4.1) of modern aero-engines; and,

(f) with aeroacoustic source terms included, in certain enterprising attempts at modelling jet noise generation and emission (Phillips, 1960 and Mani, 1976).

6. Acoustic Streaming

6.1. Streaming as a result of acoustic attenuation

Sound waves act on the air with a Reynolds stress (34) even when mean flow is absent (so that subscript f becomes subscript zero). The j-component of force acting on unit volume of air (WF, pp. 337-) is then

$$F_j = -\frac{\partial}{\partial x_i} < \rho_0 u_{si} u_{sj} >: \tag{46}$$

the force generating acoustic streaming (Lighthill, 1978a).

However, the force (46) could not produce streaming for unattenuated sound waves; indeed, their linearized equations can be used to show that

$$\text{if } p^M =< \frac{1}{2} c_0^2 \rho_0^{-1} \rho_s^2 - \frac{1}{2} \rho_0 u_{si} u_{si} > \text{ then } F_j - \frac{\partial p^M}{\partial x_j} =< \frac{\partial}{\partial t}(\rho_s u_{sj}) > \tag{47}$$

which is necessarily zero (as the mean value of the rate change of a bounded quantity). Accordingly, the fluid must remain at rest, responding merely by setting up the distribution p^M of mean pressure whose gradient can balance the force. [Note: actually, on the ray-acoustic approximation (28), p^M is itself zero, but the above argument does not need to use this approximation.]

Attenuation of sound waves takes place

(a) in the bulk of the fluid through the action of viscosity, thermal conductivity and lags in attaining thermodynamic equilibrium (Ch. 6); and

(b) near solid walls by viscous attenuation in Stokes boundary layers.

All these effects produce forces (46) which act to generate acoustic streaming. It is important to note, furthermore, that even the forces due solely to viscous attenuation – being opposed just by the fluid's own viscous resistance – generate mean motions which do not disappear as the viscosity μ tends to zero (Rayleigh, 1896; Nyborg, 1953; Westervelt, 1953; and Nyborg, 1965).

6.2. Jets generated by attenuated acoustic beams

Attenuation of type (a) produces a streaming motion u_{fj} satisfying

$$\rho u_{fi} \partial u_{fj} / \partial x_i = F_j - \partial p / \partial x_j + \mu \nabla^2 u_{fj}. \tag{48}$$

Substantial streaming motions can be calculated from this equation only with the left-hand side included (Stuart, 1966); although in pre-1966 literature it was misleadingly regarded as "a fourth-order term" and so ignored – thus limiting all the theories to uninteresting cases when the streaming Reynolds number would be of order 1 or less.

We can use streaming generated by acoustic beams to illustrate the above principles. If acoustic energy is attenuated at a rate β per unit length, then a source at the origin which beams power P along the x_1-axis transmits a distribution of

$$\text{power } Pe^{-\beta x_1}, \text{ and therefore energy per unit length } c^{-1} Pe^{-\beta x_1}; \tag{49}$$

which is necessarily the integral of energy density, and so also of the uniaxial Reynold stress (35), over the beam's cross-section. It follows by differentiation that the force per unit volume (46), integrated over a cross-section, produces (Lighthill, 1978a)

$$\text{a force } c^{-1} P \beta e^{-\beta x_1} \text{ per unit length in the } x_1\text{-direction.} \tag{50}$$

At high ultrasonic frequencies the force distribution (50) is rather concentrated, the distance of its center of application from the origin being just β^{-1} (which at 1 MHz, for example, is 24mm in air). Effectively (WF, p. 345), the beam applies at this center a total force $c^{-1}P$ (integral of the distribution (50)).

The type (WF and Lighthill, 1978a) of streaming motion generated by this concentrated force $c^{-1}P$ depends critically on the value of $\rho c^{-1} P \mu^{-2}$: a sort of Reynolds number squared, which is about $10^7 P$ in atmospheric air (with P in watts). Streaming of the low-Reynolds-number "stokeslet" type predicted (for a concentrated force) by Equation (48) with the left-hand side suppressed is a good approximation only for $P < 10^{-6} W$. For a source of power $10^{-4} W$, by contrast, the force $c^{-1}P$ generates quite a narrow laminar jet with momentum transport $c^{-1}P$, and at powers exceeding $3 \times 10^{-4}W$ this jet has become turbulent, spreading conically with semi-angle about 15° and continuing to transport momentum at the rate $c^{-1}P$. Such turbulent jets generated by sound are strikingly reciprocal to a classical aeroacoustic theme!

At lower frequencies an acoustic beam of substantial power delivers a turbulent jet with a somewhat more variable angle of spread – but one which

$$\text{at each point } x_1 \text{ carries momentum transport } c^{-1}P(1 - e^{-\beta x_1}), \tag{51}$$

generated by the total force (50) acting up to that point. This momentum transport in the jet represents the source's original rate of momentum delivery minus the acoustic beam's own remaining momentum transport (49). In summary, as acoustic power is dissipated into heat, the associated acoustic momentum transport is converted into a mean motion (which, at higher Reynolds numbers, is turbulent: Lighthill, 1978a).

6.3. Streaming around bodies generated by boundary-layer attenuation

Sound waves of frequency ω well below high ultrasonic frequencies have their attenuation concentrated, if solid bodies are present, in thin Stokes boundary layers attached to each body (WF). Then the streaming generated near a particular point on a body surface is rather simply expressed by using local coordinates with that point as origin, with the z-axis normal to the body and the x-axis in the direction of the inviscid flow just outside the boundary layer: the exterior flow. The Stokes boundary layer for an exterior flow

$$\begin{gathered}(U(x,y), V(x,y))e^{i\omega t} \text{ has interior flow} \\ (U(x,y), V(x,y))e^{i\omega t}\left[1 - e^{-z\sqrt{(i\omega\rho/\mu)}}\right].\end{gathered} \tag{52}$$

[Note that my choice of coordinates makes $V(0,0) = 0$, and that the expressions (52) become identical outside the layer.] The streaming (WF and Lighthill, 1978a) is calculated from the equation

$$F_j^{\mathrm{INT}} - F_j^{\mathrm{EXT}} + \mu \partial^2 u_{fj}/\partial z^2 = 0, \tag{53}$$

with certain differences from Equation (48) explained as follows:

(a) the first term is the force (46) generating streaming within the boundary layer;

(b) we are free, however, to subtract the second, since (see §6.1) it can produce no streaming, and conveniently, the difference is zero outside the layer;

(c) gradients in the z-direction are so steep that the third term dominates the viscous force – and, indeed, in such a boundary layer, dominates also the left-hand side of Equation (48).

The solution of Equation (53) which vanishes at $z = 0$ and has zero gradient at the edge of the layer is obtained by two integrations, and its exterior value is

$$u_{fj}^{\mathrm{EXT}} = \mu^{-1} \int_0^\infty \left(F_j^{\mathrm{INT}} - F_j^{\mathrm{EXT}} \right) z dz; \tag{54}$$

where integration extends in practice, not to "infinity," but to the edge of the layer within which the integrand is nonzero. Expression (54) for the exterior streaming is yet again (see §6.1) independent of the viscosity μ since Equation (52) makes zdz of order $\mu/\rho\omega$; and it is easily evaluated.

At $x = y = 0$ (in the coordinates specified earlier) the exterior streaming (54) has

$$x\text{-component } -U\frac{3\partial U/\partial x + 2\partial V/\partial y}{4\omega} \text{ and } y\text{-component } -U\frac{\partial V/\partial x}{4\omega}, \tag{55}$$

with zero z-component. This is a generalized form of the century-old Rayleigh law of streaming (which covers cases when V is identically zero).

For the complete streaming pattern, expressions (55) are, effectively, boundary values for its tangential component at the body surface (because the Stokes boundary layer is so thin). Therefore, any simple solver for the steady-flow Navier-Stokes equations with specified tangential velocities on the boundary allows the pattern to be determined. Important note: here, the inertial terms in the Navier-Stokes equations must not be neglected, unless the Reynolds number R_s based on the streaming velocity (55) be of order 1 or less; when, however, the corresponding streaming motions would (as in §6.2) be uninterestingly small.

In the other extreme case when R_s is rather large (at least 10^3) the streaming motion remains quite close to the body (Stuart, 1966) within a steady boundary layer whose dimension (relative to that of the body) is of order $R_s^{-1/2}$. This layer is by no means as thin as the Stokes boundary layer, but it does confine very considerably the acoustic streaming motion. Equations (55) direct this motion towards one of the exterior flow's stagnation points, whence the steady-boundary-layer flow emerges as a jet (Figure 8) – yet another jet generated by sound (Riley, 1987 and Amin and Riley, 1990).

Acknowledgements

I am warmly grateful to Professors D. G. Crighton and N. Riley for invaluable advice on the above text, as well to the Leverhulme Trust for generous support.

References

Amin, N. and Riley, N., 1990. "Streaming from a sphere due to a pulsating acoustic source," *Journal of Fluid Mechanics* **210**, pp. 459-473.

Blokhintsev, D. I., 1956. *Acoustics of Nonhomogeneous Moving Medium*, Moscow: Gostekhizdat. Available in English translation as *National Advisory Committee for Aeronautics Technical Memorandum* No. 1399, Washington.

Bretherton, F. P. and Garrett, C. J. R., 1968. "Wavetrains in inhomogeneous moving media," *Proceedings of the Royal Society A* **302**, pp. 529-554.

Candel, S. M. and Poinsot, T. J., 1988. "Interactions between acoustics and combustions," *Proceedings of the Institute of Acoustics* **10**, pp. 103-153.

Crighton, D. G., 1972. "The excess noise field of subsonic jets," *Journal of Fluid Mechanics* **56**, pp. 683-694.

Crighton, D. G., 1975. "Basic principles of aerodynamic noise generation," *Progress in Aerospace Science* **16**, pp. 31-96.

Crighton, D. G., 1981. "Acoustics as a branch of fluid mechanics," *Journal of Fluid Mechanics* **106**, pp. 261-298.

Crighton, D. G., 1984. "Long range acoustic scattering by surface inhomogeneities beneath a turbulent boundary layer," *Journal of Vibration, Stress and Reliability* **106**, pp. 376-382.

Crighton, D. G. and Leppington, F. G., 1970. "Scattering of aerodynamic noise by a semi-infinite compliant plate," *Journal of Fluid Mechanics* **43**, pp. 721-736.

Crighton, D. G. and Leppington, F. G., 1971. "On the scattering of aerodynamic noise," *Journal of Fluid Mechanics* **46**, pp. 577-597.

Curle, N., 1953. "The mechanics of edge-tones," *Proceedings of the Royal Society A* **216**, pp. 412-424.

Curle, N., 1955. "The influence of solid boundaries on aerodynamic sound," *Proceedings of the Royal Society A* **231**, pp. 505-514.

Dowell, E. H., 1975. *Aeroelasticity of plates and shells*, Noordhoff.

Dowling, A. P., Ffowcs Williams, J. E., and Goldstein, M. E., 1978. "Sound production in a moving stream," *Philosophical Transactions of the Royal Society A* **288**, pp. 321-349.

Ffowcs Williams, J. E., 1963. "The noise from turbulence convected at high speed," *Philosophical Transactions of the Royal Society A* **255**, pp. 469-503.

Ffowcs Williams, J. E. and Hall, L. H., 1970. "Aerodynamic sound generation by turbulent flow in the vicinity of a scattering half plane," *Journal of Fluid Mechanics* **40**, pp. 657-670.

Ffowcs Williams, J. E. and Hawkings, D. L., 1969. "Sound generation by turbulence and surfaces in arbitrary motion," *Philosophical Transactions of the Royal Society A* **264**, pp. 321-342.

Ffowcs Williams, J. E. and Kempton, A. J., 1978. "The noise from the large-scale structure of a jet," *Journal of Fluid Mechanics* **84**, pp. 673-694.

Ffowcs Williams, J. E. and Maidanik, G., 1965. "The Mach wave field radiated by supersonic turbulent shear flows," *Journal of Fluid Mechanics* **21**, pp. 641-657.

Goldstein, M. E., 1976. *Aeroacoustics*, McGraw-Hill.

Goldstein, M. E., 1984. "Aeroacoustics of turbulent shear flows," *Annual Reviews of Fluid Mechanics* **16**, pp. 263-285.

Howe, M. S. and Ffowcs Williams, J. E., 1978. "On the noise generated by an imperfectly expanded supersonic jet," *Philosophical Transactions of the Royal Society A* **289**, pp. 271-314.

Lighthill, J., 1972. "The propagation of sound through moving fluids," *Journal of Sound and Vibration* **24**, pp. 471-492.

Lighthill, J., 1978a. "Acoustic streaming," *Journal of Sound and Vibration* **61**, pp. 391-418.

Lighthill, J., 1978b. *Waves in Fluids*, Cambridge University Press.

Lighthill, M. J., 1952. "On sound generated aerodynamically. I. General theory," *Proceedings of the Royal Society A* **211**, pp. 564-587.

Lighthill, M. J., 1954. "On sound generated aerodynamically. II. Turbulence as a source of sound," *Proceedings of the Royal Society A* **222**, pp. 1-32.

Lighthill, M. J., 1956. "Viscosity effects in sound waves of finite amplitude," pp. 250-351 of *Surveys in Mechanics* (ed. Batchelor, G. K. and Davies, R. M.), Cambridge University Press.

Lighthill, M. J., 1962. "Sound generated aerodynamically. The Bakerian Lecture," *Proceedings of the Royal Society A* **267**, pp. 147-182.

Lighthill, M. J., 1963. "Jet noise. The Wright brothers' lecture," *American Institute of Aeronautics and Astronautics Journal* **1**, pp. 1507-1517.

Mani, R., 1976. "The influence of jet flow on jet noise, Parts 1 and 2," *Journal of Fluid Mechanics* **73**, pp. 753-793.

Möhring, W., 1978. "On vortex sound at low Mach number," *Journal of Fluid Mechanics* **85**, pp. 685-691.

Mungur, P. and Gladwell, G. M. L., 1969. "Acoustic wave propagation in a sheared fluid contained in a duct," *Journal of Sound and Vibration* **9**, pp. 28-48.

Nyborg, W. L., 1953. "Acoustic streaming due to attenuted plane waves," *Journal of the Acoustical Society of America* **25**, pp. 68-75.

Nyborg, W. L., 1965. "Acoustic streaming," Chap. 11 in *Physical Acoustics, Principles and Methods* (ed. Mason, W. P.), Academic Press.

Parry, A. B. and Crighton, D. G., 1991. "Asymptotic theory of propeller noise," *American Institute of Aeronautics and Astronautics Journal* **27**(1), 1989, pp. 1184-1190 and **29**(2), to appear.

Phillips, O. M., 1960. "On the generation of sound by supersonic turbulent shear layers," *Journal of Fluid Mechanics* **9**, pp. 1-28.

Powell, A., 1953a. "The noise of choked jets," *Journal of the Acoustical Society of America* **25**, pp. 385-389.

Powell, A., 1953b. "On edge-tones and associated phenomena," *Acustica* **3**, pp. 233-243.

Powell, A., 1960. "Aerodynamic noise and the plane boundary," *Journal of the Acoustical Society of America* **32**, pp. 982-990.

Powell, A., 1964. "Theory of vortex sound," *Journal of the Acoustical Society of America* **36**, pp. 177-195.

Pridmore-Brown, D. C., 1958. "Sound propagation in a fluid flowing through an attenuating duct," *Journal of Fluid Mechanics* **4**, pp. 393-406.

Rayleigh, Lord, 1896. *The Theory of Sound*, **II**, Second edition, Macmillan, London.

Reynolds, O., 1895. "On the dynamical theory of incompressible viscous fluids and the determination of the criterion," *Philosophical Transactions of the Royal Society A* **186**, pp. 123-164.

Ribner, H. S., 1964. "The generation of sound by turbulent jets," *Advances in Applied Mechanics* **8**, pp. 103-182.

Riley, N., 1987. "Streaming from a cylinder due to an acoustic source," *Journal of Fluid Mechanics* **180**, pp. 319-326.

Shankar, P. N., 1971. "On acoustic refraction by duct shear layers," *Journal of Fluid Mechanics* **47**, pp. 81-91.

Stuart, J. T., 1966. "Double boundary layers in oscillatory viscous flow," *Journal of Fluid Mechanics* **24**, pp. 673-687.

Westervelt, P. J., 1953. "The theory of steady rotational flow generated by a sound field," *Journal of the Acoustical Society of America* **5**, pp. 60-67.

Whitham, G. B., 1956. "On the propagation of weak shock waves," *Journal of Fluid Mechanics* **1**, pp. 290-318.

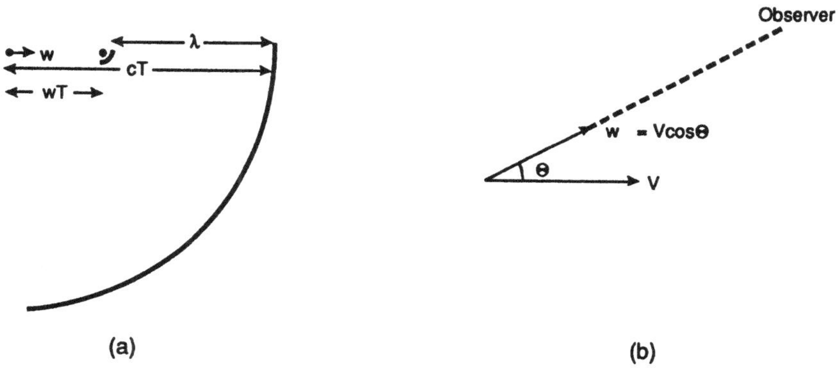

Figure 1

(a) When a source of sound of period T approaches an observer at velocity w, successive wave crests are emitted towards him with separation $\lambda = cT - wT$.

(b) Case of observer at angle θ to source's direction of motion at speed V; then velocity of approach is $w = V \cos \theta$.

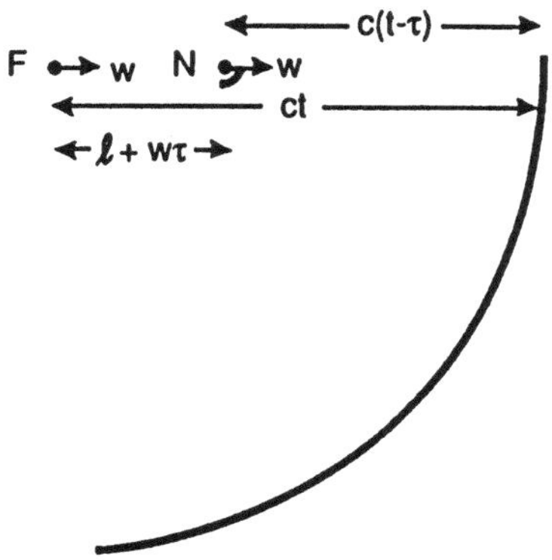

Figure 2. Sounds arriving simultaneously from a source's far side F and near side N took times t and $t - \tau$ (say) to reach the observer, while the distance FN increased from ℓ to $\ell + w\tau$ during the time τ between emissions from F and N.

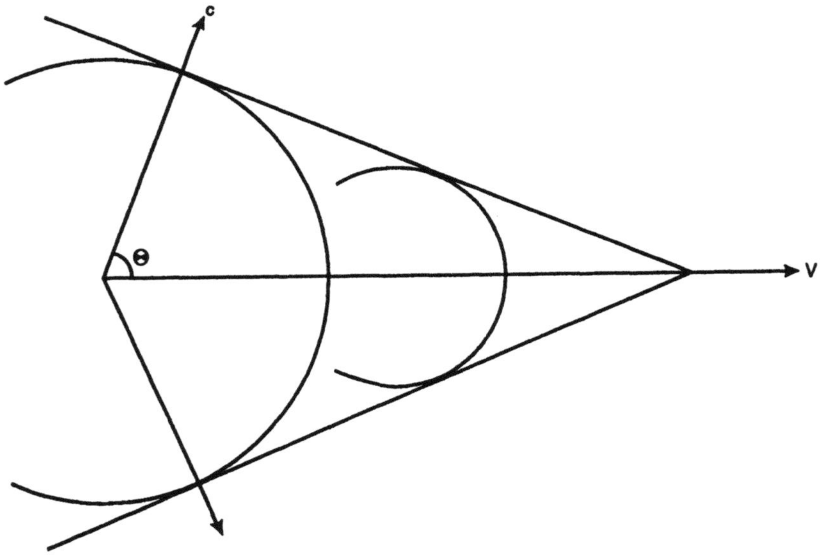

Figure 3. Supersonic source convection produces radiation along rays in the Mach direction (11).

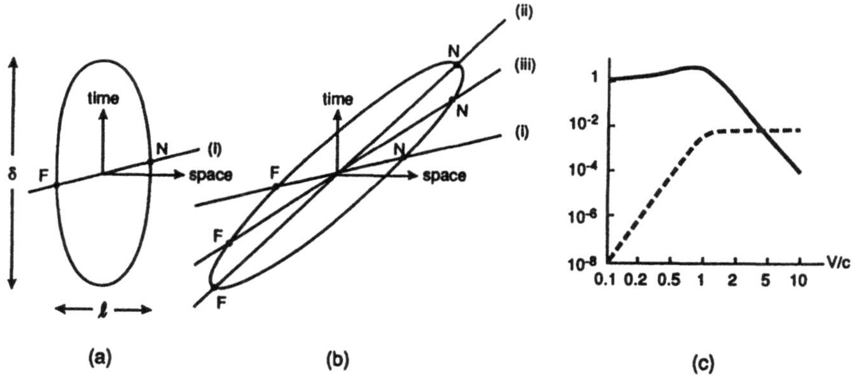

Figure 4. A uniformly valid Doppler-effect approximation.

Diagram (a) Space-time diagram for unconvected "eddies" of correlation length ℓ and duration δ.

Diagram (b) Case of "eddies" convected towards observer at velocity w; being Diagram (a) sheared by a distance w per unit time. Here, lines sloping by a distance c per unit time represent emissions received simultaneously by observer.

Case (i): w/c small. Case (ii): $w/c = 1$. Case (iii): intermediate value of w/c.

Diagram (c) ———— Average modification factor (13). - - - - - - - - Acoustic efficiency, obtained by applying this factor to a low-Mach-number "quadrupole" efficiency of (say) $10^{-3}(V/c)^5$.

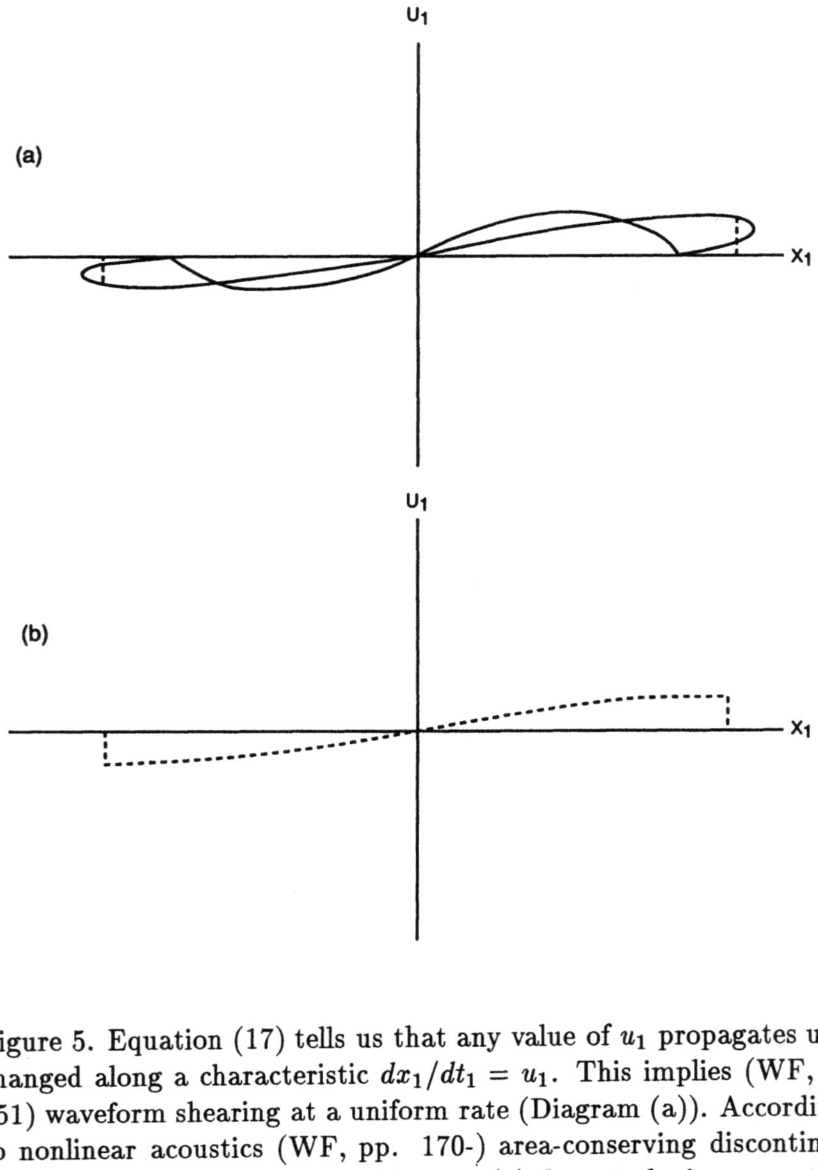

Figure 5. Equation (17) tells us that any value of u_1 propagates unchanged along a characteristic $dx_1/dt_1 = u_1$. This implies (WF, p. 151) waveform shearing at a uniform rate (Diagram (a)). According to nonlinear acoustics (WF, pp. 170-) area-conserving discontinuities, shown as broken lines in Diagram (a), have to be incorporated wherever necessary to keep the solution one-valued; leading, in the case illustrated, to the N-wave form (Diagram (b)).

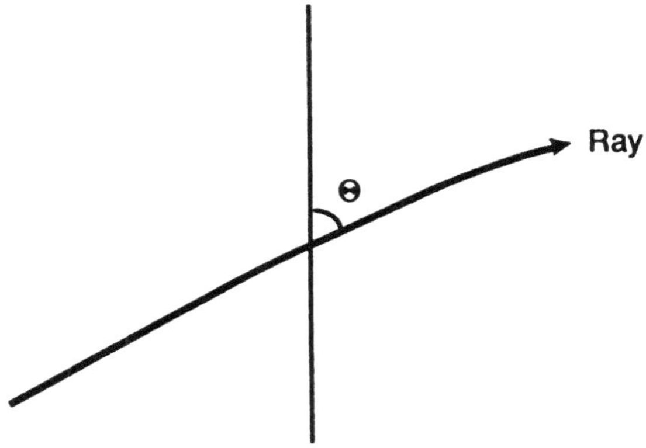

Figure 6. The angle θ between a ray and the vertical varies in accordance with the extension (41) to Snell's law.

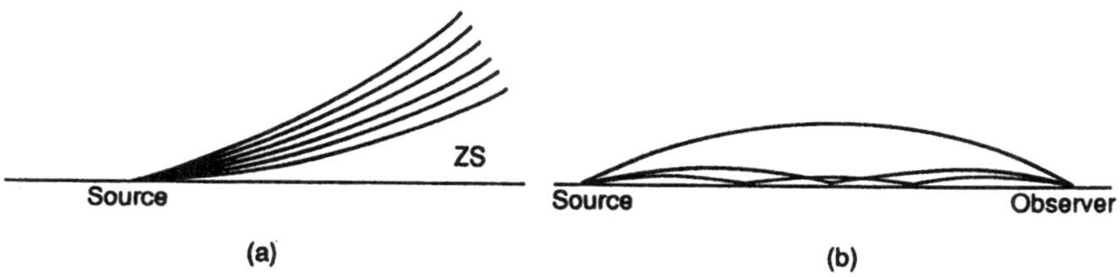

Figure 7. Effects of ray curvature (44) on propagation from a source on horizontal ground.

Diagram (a) Rays of given upward curvature (due to temperature lapse or upwind propagation) can leave a Zone of Silence (ZS) below the ray emitted horizontally.

Diagram (b) Rays of given downward curvature (due to temperature inversion or downwind propagation) can enhance received signals through multiple-path communication.

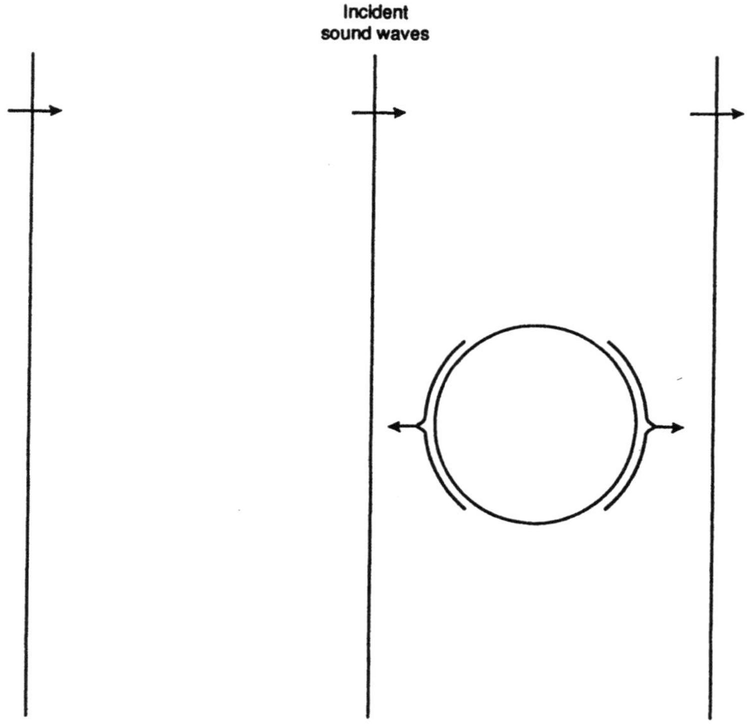

Figure 8. Plane sound waves incident on a sphere may generate a steady streaming motion (Amin and Riley, 1990), concentrated in a relatively thick boundary layer, and directed towards one of the exterior flow's stagnation points – whence it emerges as a jet.

CLASSICAL THEORETICAL APPROACHES

CLASSICAL THEORETICAL APPROACHES TO COMPUTATIONAL AEROACOUSTICS

William E. Zorumski

Mail Stop 461, NASA Langley Research Center
Hampton, Virginia 23681-0001

ABSTRACT

An assessment is given of the relation of classical theoretical approaches to the newly defined field of computational aeroacoustics. It is indicated that classical methods play two important roles. First, the classical methods supply desirable formulations of the governing equations which are suitable for computational implementation. Second, the classical methods provide boundary conditions which are essential for accurate numerical simulations within a finite computational domain.

1. Introduction

The session on Classical Theoretical Approaches contained four papers, as indicated in the contents of this volume. The paper by Crighton (1992) reviewed the principal results of theoretical methods and pointed out many of the pitfalls in direct computation. These include the problems of the extreme range of magnitude of the acoustic variables and the range of frequencies where results are desired. Lilley (1992) gave a systematic review of the classical theoretical formulations which are used in the acoustic analogy, and the partial differential equations of aeroacoustics, the major ones of which are due to Lighthill, Philips, Lilley, and Howe. Geer (1992) presented examples of the use of perturbation methods, the method of multiple scales, to the problem of acoustic radiation. Tam (1992) gave an account of his newly-developed numerical method for acoustic wave propagation. The principal feature of Tam's method is the preservation of the dispersion characteristics of the acoustic equation by the numerical simulation. Each of these papers offers unique insights into a number of important issues of computational aeroacoustics.

This overview of the classical approaches attempts to complement these papers by speculating on the ways that a symbiotic relationship may evolve between the numerical simulations of Computational Fluid Dynamics and the analytical methods of Classical Theoretical Acoustics. Included here is one attempt to answer two of the questions posed at the workshop, namely (1) "what are the fundamental problems of aeroacoustics?" and (2) "what are the appropriate boundary conditions?"

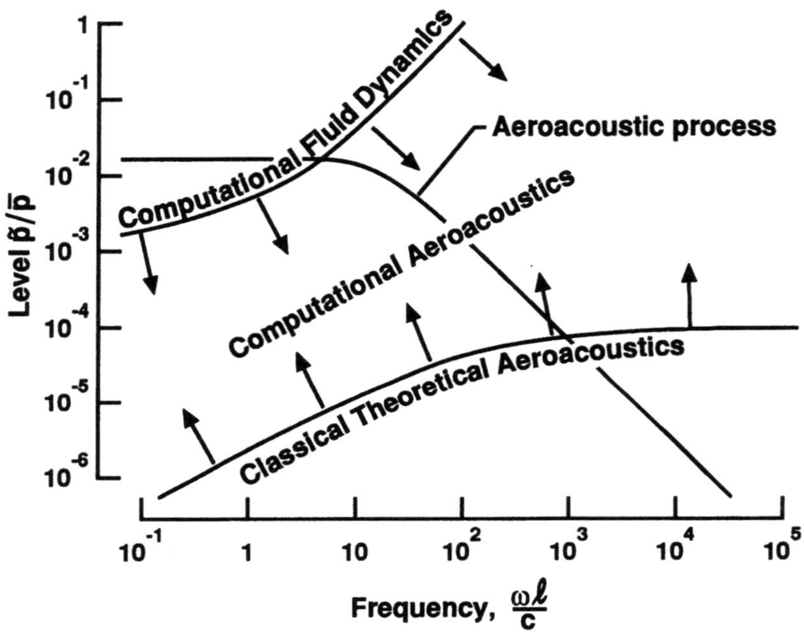

Figure 1: Levels and frequencies in Computational Aeroacoustics.

2. Fundamental Problems

One of the goals of this workshop has been to define or identify the fundamental problems of aeroacoustics. This question can be viewed from the perspective of analysis and computation or from the perspective of physics. Figure 1 serves to illustrate the first perspective, and relates to the issues of small variables and high frequencies discussed more fully by Crighton (1992). Here, $\bar{p}$ is the mean pressure and $\tilde{p}$ is the fluctuation about it, while $\omega = 2\pi f$ is the circular frequency, c is the sound speed, and l is a characteristic length. The range of magnitude of acoustic variables is very large, as is the range of frequencies and wavelengths of interest. An aeroacoustic process such as jet noise has pressure fluctuations at some frequencies which are comparatively large, say $O(10^{-2})$ of the local average pressure, but there is also interest, because of the response characteristic of the human ear, in low level fluctuations as small as $O(10^{-6})$ at high frequencies as shown on Figure 1. Computational Fluid Dynamics is typified by low frequencies, even zero frequency or steady state, and high levels where the dependent variables are $O(1)$. Classical acoustics is typified by low levels, $< O(10^{-4})$, and high frequencies, $O(10^2)$ to $O(10^5)$. Researchers in CFD are pushing to lower levels (with greater accuracy) and higher frequencies, while researchers in aeroacoustics are pushing to higher levels where nonlinearities become important. The field of Computational Aeroacoustics may be defined in this sense as the domain between Computational Fluid Dynamics and Classical Acoustics.

The physical perspective on the fundamental problems of Computational Aeroacoustics suggests two general categories. These are acoustic sources, or sound generation, and acoustic propagation. In the source category, there are three basic problems related to turbulent flows. These are Jet Noise, Boundary Layer Noise, and Shock/Turbulence Interaction Noise. Jet Noise is the ultimate canonical problem of aeroacoustics. The field began with Jet Noise and will probably end with Jet Noise. Boundary Layer Noise is similar, but is markedly influenced by the presence of a surface. Shock/Turbulence Interaction Noise can occur within either jets or boundary layers, but can also be studied in isolation. Another general category of source noise is the sound generated by rotating machinery. This includes propellers, helicopter rotors, and aircraft engine fans, or turbofans. In the propagation category, there are the areas of propagation within waveguides, such as aircraft engine ducts, and atmospheric sound propagation. The physical boundary conditions, typically nonlinear, make the duct propagation problem challenging. Both duct propagation and atmospheric propagation involve inhomogeneous media, which is a significant complication.

3. Boundary Conditions

A second goal of the workshop was to define appropriate boundary conditions for Computational Aeroacoustics. In one sense, this is trivial. The boundary conditions for far-field problems are that there

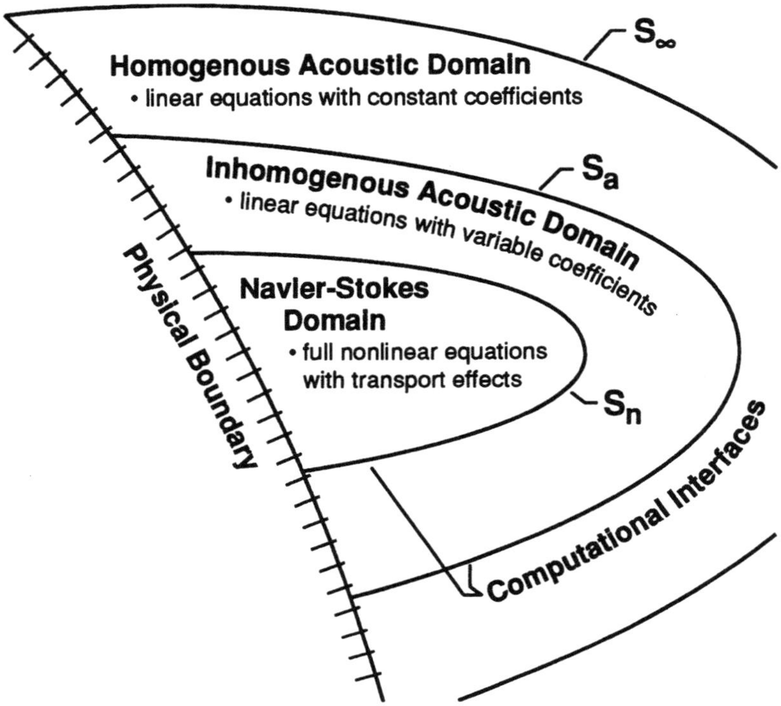

Figure 2: Domains of Computational Aeroacoustics.

are only radiated waves at infinity. But this is a computationally impossible condition to implement, because no mesh can extend to infinity. Morris (1992) has given an analysis of the grid requirements of Computational Aeroacoustics which shows that present computers are inadequate even for finite computational domains. The massively parallel systems envisioned will be stretched to their limit by the problems of Computational Aeroacoustics. It would seen reasonable then to pursue methods which minimize the size of the computational domain and the number of grid points. Advancements in the area of boundary conditions may provide a means of reducing the size of the computational domain to one suitable for present day computers or even workstations.

The term "boundary conditions" is not entirely appropriate for the surface of a computational domain because no physical boundary need exist there. The computational domain surface may be an interface, defined somewhat arbitrarily, across which solutions must be matched. One typically thinks of a single bounding surface, or interface, and the boundary condition on this surface is taken to be some condition representing "outgoing waves." Tam (1992) shows that the accuracy of his boundary condition is of order $r^{-5/2}$, where r is the distance from the source. When the computational domain becomes small, then this boundary condition becomes less accurate. Boundary conditions of this sort depend on knowing which direction the wave is travelling, but in a general noise generation problem, there are many waves travelling in many directions. Unless, of course, the bounding surface is so far from the source that the direction becomes radial. The "outgoing wave" boundary condition is suitable in general only for large computational domains. There may be particular problems, however, where the wave direction is known and this boundary condition will suffice.

The accuracy of the boundary condition may be improved by matching the computational solutions within a given domain to a near-field asymptotic formula for the radiated waves (Geer, 1992). In contrast to the idea of a single bounding surface, or interface, it may be advantageous to consider matching solutions across interfaces between several domains. Figure 2 illustrates one possible scheme for implementing this idea. The computational domains are defined by any physical boundaries and a nested set of surfaces, each one enclosed fully by another. The innermost domain might be a Navier-Stokes domain, where the full equations of fluid dynamics are simulated by available CFD methods. This domain is enclosed by a bounding surface S_n. It may be advantageous to enclose this domain within a buffer domain where the Navier-Stokes equations are decomposed into a set of nonlinear steady flow equations and linear equations for the unsteady parts. The acoustic equations within this buffer domain have spatially-variable coefficients, so that it could be called the inhomogeneous acoustic domain. The buffer domain is enclosed by the surface S_a, and outside of this surface is the homogeneous acoustic domain, extending to infinity. In the homogeneous acoustic domain, the equations would have constant coefficients. The acoustic field here would be represented by the classic solutions of theoretical acoustics. Each domain would utilize a computational method optimized for that domain, hopefully avoiding wasted grid points and computations. These potential savings would come at the cost of more interfacing operations, and the merits of the trade remain to be determined.

Whatever scheme of domains is adopted, a primary goal would be to make the computational domain within S_a as small as possible. In addition to minimizing grid points, the buffer domain could be useful in alleviating the problem of small acoustic variables, since these variables would be defined separately in the buffer and in the homogeneous acoustic domain.

The homogeneous acoustic domain can be represented entirely by the Kirchhoff-Helmholtz formula (Pierce, 1981). If this surface is moving, a more general form of this formula is available (Farassat and Meyers, 1988). The form given by Pierce is

$$\begin{aligned} \tilde{p}(\vec{x},t) &= \frac{\bar{\rho}}{4\pi}\int\int_{S_a} \frac{\dot{v}_n(\vec{x}_s,\tau)}{r}\,dS \\ &+ \frac{1}{4\pi c}\int\int_{S_a} \vec{e}_r \cdot \vec{n}_s \left(\frac{\partial}{\partial t}+\frac{c}{r}\right)\frac{\tilde{p}(\vec{x}_s,\tau)}{r}\,dS \end{aligned} \tag{1}$$

where $\tau = t - r/c$ is the retarded time, or source time, and r is the distance between the source point $\vec{x}_s$ and the receiver point $\vec{x}$. The other variables are the acoustic pressure p, the constant ambient density $\bar{\rho}$, and the constant ambient sound speed c.

The elements of the formula and its vectors are depicted in Figure 3. The unit vector $\vec{n}_s$ is normal to the surface S_a, and the unit vector $\vec{e}_r$ points from the source to the receiver position. The acoustic velocity normal to the surface is v_n and its derivative is indicated by the dot. All variables under the integral signs are evaluated at retarded time τ.

The Kirchhoff formula holds for all points in the homogeneous acoustic domain including, in the limit, points on the surface S_a. When the observer position is on the surface, the Kirchhoff formula becomes an integral equation for the surface pressure. This integral equation is an exact boundary condition for the computational domain enclosed by S_a and emphasizes the fact that the proper radiation boundary condition is multi-dimensional, not local. Any combination of computational methods and boundary conditions which produce results inconsistent with this integral equation is incorrect.

Since the Kirchhoff formula is an exact boundary condition, an attempt should be made to implement it in parallel with methods

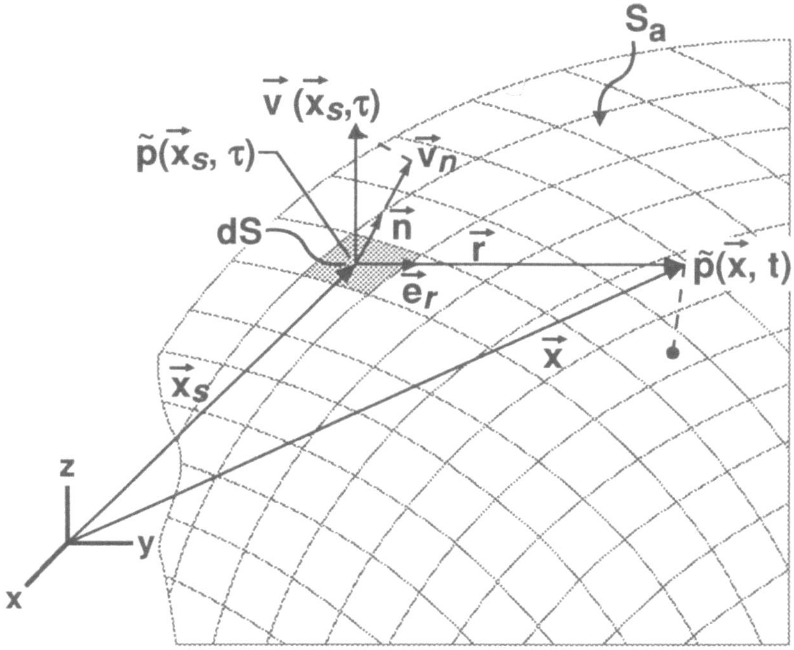

Figure 3: Elements of the Kirchhoff-Helmholtz radiation formula.

within the computational domain. This is not a new idea (Pierce, 1990), but the development of computational technology may make it feasible now. Many variations on this theme are possible, because the Kirchhoff formula relies on a Green's function. The one given above has the free-space Green's function as its basis. Other Green's functions tailored for the homogeneous acoustic domain may provide a superior computational capability. This task lies squarely within the realm of Classical Theoretical Acoustics.

4. Conclusions

Classical theoretical approaches to aeroacoustics supply formulations of the governing equations of fluid motion which are well suited as the basis of Computational Aeroacoustics. The success of the acoustic analogy demonstrates that these formulations contain at least the relevant physics of aeroacoustic phenomena and much of the quantitative information as well. Limits on the success of these formulations can be attributed to the fact that the source terms are functions of the dependent variables. A "solution" to a Lighthill's equation or a Lilley's equation then simply produces an integral equation for the aeroacoustic variable. But this solution, that is, the construction of a Green's function for the partial differential equation, is still a very important step. Its completion makes possible the implementation of acoustic analogies and a direct computational attack on aeroacoustics via the integral equations.

References

Crighton, David, 1992. "Computational aeroacoustics for low Mach number flows in ducts," Proceedings of the ICASE/NASA Langley Workshop on Computational Aeroacoustics, Hampton, Virginia.

Farassat, F. and Meyers, M. K., 1988. "Extension of Kirchhoff's formula to radiation from moving surfaces," *Journal of Sound and Vibration* **123**(3), pp. 451–460.

Geer, James, 1992. "A multiple scales approach to sound generation by vibrating bodies," Proceedings of the ICASE/NASA Langley Workshop on Computational Aeroacoustics, Hampton, Virginia.

Lilley, Geoffrey, 1992. "On the noise radiated from an aerodynamically unsteady fluid flow," Proceedings of the ICASE/NASA Langley Workshop on Computational Aeroacoustics, Hampton, Virginia.

Morris, Philip, 1992. "Validation of computational aeroacoustics algorithms," Proceedings of the ICASE/NASA Langley Workshop on Computational Aeroacoustics, Hampton, Virginia.

Pierce, Allan D., 1981. *Acoustics: An Introduction to Its Physical Principles and Applications*, McGraw-Hill Book Company, New York.

Pierce, Allan D., 1990. "The Helmholtz-Kirchhoff integral relation as a framework for developing algorithms for sound propagation through inhomogeneous moving media," Computational Acoustics: Ocean-Acoustic Models and Supercomputing, Proceedings of the Second IMACS Symposium on Computational

Acoustics, Princeton, NJ, USA, Volume I, Edited by D. Lee, A. Cakmak, and R. Vichnevetsky, North-Holland, New York.

Tam, Christopher, 1992. "A study of the short-wavelength components in direct numerical simulations of aeroacoustic problems," Proceedings of the ICASE/NASA Langley Workshop on Computational Aeroacoustics, Hampton, Virginia.

COMPUTATIONAL AEROACOUSTICS FOR LOW MACH NUMBER FLOWS

D. G. Crighton

Department of Applied Mathematics and Theoretical Physics,
University of Cambridge, Cambridge CB3 9EW
UNITED KINGDOM

1. General Remarks

This is the text of an introductory lecture intended to provoke discussion of some of the problems likely to face participants in the Workshop as they develop the subject of Computational Aeroacoustics (CAA). Many of the points were already made by the author in a lecture at the 1st IMACS Symposium on Computational Acoustics (Crighton, 1988). Nonetheless, it is appropriate to repeat and amplify them in a Workshop devoted specifically to CAA and not to the wider problems of acoustics – and in particular, in a Workshop in which it is hoped that interest and expertise will have been brought together in more than the critical mass needed to really make CAA take off as a significant tool in aeroacoustics investigations.

First, why is there interest in low Mach number aeroacoustics at all, whether analytical, experimental or computational? Might this not turn out to be a difficult area, and even irrelevant, given that most flows of aeronautical interest have velocities in the moderate to high subsonic range? Is there a danger that we are tackling a problem which is apparently simplified, but in fact notoriously intractable and of less than major applicability – much as has turned out to be the case with homogeneous turbulence?

The answer is in the negative, and largely for the reasons given in his lecture at this Workshop by Sir James Lighthill; unless the mean flow velocity exceeds roughly *twice* the ambient speed of sound (in which case the eddy convection velocity exceeds the ambient sound speed and Mach wave emission takes place, determined principally by the spatial turbulence structure, but limited in amplitude by temporal correlations) the radiation is determined primarily by the slow temporal evolution of convected eddies; and many of the effects of eddy convection, mean flow fluid shielding, and refraction, can be represented as multiplicative corrections to the radiation of essentially low-Mach-number eddies. Thus, for example, convection is equivalent to a factor $(1 - M_c \cos\theta)^{-5}$ on intensity in any direction

θ, with M_c the convection Mach number (Ffowcs Williams, 1963), fluid shielding to a factor $(1 - M_J \cos\theta)^{-4}$ at low frequencies, with M_J the flow Mach number based on ambient sound speed (Mani, 1976), etc. Most not-so-small Mach number effects can be accounted for subsequently in such ways, but one always has ultimately to deal with the low-Mach-number aeroacoustics problem.

Second, why are we making a fuss about *computational* aeroacoustics? After all, we have many computational developments of other fields within mechanics, such as aerodynamics and structural mechanics. But in computational aeroacoustics we have *qualitatively* new issues which make *computational aeroacoustics* a significant scientific issue, in which quite new constraints and demands are placed on computational schemes. The problem is that *all numerical analysis procedures are inherently noisy* (in the acoustic sense). Discretization on meshes or finite elements, spectral cut-off, confinement to a box, enforcement of periodicity – all these and other computational procedures correspond to the application to the fluid of certain acoustic source mechanisms which may be small in strength – if the numerical procedure is "accurate" – but which may be acoustically efficient; far more so, in fact, in some cases, than the hydrodynamic flow under discussion itself. Then the numerical procedure may actually be "noisier" than the flow! There is thus a crucial need in all studies to *preserve the multipole structure of the flow.* The dangers of not doing so have long been appreciated in analytical studies (think of the controversy in the 1960's over dipole sound from turbulence over plane homogeneous boundaries); now we need to hear these warnings in the computational context.

A final remark in this section, provoked by Sir James Lighthill's reference to jet noise levels in the 1950's as "terrifyingly high"; here one needs to have in mind the extraordinary disparity between energies in the flow (e.g., a jet flow) and in the sound field, and the extraordinary dynamic and frequency range of relevance to *people.* Dynamic levels range from around 0 dB at the threshold of audibility to 140 dB at the threshold of pain, equivalent to an intensity variation by a factor 10^{14}; while frequencies perceived by the human ear range from 20 Hz to 20 000 Hz – 10 octaves (to be contrasted with visual sensitivity over less than one octave of light). Moreover, the ear is much more sensitive to sound (especially sound with a strong tonal content) in the 2000–4000 Hz range, and sound in this range is heavily penalized by the aircraft noise certification procedure – but such sound is particularly hard to predict, calculate or measure, because it lies four or five octaves higher than the peak jet noise (which may lie in the 200–400 Hz range).

These large frequency and dynamic ranges are found nowhere else in normal experience. They have amusing implications in popular terms; an early-generation 4-jet transport (Boeing 707) at take-off produces as much sound intensity as the whole world's population shouting (in phase) together; yet 30 years later the Boeing 767 (with roughly double the range, double the payload, four times as much thrust per engine) produces only the sound intensity of the inhabitants of a large city (London, New York) shouting coherently together! But nevertheless, in the 45 seconds of take-off roll of the "terrifyingly loud" Boeing 707, the total energy radiated as sound is only about enough to cook one egg!

In aeroacoustics we are dealing with the *minutest* energy levels, compared with the whole flow – and moreover, with those energies on length scales $O(M^{-1})$ larger than the scales of the "energy-containing eddies". Calculations of these lie, in energy and scale, far outside the range of conditions adequately resolved by standard CFD procedures. This is what makes *computational aeroacoustics* at low Mach number not merely a *technologically*, but also a *scientifically*, important problem.

2. Low Mach Number Aeroacoustics

Here we have the following agenda; "given the flow field" (for example, given the vorticity $\boldsymbol{\omega}(\mathbf{x}, t)$ confined to some bounded domain V), determine the acoustic field far from V, from

(i) an acoustic analogy, together with the free-space Green's function or an exact Green's function appropriate to the particular geometrical and mechanical configuration; or

(ii) evaluation of a Kirchhoff-Helmholtz integral over a surface S enclosing all sources; or

(iii) Fourier integrals and stationary phase for data given on a planar or cylindrical surface outside the flow; or

(iv) asymptotic matching (as $M \to 0$) of an inner rotational core with length scale l to an outer acoustic field with length scale lM^{-1}; or

(v) directly, along with the flow field, by spectral, difference or element methods.

Basic difficulties with this program are the following:

(a) the numerical analysis must preserve the *multipole* structure of the basic acoustic source. This applies both to volume sources representing the turbulent eddies themselves, and sources on any boundary surfaces in the flow. Replacement of a quadrupole by a dipole will grossly overestimate the direct radiation (unless the dipole radiation integral is carefully evaluated with full account of retarded-time variations);

(b) there are large length scale disparities between the eddy scale l and the wavelength $\lambda \sim a_0/(u/l) \sim lM^{-1}$ of sound generated by eddies of velocity u and lifetime l/u;

(c) there are gross energy density disparities between the hydrodynamic near field and the acoustic far field. Even when spherical spreading effects have been removed, the energies in the far field are smaller than those in the flow itself by $O(M^4)$;

(d) what is important for certification purposes is not necessarily the total acoustic energy – dominated by low frequency components of little subjective importance (frequencies $\sim$ 200 Hz) – but the subjectively weighted EPNdB measure, which places heavy penalties on noise (and especially tones) in the 2-4 kHz band. It is vital, therefore, to control not the energy-containing scales, but rather those corresponding to frequencies four to five octaves higher, where the energy levels may be 10–20 dB lower than those at the peak frequency;

(e) in applications (the certification process) we are concerned with propagation distances, of say 300 m, over which there may be significant nonlinear pumping of energy to higher frequencies. A typical flow scale – the jet radius – may be of order 1 m, the sources may be distributed over an axial extent of 10 m, and in laboratory or rig experiments it is likely that the range to a far-field observer will be no more than 30 m. However, the real needs (which are difficult to achieve in carefully-controlled experiments) – whether for theory, experiment or computation – involve significantly larger physical domains than the usual linear far field.

Many of these issues pose formidable problems for computation – and indeed, have not always yet been addressed in either theory or experiment – but taken together they define an area of study of particular intricacy for which a computational solution would represent a major advance.

3. Acoustic Analogies

This term refers to the recasting of the exact equations of fluid motion in the form of an inhomogeneous wave equation appropriate to the fact that in the far field (and ignoring here nonlinear waveform distortion) pressure fluctuations propagate through still fluid at the ambient sound speed a_0. Thus

$$\left(\nabla^2 - \frac{1}{a_0^2}\frac{\partial^2}{\partial t^2}\right) h = q \,, \tag{3.1}$$

for some h equivalent in the far field to $p - p_0$. The right side $q(\mathbf{x}, t)$ acts as a source for h – and also includes all propagation mechanisms such as scattering by turbulent eddies and temperature inhomogeneities, refraction in the mean velocity and temperature fields, fluid shielding of each eddy by the local (moving) fluid environment, nonlinear self-distortion of propagating sound waves, etc.

There are three analogies of the form (3.1), one exact and due to Lighthill (1952), with

$$q = -\frac{\partial^2 T_{ij}}{\partial x_i \partial x_j} \,, \quad T_{ij} = \rho u_i u_j + p_{ij} - a_0^2 \rho \delta_{ij} \,, \tag{3.2}$$

the other two asymptotically equivalent for $M \to 0$, and due to Powell (1964) and Howe (1975), with

$$q = -\rho_0 \frac{\partial L_i}{\partial x_i} \,, \quad L_i = (\boldsymbol{\omega} \wedge \mathbf{u})_i \,, \tag{3.3}$$

and to Ribner (1962) with

$$q = \frac{1}{a_0^2}\frac{\partial^2 p^{(0)}}{\partial t^2} \,, \quad \nabla^2 p^{(0)} = -\frac{\partial^2 \rho_0 u_i u_j}{\partial x_i \partial x_j} \,. \tag{3.4}$$

For (3.2) the sources are quadrupole with a double integral vanishing property; since T_{ij} vanishes far from the flow we have

$$\int q(\mathbf{x}, t) d^3\mathbf{x} = 0 \,, \tag{3.5}$$

$$\int x_m q(\mathbf{x},t) d^3\mathbf{x} = 0\,. \tag{3.6}$$

The same must hold for the other two representations of q, but in the case of (3.3) with the Lamb vector $\mathbf{L}$ as a dipole source we are guaranteed only (3.5), and (3.6) will hold only if special care is taken, while in the case of (3.4), with a monopole source related to the "pseudo-sound" pressure field generated within the flow, neither (3.5) nor (3.6) will automatically be satisfied.

Although the quadrupole formulation is least prone to lead to overestimation of the sound levels, the quadrupole strength is not easily measured, it decays only algebraically away from a localized vorticity distribution ($T_{ij} \sim |\mathbf{x}|^{-6}$), and it does not emphasize the primary feature of the source field – the vorticity $\boldsymbol{\omega}$. The Powell-Howe vortex dipole is attractive in being precisely localized on the vorticity. This has made it very popular for the solution of model problems involving line and ring concentrations of vorticity and their interaction with solid bodies – some of these problems having been also studied experimentally by Kambe and his group (see later). Ribner's monopole is only very weakly localized ($p^{(0)} \sim |\mathbf{x}|^{-3}$) away from the flow, and integrals involving $p^{(0)}$ must be taken over an extensive region – of acoustic wavelength size – with account taken of retarded-time variations, in order to get results consistent with those of the quadrupole formulation. Such features have not been conducive to the development of a broad aeroacoustic theory (including, in particular, effects of flow interaction with solid bodies) based on the Ribner monopole idea.

We must emphasize the dangers of using convenient forms of acoustic analogy which do not (in free space) preserve the underlying quadrupole structure. The far-field solution of the Lighthill equation in free space is

$$(p - p_0)(\mathbf{x},t) = \frac{1}{4\pi a_0^2 |\mathbf{x}|} \hat{x}_i \hat{x}_j \frac{\partial^2}{\partial t^2} \int T_{ij}\left(\mathbf{y}, t - \frac{|\mathbf{x}-\mathbf{y}|}{a_0}\right) d^3\mathbf{y}\,, \tag{3.7}$$

with $\hat{x}_i = x_i/|\mathbf{x}|$ the unit vector in the radiation direction. In general there are no grounds for supposing that $\int T_{ij}(\mathbf{y},t) d^3\mathbf{y} = 0$ (i.e. no grounds for supposing that T_{ij} is in fact an octupole, $T_{ij} = \partial W_{ijk}/\partial x_k$ with W_{ijk} vanishing at infinity) and so, since the maximum variation in retarded time $|\mathbf{y}|/a_0$ is l/a_0 and is small on the time scale l/u of low Mach number eddies, we can write

$$(p - p_0)(\mathbf{x}, t) = \frac{1}{4\pi a_0^2 |\mathbf{x}|} \hat{x}_i \hat{x}_j \frac{\partial^2}{\partial t^2} S_{ij}(t - |\mathbf{x}|/a_0) , \tag{3.8}$$

with $S_{ij}(t) = \int T_{ij}(\mathbf{y}, t) d^3\mathbf{y}$ the total instantaneous quadrupole strength. The naive form of the Powell-Howe theory, after similar neglect of retarded-time variations, gives an apparent dipole,

$$(p - p_0)(\mathbf{x}, t) = \frac{\rho_0}{4\pi a_0 |\mathbf{x}|} \hat{x}_i \frac{\partial}{\partial t} \int L_i \left(\mathbf{y}, t - \frac{|\mathbf{x}|}{a_0}\right) d^3\mathbf{y} , \tag{3.9}$$

which actually vanishes identically, because for incompressible flow

$$\int L_i(\mathbf{y}, t) d^3\mathbf{y} = \int \left[\frac{\partial}{\partial y_j}(u_i u_j) - \frac{\partial}{\partial y_i}\left(\tfrac{1}{2} u_j u_j\right) \right] d^3\mathbf{y} = 0 . \tag{3.10}$$

Computationally, however, (3.10) may not be precisely respected, and in that case there will be a spurious, but powerful, dipole radiation according to (3.9). The ratio of (3.9) to (3.8) will then be $O(M^{-1}\Delta s)$, where Δs is the normalized discretization length, and for small Mach numbers ($M \sim 10^{-5}$) typical of underwater turbulence the numerical dipole will dominate the genuine flow quadrupole unless the step size is extremely small. The author has seen this occur in computations, via (3.9), of the radiation from two counter-rotating line vortices approaching a rigid plane. This is a simple problem involving four vortices, by image arguments, the radiation is of quadrupole type, and the problem can be solved analytically by several methods. It can be shown that the total energy E radiated as sound, over the whole process in which the vortices approach the plane and diverge over it, is proportional to U^6, with U the velocity of approach of the vortices to the plane at large distances. Computation according to (3.9) gave $E \propto (\Delta s)^2 U^4$, a two-dimensional dipole result; the numerical force giving rise to the non-zero dipole is that which is needed to make the vortices travel along small rectilinear line segments Δs, rather than along the smooth path they would actually follow. If the rigid plane is replaced by a homogeneous flexible surface on which an impedance condition holds, then the exact proportionality $E \propto U^6$ still holds, but numerical calculations now gave $E \propto (\Delta s)^2 U^2$, different by a factor $(\Delta s)^2 M^{-4}$ which can be large, even if Δs is small, for very small Mach numbers $M = U/a_0$. The numerical radiation this time is monopole in character, associated with oscillations of the surface. In the exact solution these cancel precisely and leave no residual monopole, but in the computation, discretization and confinement of the system to a large box led to imperfect cancellation and an efficient numerical monopole.

More powerful dipole radiation arises when vortices pass out of the computational domains, as in the many simulations of turbulent shear layers by line and ring vortices (see, e.g., Morfey, 1979; Acton, 1980). The computations described by Morfey (1979, Fig. 5) give a broadband sound spectrum more than 30dB higher in level than the radiation measured from axisymmetric turbulent jets simulated by the ring vortex method, even though (Acton, 1980) that method gives mean velocity and Reynolds stress distributions in remarkable agreement with flow measurements.

To retain the advantages of the Powell-Howe form we must recognize (3.10) explicitly, and expand the retarded-time argument of $L_i(\mathbf{y}, t - |\mathbf{x} - \mathbf{y}|/a_0)$ to get

$$(p - p_0)(\mathbf{x}, t) = \frac{\rho_0}{4\pi a_0^2 |\mathbf{x}|} \hat{x}_i \hat{x}_j \frac{\partial^2}{\partial t^2} \int y_j L_i \left(\mathbf{y}, t - \frac{|\mathbf{x}|}{a_0} \right) d^3\mathbf{y} , \quad (3.11)$$

which is clearly of quadrupole form, yet has the advantage of being localized on the support of the vorticity distribution.

In this general category – and especially with the possibilities for computation of the time-dependent vorticity by CFD codes in mind – we should include Möhring's (1978) theory of vortex sound. This starts from the radiation integral of the Powell-Howe theory (convolution of the divergence of the Lamb vector with the scalar Green's function $G(\mathbf{x}, t)$ for the wave equation), introduces a vector Green's function $\mathbf{G}(\mathbf{x}, t)$ via $\nabla G = \text{curl } \mathbf{G}$, and then uses the Helmholtz dynamical vorticity law $\partial \boldsymbol{\omega}/\partial t + \text{ curl } \mathbf{L} = 0$ to finally write the radiation explicitly in terms of the vorticity alone; thus

$$h(\mathbf{x}, t) = \rho_0 \frac{\partial}{\partial t} \int \mathbf{G}(\mathbf{x}, t; \mathbf{y}, \tau) \cdot \boldsymbol{\omega}(\mathbf{y}, \tau) d^3\mathbf{y} d\tau . \quad (3.12)$$

This approach has some technical difficulties, among them the fact that in general $\mathbf{G}$ does not exist, because the integrability condition for its existence is $\nabla^2 G = 0$ and is not generally satisfied. However, since G satisfies the wave equation, the integrability condition is "nearly satisfied" at low Mach number (!), and in fact vector Green's functions with the necessary properties have been found by Möhring and others for free space and with the inclusion of certain types of boundary. For the sound of vorticity in free space one has

$$(p - p_0)(\mathbf{x}, t) = \frac{\rho_0}{12\pi a_0^2 |\mathbf{x}|^3} \frac{\partial^3}{\partial t^3} \int (\mathbf{x} \cdot \mathbf{y}) \mathbf{y} \cdot \left[\boldsymbol{\omega}(\mathbf{y}, t - |\mathbf{x}|/a_0) \wedge \mathbf{x} \right] d^3\mathbf{y} \,, \tag{3.13}$$

which should be computable by CFD, provided there is careful treatment of the way in which vorticity escapes (by convection and induction downstream) from the computational domain. This aspect has not received due attention in the West, but is the subject of a nice paper by Fedorchenko (1986). In view of this it may now be worthwhile to calculate vector Green's functions for other flow and geometrical configurations, and in particular to formulate problems of the scattering of near-field pressure fluctuations into far-field sound in terms of the vector Green's functions. Kambe (1986, 1992) has effectively done this for certain simple but important scatterers, including the cylinder and half-plane.

4. Surface Integral Formulations

Suppose we have computations in some domain V bounded by a surface S, with S sufficiently far from the flow for the linear wave equation to hold outside S. Then we can calculate the pressure anywhere outside S, and in the far field in particular, from the Kirchoff-Helmholtz integral

$$p = \int_S \left(p \frac{\partial G}{\partial n} - G \frac{\partial p}{\partial n} \right) dS \,. \tag{4.1}$$

This is the basis of the aeroacoustic scheme proposed by Liepmann (1954), and recently revived and developed (under the name of a "theory of displacement thickness fluctuations") by Howe (1981) for various rather specific problems associated with boundary layer noise. However, the scheme is not free of difficulties. For example, if it is assumed that eddies have a finite correlation volume l^3, then from Lighthill's theory the predicted radiated power from a volume V goes up as V; but if the fluctuations over S have a finite correlation area l^2, then the radiated power from a prescription like (4.1) goes up only as $S \sim V^{2/3}$. Evidently, then, small-scale volume correlations must imply surface correlations of much larger scale, and although retarded-time variations may be negligible for a suitable set of volume sources, they may not be for surface sources. It has not been straightforward to reconcile the predictions of Liepmann's theory (4.1) with those of Lighthill's theory, though a determined and largely successful attempt for jet noise has been made by Laufer, Ffowcs Williams and Childress (1964).

In other applications there may be other constraints to be incorporated into (4.1). Thus, for incompressible inviscid flow over a homogeneous plane boundary we have the Kraichnan (1956) – Phillips (1956) theorem

$$\int_{\text{PLANE}} p(\mathbf{y},t)d^2\mathbf{y} = 0\,, \tag{4.2}$$

or equivalently, for turbulent flow over a plane boundary, the correlation *area* of pressure fluctuations at any frequency must *vanish.* In recent computations of such turbulent flow, intense localized *bursts* of pressure have been seen, as a characteristic and dominant feature of surface pressure fluctuations induced by a turbulent boundary layer. It is tempting to fit a convenient *one-signed* analytical profile to these (a Gaussian, say) and to work out the predicted sound using (4.1) with an "exact" Green's function having $G = 0$ on the boundary. That would not respect (4.2), however, and one must recognize that surrounding a highly localized pressure spike there are regions of pressure fluctuation of the opposite sign which are not conspicuous in the computed pressure results, but are needed to comply with (4.2); see Domaradzki, Shah and Crighton (1992) for discussion of this.

Notwithstanding these pitfalls, it would probably pay the aeroacoustics community to consider schemes based on (4.1) more fully, with a view to determining how data on S can be computed, given that S must be outside "all sources" and that outside S the wave equation must hold.

Some further remarks are in order for the (important) cases in which S is a plane or a cylinder. Then at each frequency ω, (4.1) can be reduced (for a 2D problem to be specific, with $y = 0$ as the plane) to the Fourier integral

$$p(x,y,\omega) = \int_{-\infty}^{+\infty} F(k)\exp\left[ikx + i(k_0^2 - k^2)^{1/2}y\right]dk\,, \tag{4.3}$$

with $\exp(-i\omega t)$ understood and with $k_0 = \omega/a_0$ the acoustic wavenumber at frequency ω. In the far field, with $x = r\cos\theta$, $y = r\sin\theta$, we have

$$p \sim \left(\frac{2k_0\pi}{r}\right)^{1/2} \sin\theta F(k_0\cos\theta)e^{ik_0 r - \pi i/4}, \tag{4.4}$$

with an analogous expression for cylindrical geometry and data specified on a cylinder $r = a$. Equation (4.4) shows the *filtering action* of the acoustic far field. Of all the spectral components $F(k)$ present in the source distribution $f(x)$ on $y = 0$, only the single component $F(k_0\cos\theta)$ reaches the observer at (r, θ) – and that is the component corresponding to a plane acoustic wave propagating in direction θ. The far field thus comprises the elements of $F(k)$ for $|k| < \omega/a_0$ – precisely the "sound wave" parts of $f(x)$. Such parts lie, for low Mach numbers, $M \ll 1$, far in wavenumber from the spectral region $k \sim l^{-1} \sim \omega/u$ where the energy resides and which is well predicted by CFD codes.

In some cases even greater demands may be placed on CFD. Laufer and Yen (1983) conducted experiments on a low Mach number jet flow (mean flow exit Mach number $M_J = 0.1$) and found very unusual features of the acoustic field directivity. Specifically, wave packets corresponding to various ring vortex evolution and merging processes in the jet shear layers were found to give far-field directivity fitted by $\exp(90M_c\cos\theta)$, with $\theta = 0$ on the downstream jet axis and $M_c(\sim 0.6M_J)$ the convection Mach number. Such a variation (called *superdirective* by Crighton and Huerre, 1990) could not arise from any "compact" source (small compared with the acoustic wavelength). But the wave packet on the jet shear layers was found to be fitted by the form $\exp[i\kappa(x - U_c t) - \epsilon^2\kappa^2x^2]$, with different values of wavenumber κ for three different prominent frequencies, with the same convection speed $U_c = a_0M_c$ for all three, and the same envelope width parameter $\epsilon \sim 0.16$ for all three. Thus the largest apparent length in the source is the envelope scale $l \sim (\epsilon\kappa)^{-1}$, while the wavelength of sound is $\lambda \sim 2\pi(M_c\kappa)^{-1} \gg l$, and so it is not clear that such a small source can produce a superdirective field. What matters for the acoustic field is, however, the *integral* scale L, defined not by the envelope width (within which there is substantial cancellation through rapid oscillations $\exp(i\kappa x)$), but rather by $L = \int_{-\infty}^{+\infty} \exp(i\kappa x - \epsilon^2\kappa^2x^2)dx$, which gives $L \sim \kappa^{-1}\epsilon^{-2}$ for $\epsilon \ll 1$. In the case of the Laufer and Yen experiments we find that L and λ are very close, and this explains (Crighton and Huerre, 1990) how an apparently small source can give rise to a superdirective field; the integral scale – which is what matters acoustically – is much larger than the envelope size, and is determined by the tail of the envelope rather than by the main part of the envelope. To get the superdirective field right, computation would have to correctly analyze the wave packet far into the tails, $\kappa x = O(\epsilon^{-2})$, of a packet which is exponentially confined to the much smaller range $\kappa x = O(\epsilon^{-1})$.

Such extreme features are probably exceptional, but they do illustrate the crucial near-total-cancellation aspects of low Mach number aeroacoustics – and they have indeed been clearly seen in experiment. A rather different phenomenon, illustrating the extreme sensitivity of jet flows and their acoustic fields to external forcing, is that of *broadband amplification*, discovered independently by Bechert and Pfizenmaier (1975) and Moore (1977). These authors perturbed round jets ($M_J \sim 0.4$) with very-low-amplitude plane wave forcing generated by a loudspeaker in the settling chamber, and found that even very small levels ($u'/U_J < 0.1\%$) of coherent tonal forcing could amplify the sound field almost uniformly over several octaves of frequency and over almost the whole angular range by up to 8dB. This very dramatic effect has since been detected in experiments on high speed jets, high temperature jets, and with broadband forcing, and there is some evidence for its presence in full-scale jet engines. There is no theory available to indicate the mechanism involved, or to explain the large and uniform magnitude of the effect. For CAA it serves as a warning that perturbations at the boundaries of computational domains may have unexpectedly large effects on the sound field.

5. Asymptotic Matching

The idea that one can match the outer asymptotics, for $|\mathbf{x}| \gg l$, of an inner rotational incompressible core scaled on an eddy size l, to the inner asymptotics, for $|\mathbf{x}| \ll \lambda = lM^{-1}$, of an outer irrotational acoustic field scaled on the wavelength $\lambda = lM^{-1}$, has been put forward by several authors (Möhring, Müller and Obermeier, 1969; Crow, 1970). This would provide an effective means to find the aeroacoustics of numerically-determined low Mach number flows, if it were free of pitfalls, which it is not. Crow gives a particularly illuminating discussion of some of these. First, he makes it clear that there can be only one small parameter M in the problem, and another parameter $\Lambda = l/D$ presents itself, where l is the eddy scale and D a measure of the whole source region, so that Λ^{-3} is the number of eddies present in the field. Even if $\Lambda = O(1)$ – a few eddies – and there are no boundaries present, the problem is not entirely straightforward.

Scale $\mathbf{x}$ on l, t on l/u, and pressure $p - p_0$ on $\rho_0 u^2$. Then we can construct an inner expansion

$$P(\mathbf{X}, t, M) = P_0(\mathbf{X}, t) + MP_1(\mathbf{X}, t) + \ldots, \tag{5.1}$$

where the incompressible pressure P_0 satisfies Poisson's equation

$$\nabla^2_{\mathbf{X}} P_0 = -\frac{\partial^2 V_i V_j}{\partial X_i \partial X_j}, \tag{5.2}$$

with $\mathbf{V}$ the scaled velocity, incompressible and determined entirely by the vorticity. As $\mathbf{X} = \mathbf{x}/l \to \infty$, P_0 is quadrupole,

$$P_0 \sim B_{ij}(t) \frac{\partial^2}{\partial X_i \partial X_j}\left(\frac{1}{|\mathbf{X}|}\right), \tag{5.3}$$

$$B_{ij}(t) = \frac{1}{4\pi} \int (V_i V_j)(\mathbf{Y}, t) d^3\mathbf{Y}\,. \tag{5.4}$$

Now scale $\mathbf{x}$ on the wavelength lM^{-1}, t on l/u again, and $p - p_0$ on $\rho_0 u^2$. Then using lower case symbols we construct an outer wave-field expansion

$$p(\mathbf{x}, t, M) = M^3 p_0(\mathbf{x}, t) + \ldots, \tag{5.5}$$

where M^3 is needed because $P_0 = O(|\mathbf{X}|^{-3})$ as $|\mathbf{X}| \to \infty$. All terms satisfy the homogeneous wave equation, and in a multipole expansion

$$\begin{aligned} p_0 &= \frac{A_0(t - |\mathbf{x}|)}{|\mathbf{x}|} + \frac{\partial}{\partial x_i}\left(\frac{A_i(t - |\mathbf{x}|)}{|\mathbf{x}|}\right) \\ &+ \frac{\partial^2}{\partial x_i \partial x_j}\left(\frac{A_{ij}(t - |\mathbf{x}|)}{|\mathbf{x}|}\right) + \ldots \end{aligned} \tag{5.6}$$

we must stop at the indicated quadrupole A_{ij} and exclude octupole and higher-order multipoles which are too singular, as $|\mathbf{x}| \to 0$, to be matched to (5.3). It is tempting to match (5.6), as $|\mathbf{x}| \to 0$, to (5.3) by taking

$$A_{ij}(t) = B_{ij}(t)\,,\ A_i \equiv 0\,,\ A_0 \equiv 0\,, \tag{5.7}$$

and then we would have the Lighthill result

$$p \sim \frac{M^3}{4\pi|\mathbf{x}|} \frac{\partial^2}{\partial x_i \partial x_j} \int (V_i V_j)(\mathbf{Y}, t - |\mathbf{x}|) d^3\mathbf{Y} \tag{5.8}$$

in the far field, $|\mathbf{x}| \to \infty$.

However, strict asymptotic matching allows us to conclude only that the most singular term in (5.6), namely

$$M^3 \frac{\partial^2}{\partial x_i \partial x_j}\left(\frac{A_{ij}(t)}{|\mathbf{x}|}\right) \tag{5.9}$$

should match the leading term (5.3) from the inner field, so that $A_{ij}(t) = B_{ij}(t)$ is correct, but it cannot be concluded that the dipole A_i and the monopole A_0 vanish. Those less singular wave fields must be determined by matching to a better approximation to the inner field than P_0. Therefore, in the absence of any knowledge of P_1, which is determined by compressibility effects *in the source region* – we cannot determine the leading order wave field. Crow (1970) does in fact determine compressibility corrections to P_0 and shows from higher-order matching that the dipole and monopole coefficients vanish, and that (5.7) holds.

In the presence of boundaries, however, or of material flow inhomogeneities that do generate monopoles and dipoles, the position is not clear. The following simple example makes the point very plainly and should serve as a test case for computational aeroacoustics.

Consider time-harmonic oscillations of a sphere of radius R in the compact limit $\omega R/a_0 = \epsilon \ll 1$. The inner field, scaled on R, is incompressible, and for $|\mathbf{x}| \gg R$ has a dipole form

$$\phi \sim B_1 \frac{\partial}{\partial X_1}\left(\frac{1}{|\mathbf{X}|}\right) \tag{5.10}$$

which is exact for a sphere oscillating along OX_1. For the outer field, matching to a dipole allows only monopoles and dipoles, so that

$$\phi = \epsilon^2 C_0 \left(\frac{\exp ik_0|\mathbf{x}|}{|\mathbf{x}|}\right) + \epsilon^2 C_1 \frac{\partial}{\partial x_1}\left(\frac{\exp ik_0|\mathbf{x}|}{|\mathbf{x}|}\right) + o(\epsilon^2) . \tag{5.11}$$

Matching, with $\mathbf{x} = \epsilon\mathbf{X}$, gives $C_1 = B_1$ and leaves the monopole C_0 to be determined by matching to the second term of the inner expansion; one finds $C_0 = 0$ and the far-field is pure dipole, with directivity $\cos\theta$.

Now hold the sphere fixed, and irradiate it with plane waves $\exp(ik_0x_1)$. To leading order the generation problem (above) and the problem here for the scattered field are formally identical in the inner region, and so

$$\phi \sim D_1 \frac{\partial}{\partial X_1}\left(\frac{1}{|\mathbf{X}|}\right) . \tag{5.12}$$

Matching to an outer wave field

$$\phi = \epsilon^2 E_0 \left(\frac{\exp ik_0|\mathbf{x}|}{|\mathbf{x}|}\right) + \epsilon^2 E_1 \frac{\partial}{\partial x_1}\left(\frac{\exp ik_0|\mathbf{x}|}{|\mathbf{x}|}\right) + o(\epsilon^2) \tag{5.13}$$

gives, as before, $E_1 = D_1$, but one cannot conclude that $E_0 = 0$, and in fact determination of the next term in the inner expansion this time gives $E_0 \neq 0$. There is a non-zero monopole at leading order in the sound field, determined by second-order (compressibility) effects in the source region. The monopole and dipole are strictly comparable, and the far-field directivity is

$$\phi \propto \epsilon^2(2 - 3\cos\theta) . \tag{5.14}$$

This must serve as a warning to the effect that – except in simple cases where it is known, for example, that there are only quadrupoles at leading order – asymptotic matching of the results of a strictly incompressible calculation to a wave field may not be as straightforward as one would hope.

If $\Lambda \neq O(1)$ but is large, say $\Lambda = O(M^{-1})$ or larger, as could easily be the case in many applications (certainly in astrophysics, but also in many terrestrial cases), then as Crow (1970) points out, many new mechanisms may need to be incorporated explicitly. These include the scattering of sound of one eddy by the many other eddies through which the sound must pass, the stretching of vorticity by the aerodynamically generated sound waves, and even steady and unsteady effects associated with cumulative nonlinear distortion of the sound waves themselves. Such processes have hardly begun to be studied, and the extent to which they are taken into account by a Lighthill integral

$$(p - p_0)(\mathbf{x}, t) = \frac{1}{4\pi|\mathbf{x}|}\frac{\partial^2}{\partial x_i \partial x_j}\int T_{ij}\left(\mathbf{y}, t - \frac{|\mathbf{x}-\mathbf{y}|}{a_0}\right) d^3\mathbf{y} \tag{5.15}$$

remains an open question.

6. Conclusions

Although this talk has emphasized possible difficulties for computational aeroacoustics (CAA), several positive points must be made. First, the way is now open to calculation of relatively simple low Mach number aeroacoustic fields through the evaluation of radiation integrals with integrands specified by CFD. For free-space flows one has any of the prescriptions (3.8), (3.11) and (3.13), though no attempt seems to have yet been made to assess the relative merits of these three. There are similar prescriptions, using various forms of exact Green's function, for aeroacoustic sources (vorticity) in the vicinity of a circular cylinder, the edge of a half-plane, and a circular hole in a plane. Second, for simple ring vortex flows these radiation integrals have been evaluated analytically and compared with a long series of experiments by Kambe and his co-workers. The agreement between theory and experiment is quite remarkable, absolute levels and directivities being predicted and measured with differences of less than 1dB over most of the range of the variables; for reviews of some of this work, see Kambe (1986, 1992). In Kambe (1992) there is even evidence for the detection of a sound field from the process of vortex line re-connection following the oblique impact of two vortex rings. This sound field is nominally of octupole type and would be expected to be insignificant at the low Mach numbers ($M \sim 0.1$) of the experiment. However, large accelerations in the re-connection process lead to a numerically large octupole coefficient, and the octupole is clearly distinguishable from the quadrupole which does not reflect changes in the vortex line topology.

Whether the matched asymptotic expansion has anything to offer remains to be seen. Certainly, it requires computation over a volume significantly larger than the region containing vorticity, though not as large as the wavelength scale needed for direct computation of the sound along with the flow. But if an appropriate integral expression is unavailable, matched expressions (in which the wave field to be matched to the inner computationally-determined flow is guessed analytically, up to constants to be determined by matching) may offer an advantage over full direct computation of flow and sound.

In any event, there are now a number of *canonical* aeroacoustics problems which have been analyzed from several points of view, including the experimental studies noted above, and which should now serve as test cases for all possible CAA schemes including, of course, direct computation of the whole flow and acoustic field. These problems include the collision of line and ring vortices, oblique collision of ring vortices, interaction of line and ring vortices with cylinders and half-planes, and scattering by sound-vortex interaction.

References

Acton, E., 1980. "A modelling of large eddies in an axisymmetric jet," *J. Fluid Mech.* **98**, pp. 1–31.

Bechert, D. W. and Pfizenmaier, E., 1975. "On the amplification of broad band jet noise by a pure tone excitation," *J. Sound Vib.* **43**, pp. 581–587.

Crighton, D. G., 1988. "Goals for computational aeronautics," In *Computational Acoustics: Algorithms and Applications* (eds. D. Lee, R. L. Sternberg, M. H. Schultz), Elsevier-North Holland, pp. 3–20.

Crighton, D. G. and Huerre, P., 1990. "Shear layer pressure fluctuations and superdirective acoustic sources," *J. Fluid Mech.* **220**, pp. 355–368.

Crow, S. C., 1970. "Aerodynamic sound emission as a singular perturbation problem," *Stud. Appl. Math.* **49**, pp. 21–44.

Domaradzki, J. A., Shah, K., and Crighton, D. G., 1992. "Scattering of hydrodynamic pressure pulses by coating inhomogeneities," *J. Acoust. Soc. Am.* (in press).

Fedorchenko, A. T., 1986. "On vortex outflow through the permeable boundary of the computational domain of non-stationary subsonic flow," Zh. Vychisl. Mat. i Mat. Fiz. (USSR Computational Maths. Math. Phys.) **26** (1), pp. 114–129.

Ffowcs Williams, J. E., 1963. "The noise from turbulence convected at high speed," *Phil. Trans. R. Soc. Lond. A.* **255**, pp. 469–502.

Howe, M. S., 1975. "Contributions to the theory of aerodynamic sound, with application to excess jet noise and the theory of the flute," *J. Fluid Mech.* **71**, pp. 625–673.

Howe, M. S., 1981. "The role of displacement thickness fluctuations in hydroacoustics, and the jet-drive mechanism of the flue organ pipe," *Proc. R. Soc. Lond. A.* **374**, pp. 543–568.

Kambe, T., 1986. "Acoustic emission by vortex motions," *J. Fluid Mech.* **173**, pp. 643–666.

Kambe, T., 1992. "Observed and computed waves of aerodynamic sound," Proc. XVIII Intl. Cong. Theoretical and Appl. Mech. Haifa.

Kraichnan, R., 1956. "Pressure fluctuations in turbulent flow over a flat plate," *J. Acoust. Soc. Am.* **28**, pp. 378–390.

Laufer, J., Ffowcs Williams, J. E., and Childress, S., 1964. "Mechanism of noise generation in the turbulent boundary layer," *AGARDograph* **90** (NATO).

Laufer, J. and Yen, T. C., 1983. "Noise generation by a low-Mach-number jet," *J. Fluid Mech.* **134**, pp. 1–31.

Liepmann, H. W., 1954. "On acoustic radiation from boundary layers and jets," GALCIT Report, Calif. Inst. of Tech.

Lighthill, M. J., 1952. "On sound generated aerodynamically. I. General theory," *Proc. R. Soc. Lond. A.* **211**, pp. 564–587.

Mani, R., 1976. "The influence of jet flow on jet noise. Part I. The noise of unheated jets," *J. Fluid Mech.* **73**, pp. 753–778.

Möhring, W. F., 1978. "On vortex sound at low Mach number," *J. Fluid Mech.* **85**, pp. 685–691.

Möhring, W. F., Müller, E.-A., and Obermeier, F., 1969. "Schallerzeugung durch instationäre Strömung als singuläres Störungsproblem," *Acustica* **21**, pp. 184–188.

Moore, C. J., 1977. "The role of shear layer instabilities in jet exhaust noise," *J. Fluid Mech.* **80**, pp. 321–367.

Morfey, C. L., 1979. "Sound radiated from a simulated jet flow," In *Mechanics of Sound Generation in Flows*, ed. E.-A. Müller, Springer-Verlag, New York, pp. 12–18.

Phillips, O. M., 1956. "On the aerodynamic surface sound from a plane turbulent boundary layer," *Proc. R. Soc. Lond. A.* **234**, pp. 327–335.

Powell, A., 1964. "Theory of vortex sound," *J. Acoust. Soc. Am.* **36**, pp. 177–195.

Ribner, H. S., 1962. "Aerodynamic sound from fluid dilatations," *U.T.I.A. Rep.* **86** (AFOSR TN3430) (see also *Advances in Applied Mechanics* **8**, pp. 103–182, 1964).

A MULTIPLE SCALES APPROACH TO SOUND GENERATION BY VIBRATING BODIES

James F. Geer

Watson School of Engineering and Applied Science
State University of New York
Binghamton, New York 13902

Dennis S. Pope

Lockheed Engineering and Sciences Company
144 Research Drive
Hampton, Virginia 23666

ABSTRACT

The problem of determining the acoustic field in an inviscid, isentropic fluid generated by a solid body whose surface executes prescribed vibrations is formulated and solved as a multiple scales perturbation problem, using the Mach number M based on the surface velocity as the perturbation parameter. Following the idea of multiple scales, new "slow" spacial scales are introduced, which are defined as the usual physical spacial scale multiplied by various powers of M. The governing nonlinear differential equations lead to a sequence of linear problems for the perturbation coefficient functions. However, it is shown that the higher order perturbation functions obtained in this manner will dominate the lower order solutions unless their dependence on the slow spacial scales is chosen in a certain manner. In particular, it is shown that the perturbation functions must satisfy an equation similar to Burgers' equation, with a slow spacial scale playing the role of the time-like variable. The method is illustrated by a simple one-dimensional example, as well as by the pulsating sphere. The results are compared with solutions obtained by purely numerical methods and some insights provided by the perturbation approach are discussed.

1. Introduction

We wish to describe the acoustic field in an inviscid, isentropic fluid generated by a solid body whose surface executes prescribed vibrations. When the Mach number M based on the maximum surface velocity is small, it is natural to consider a perturbation approach to the problem. Crow (1970) carefully analyzed a general class of problems of this type and showed how it can be treated as a singular perturbation problem by the method of inner and outer asymptotic expansions. Whitham (1956, 1974) examined this problem in the high frequency limit by using a perturbation approach similar to Lighthill's (1949, 1961) method of strained coordinates. He essentially "strained" the time coordinate and obtained expressions for the acoustic field valid in the far field for large frequency excitations. The approach we shall use is not restricted by the high frequency assumption, although our results do reduce to Whitham's results (to leading order in M) in the far field and at high frequencies. In addition, the analysis we shall present, being based on multiple *spacial scales*, appears to provide more insight into the *effects* of the nonlinearity on the form of the solution, as well as insights into the interaction of the value of the Mach number and the angular dependence of the solution (see Section 6).

In Section 2 we formulate the mathematical problem we wish to solve and then outline the multiple scales approach in Section 3. We apply our method to a simple one-dimensional example in Section 4 and then to the problem of the pulsating sphere in Section 5. We discuss our results in Section 6. Due to space limitations, many of the details of the applications of our method, especially to the three-dimensional case, are not presented here. These details, along with other three-dimensional examples, will be presented and discussed elsewhere.

2. Problem Formulation

In terms of a Cartesian coordinate system (x_1, x_2, x_3), with the origin fixed at some convenient location inside the body, the equations of conservation of mass and momentum become

$$\partial\rho/\partial t + \vec{\nabla}\cdot(\rho\,\vec{u}) = 0, \tag{2.1}$$

$$\partial p/\partial x_i + \rho\left(\partial u_i/\partial t + (\vec{u}\cdot\vec{\nabla})u_i\right) = 0,\ i = 1,2,3. \tag{2.2}$$

Here ρ, u_i, p, and t represent the fluid density, the fluid velocity component in the positive x_i direction, the fluid pressure, and the time coordinate, respectively. Also, $\vec{u} = u_1\,\vec{i}_1 + u_2\,\vec{i}_2 + u_3\,\vec{i}_3$ and

$\vec{\nabla} = \vec{i}_1 (\partial/\partial x_1) + \vec{i}_2 (\partial/\partial x_2) + \vec{i}_3 (\partial/\partial x_3)$, where $\vec{i}_j$ is a unit vector in the positive x_j direction. We also let $p = k\rho^\gamma$ and define

$$c^2(\rho) = dp/d\rho = k\gamma(\rho)^{\gamma-1}. \tag{2.3}$$

We let ρ_0, U, and L be typical (constant) values for the density, velocity, and length scales, respectively, associated with the flow, and define nondimensional variables (denoted by a "^" above the quantity) by $\hat{\rho} = \rho/\rho_0$, $\hat{\vec{u}} = \vec{u}/U$, $\hat{p} = (p - p_0)/(\rho_0 U^2)$, $\hat{x}_j = x_j/L (j = 1, 2, 3)$, and $\hat{t} = t/(L/c_0)$, where $p_0 = k\rho_0^\gamma$ and $c_0 = c(\rho_0)$. We now rewrite Equations (2.1) and (2.2) in terms of these variables and then omit the "^" above the various quantities to obtain

$$\partial\rho/\partial t + M \vec{\nabla} \cdot (\rho \vec{u}) = 0, \tag{2.4}$$

$$\partial p/\partial x_i + \rho^{2-\gamma}(M\partial u_i/\partial t + M^2(\vec{u} \cdot \vec{\nabla})u_i) = 0, \; i = 1, 2, 3, \tag{2.5}$$

where $M = U/c_0$ is the Mach number of the flow.

We now seek approximate solutions to Equations (2.4) and (2.5), subject to appropriate boundary and initial conditions, which will be formally valid for small values of M.

3. A Multiple Scales Perturbation Solution Method

Following the method of multiple scales (see e.g. Nayfeh (1973)), we introduce the spacial scales (variables) $\vec{x}^{(k)}$, $k = 0, 1, 2, \ldots$, related to $\vec{x}$ by

$$\vec{x}^{(k)} = M^k \vec{x}, \; k = 0, 1, 2, \ldots . \tag{3.1}$$

Thus $\vec{x}^{(0)} = \vec{x}$, $\vec{x}^{(1)} = M \vec{x}$, $\vec{x}^{(2)} = M^2 \vec{x}$, etc. We then assume that $\vec{u}$ and ρ are functions of these scales, as well as functions of t and M, i.e. $\rho = \rho(t, \vec{x}^{(0)}, \vec{x}^{(1)}, \vec{x}^{(2)}, \ldots, M)$ and $\vec{u} = \vec{u}(t, \vec{x}^{(0)}, \vec{x}^{(1)}, \vec{x}^{(2)}, \ldots, M)$.

The method of multiple scales now treats the scales $\{x_i^{(k)}\}$ as *independent* variables and seeks to determine ρ and $\vec{u}$ as functions of these variables. In particular, for "small" values of M, we look for solutions in the form

$$\rho = 1 + M\rho^{(1)} + M^2\rho^{(2)} + \ldots, \; \vec{u} = \vec{u}^{(0)} + M \vec{u}^{(1)} + M^2 \vec{u}^{(2)} + \ldots . \tag{3.2}$$

Here each of the coefficient functions $\rho^{(j)}$ and $\vec{u}^{(j)}$ is independent of M, but may depend upon t and the scales $x_i^{(k)}$, i.e. $\rho^{(j)} = \rho^{(j)}(t, \vec{x}^{(0)}, \vec{x}^{(1)}, \ldots)$, and $\vec{u}^{(j)} = \vec{u}^{(j)}(t, \vec{x}^{(0)}, \vec{x}^{(1)}, \ldots)$.

To determine these coefficient functions, we substitute the expansions (3.2) into Equations (2.4)-(2.5) and use the relation

$$\vec{\nabla} = \vec{\nabla}^{(0)} + M\, \vec{\nabla}^{(1)} + M^2 \nabla^{(2)} + \ldots, \tag{3.3}$$

where $\vec{\nabla}^{(k)} \equiv \vec{i}_1 (\partial/\partial x_1^{(k)}) + \vec{i}_2 (\partial/\partial x_2^{(k)}) + \vec{i}_3 (\partial/\partial x_3^{(k)})$, to express the left side of each equation as a power series in M. We then equate to zero the coefficient of each power of M, since the right side of each equation is zero. In this way, we find that the perturbation coefficient functions satisfy the system of equations

$$\partial \rho^{(1)}/\partial t + \vec{\nabla}^{(0)} \cdot \vec{u}^{(0)} = 0 \tag{3.4a}$$

$$\partial \rho^{(1)}/\partial x_i^{(0)} + \partial u_i^{(0)}/\partial t = 0, \; i = 1, 2, 3, \tag{3.4b}$$

and

$$\partial \rho^{(k+1)}/\partial t + \vec{\nabla}^{(0)} \cdot \vec{u}^{(k)} = F^{(k)} \tag{3.5a}$$

$$\partial \rho^{(k+1)}/\partial x_i^{(0)} + \partial u_i^{(k)}/\partial t = G_i^{(k)}, \; i = 1, 2, 3, \text{ for } k \geq 1. \tag{3.5b}$$

Here the functions $F^{(k)}$ and $G_i^{(k)}$ depend only upon $\rho^{(j)}$ with $j < k+1$ and $\vec{u}^{(j)}$ with $j < k$. In particular, we find

$$F^{(1)} = -(\vec{\nabla}^{(0)} \cdot (\rho^{(1)}\, \vec{u}^{(0)}) + \vec{\nabla}^{(1)} \cdot \vec{u}^{(0)}), \tag{3.6a}$$

$$G_i^{(1)} = -((\vec{u}^{(0)} \cdot \vec{\nabla}^{(0)}) u_i^{(0)} + (2-\gamma)\rho^{(1)} \partial u_i^{(0)}/\partial t + \partial \rho^{(1)}/\partial x_i^{(1)}). \tag{3.6b}$$

The general solution to Equations (3.4) can be expressed as

$$\rho^{(1)} = -\rho \phi/\partial t \text{ and } \vec{u}^{(0)} = \vec{\nabla}^{(0)} \phi, \tag{3.7}$$

where ϕ is a solution of the linear wave equation

$$\partial^2 \phi/\partial t^2 = ((\partial/\partial x_1^{(0)})^2 + (\partial/\partial x_2^{(0)})^2 + (\partial/\partial x_3^{(0)})^2)\phi. \tag{3.8}$$

From the structure of the system of Equations (3.5), we see that these equations can (in principle) be solved recursively, starting with

$k = 1$. In particular, (3.5) is a system of *linear* equations for the unknowns $\rho^{(k+1)}$ and $\vec{u}^{(k)}$. Consequently, we can express the solution as the superposition of a particular solution and a homogeneous solution. The homogeneous solution has the same form as Equations (3.7) and (3.8). The particular solution will, of course, depend on the index k, as well as on the solutions for the lower order perturbation coefficients. In general, these particular solutions have the property (as we shall demonstrate explicitly in the following sections) that they tend to *decay more slowly in magnitude as* $|\vec{x}| \to \infty$ *than* $\rho^{(1)}$ *and* $\vec{u}^{(0)}$. Consequently, if this decay were to be left unchecked, the perturbation expansions (3.2) would become invalid as $|\vec{x}| \to \infty$. However, as we shall show, it is possible to satisfy the requirement that $\rho^{(k+1)}$ and $\vec{u}^{(k)}$ should *decay at least as fast as* $\rho^{(1)}$ and $\vec{u}^{(0)}$ as $|\vec{x}| \to \infty$ by properly choosing the dependence of $\rho^{(k+1)}$ and $\vec{u}^{(k)}$ on the "slower" spacial scales $\vec{x}^{(1)}$, $\vec{x}^{(2)}$, etc. Thus, this method will yield an expansion (3.2) which will be uniformly valid as $|\vec{x}| \to \infty$ and, consequently, will more accurately represent the far field behavior of the acoustical waves.

In the following section we shall demonstrate some of these ideas with a simple, one-dimensional example, and then proceed to a three dimensional problem in Section 5.

4. A One-Dimensional Example

To illustrate the general ideas of our approach, we consider first a flow which varies in only one spacial coordinate. We let this one spacial coordinate be denoted by $x_1 = x$. Then, Equations (2.4)-(2.5) become

$$\partial\rho/\partial t + M(\partial/\partial x)(\rho u) = 0, \tag{4.1}$$

$$\partial\rho/\partial x + \rho^{2-\gamma}(M\partial u/\partial t + M^2 u(\partial u/\partial x)) = 0. \tag{4.2}$$

We shall assume that u is a specified function, say $\tilde{f}(t)$, at $x = 0$ for all $t \geq 0$, and that both u and $\rho - 1$ are specified to be zero at $t = 0$ for all $0 < x < \infty$.

Following the method outlined above, we define the spacial scales $x^{(i)} = M^i x$, $i = 0, 1, 2, \ldots$. We also find it convenient to define the differential operators D_i by

$$D_i \equiv \partial/\partial x^{(i)}, \quad i = 0, 1, \ldots. \tag{4.3}$$

We then look for solutions for ρ and u in the form of (3.2), where $\rho^{(1)}$ and $u^{(0)}$ satisfy Equations (3.4). Then the general solution to Equation (3.8) is given by $\phi = F(t-x) + G(t+x)$, where F and G are arbitrary functions of their arguments and may also depend on the spacial scales $x^{(1)}, x^{(2)}, \ldots$. Since we shall restrict our attention to solutions which propagate only in the positive x direction, we set $G \equiv 0$ and write

$$\rho^{(1)} = u^{(0)} = f(t-x, x^{(1)}, x^{(2)}, \ldots). \tag{4.4}$$

The exact form of the function f will be determined from the boundary and initial conditions of the problem.

Using the solutions (4.4), the solution to equations (3.5) with $k = 1$ for $\rho^{(2)}$ and $u^{(1)}$ becomes

$$\begin{aligned}\rho^{(2)} &= -x(D_1 f + ((1+\gamma)/4)D_0(f^2)) + f_1(t-x, x^{(1)}, x^{(2)}, \ldots),\\ u^{(1)} &= -x(D_1 f + ((1+\gamma)/4)D_0(f^2))\\ &\quad +((\gamma-3)/4)(f^2) + f_1(t-x, x^{(1)}, x^{(2)}, \ldots).\end{aligned} \tag{4.5}$$

In (4.5), f_1 is an arbitrary function of its arguments, which will eventually be determined by the initial and boundary conditions of the problem.

In order for $\rho^{(2)}$ and $u^{(1)}$ to grow no faster that $\rho^{(1)}$ and $u^{(0)}$, respectively, as x becomes large, we must require that the term in brackets in these equations vanishes, i.e.

$$D_1 f + ((1+\gamma)/4)D_0(f^2) = 0. \tag{4.6}$$

Condition (4.6) is an equation to determine the "$x^{(1)}$ behavior" of f and, hence, the $x^{(1)}$ behavior of both $\rho^{(1)}$ and $u^{(0)}$. Equation (4.6) can be expressed in terms of Burgers' equation, with $((1+\gamma)/4)x^{(1)}$ playing the role of the "time" variable and $x^{(0)} = x$ playing the role of the spacial variable. In particular, the solution to (4.6) which satisfies the "initial" condition that $f = \tilde{f}(t-x)$ when $x^{(1)} = 0$ is

$$f = \tilde{f}(\tau), \text{ where } \tau = t - x + ((\gamma+1)/2)x^{(1)}\tilde{f}(\tau). \tag{4.7}$$

For this one-dimensional case, the solution (4.7) is the same (to leading order in M) as that obtained by Whitham (1956), based on mostly physical arguments, and by Whitham (1974), using both a strained coordinates approach and a modified characteristic approach.

As an application of these results, we consider flow in a semi-infinite tube, which is generated by the velocity of the end of the tube at $x = 0$ being prescribed as $u = \sin(t)$ for $t \geq 0$. Then the initial and boundary conditions for the flow become

$$\left.\begin{array}{l} u = 0 \\ \\ \rho = 1 \end{array}\right\} \text{at } t = 0, \text{ for all } x \geq 0 \text{ (initial conditions)}, \tag{4.8}$$

$$u = \sin(t), \text{at } x = 0, \text{ for all } t > 0 \text{ (boundary condition)}. \tag{4.9}$$

From Equations (4.4), we find $u^{(0)} = \rho^{(1)} = f(t - x, x^{(1)}, x^{(2)}, ...)$. If we were to terminate our perturbation expansion at this point and treat $x^{(i)}$ for $i \geq 1$ as constants (in particular, if we were to set them all equal to zero), then the conditions (4.8)-(4.9) yield the solution

$$u^{(0)} \equiv \rho^{(1)} = \begin{cases} \sin(t - x), & \text{for } 0 \leq x \leq t, \\ \\ 0, & \text{for } x < t. \end{cases} \tag{4.10}$$

This, of course, is just the classical wave solution of linear acoustics. However, if we continue the perturbation solution outlined above and use the solution (4.7), we find

$$u^{(0)} = \rho^{(1)} = \begin{cases} \sin(\tau), & \text{for } 0 \leq x \leq t, \\ \\ 0, & \text{for } x > t, \end{cases} \tag{4.11a}$$

where

$$\tau = t - x + ((\gamma + 1)/2)x^{(1)} \sin(\tau). \tag{4.11b}$$

With $\rho^{(1)}$ determined from (4.11), we use (4.5) and the conditions (4.8)-(4.9) to write

$$\begin{aligned} \rho &= 1 + M \sin(\tau) + M^2((3 - \gamma)/4) \sin^2(t - x) + O(M^3) \\ &\quad + O(M^4 x), \\ u &= \sin(\tau) + M((\gamma - 3)/4)(\sin^2(\tau) - \sin^2(t - x)) \\ &\quad + O(M^2) + O(M^3 x), \end{aligned} \tag{4.12}$$

for $0 \leq x \leq t$, and $\rho = 1$ and $u = 0$ for $x > t$. Here τ is defined by Equation (4.11b).

To illustrate these results, in Figure 1 we have plotted approximations to $(\rho - 1)$ as a function of distance from the source at a fixed time. These approximations were obtained using the first three terms on the right side of the first equation in (4.12). In Fig. 1(a), $M = 0.01$ and the wave is nearly linear, while in Fig. 1(b) M has been increased to $M = 0.04$. For this case, the gradual steepening of the wave as x increases is evident, with the wave approaching an "N-wave" as x increases. In these figures we have also plotted the linearized solution based on (4.10) and a solution to Equations (4.1)-(4.2) obtained by purely numerical means. The numerical solution was obtained by writing these equations in characteristic form and then using a MacCormack predictor-corrector technique. (We thank Dr. Willie R. Watson of the NASA Langley Research Center for carrying out these computations for us.) As the figures illustrate, there is very good agreement between the multiple scales solution and the numerical solution. We shall comment further on these results in Section 6.

5. Three-Dimensional Flow - The Pulsating Sphere

We now wish to describe the acoustic field outside a spherical region, when the normal component of the fluid velocity is specified on the surface of the sphere. We let the origin of a nondimensional Cartesian coordinate system (x, y, z) coincide with the average center of the sphere, whose average radius is a constant a, and then introduce spherical coordinates (r, θ, ψ) related to (x, y, z) by $x = r\sin(\theta)\cos(\psi)$, $y = r\sin(\theta)\sin(\psi)$, $z = r\cos(\theta)$. (Here all lengths have been nondimensionalized by using a as the typical length scale.)

Following the method of multiple scales, we introduce the spacial scales r_i, $i = 0, 1, 2, ...$, related to r by

$$r_0 = r,\ r_i = M^i r,\ i = 1, 2, \dots . \tag{5.1}$$

Thus $r_1 = Mr$, $r_2 = M^2 r$, etc. Then, for "small" values of M, we look for solutions for the density and velocity in the form

$$\rho = 1 + M\rho^{(1)} + M^2\rho^{(2)} + ...,\ u_r = u_r^{(0)} + Mu_r^{(1)} + M^2 u_r^{(2)} + ... \tag{5.2}$$

with analogous expressions holding for u_θ and u_ψ. (Here u_r, u_θ, and u_ψ are the velocity components in the directions of increasing r, θ, and ψ, respectively.) In (5.2) each of the coefficient functions $\rho^{(j)}, u_r^{(j)}$, etc., is independent of M, but may depend upon t and the spacial scales r_i, i.e. $\rho^{(j)} = \rho^{(j)}(t, \theta, \psi, r_0, r_1, r_2, ...)$, $u^{(j)} = u_r^{(j)}(t, \theta, \psi, r_0, r_1, r_2, ...), ...$. To determine these functions, we substitute the expansions (5.2) into Equations (2.4)-(2.5) and use the relation

$$\partial/\partial_r = D_0 + MD_1 + M^2 D_2 + ..., \text{ where } D_j \equiv \partial/\partial r_j. \tag{5.3}$$

We then collect coefficients of like powers of M and equate to zero the coefficient of each power of M, since the right side of each equation is zero. In this way, we are led to a system of linear equations satisfied by the coefficient functions $\rho^{(j)}, u_r^{(j)}, u_\theta^{(j)}$, and $u_\psi^{(j)}$, similar in structure to Equations (3.4)-(3.5). Although the derivation and detailed form of the solutions to these equations will be presented elsewhere, a key result is the equation (corresponding to Equation (4.5)) which describes the r_1 behavior of the solution. This equation and its solution can be expressed as

$$D_1(g)+((\gamma+1)/4)P_n^m(\cos(\theta))\cos(m\psi-\beta)(1/r^2)D_0(r^2g^2) = 0, \tag{5.4}$$

where $g = f^{(n+1)}(\tau)/r$ and τ is defined implicitly by

$$\tau = t - r + 1 + ((\gamma+1)/2)F_n(r_1, \tau)MP_n^m(\cos(\theta))\cos(m\psi - \beta),$$

$$F_n(r_1, \tau) = f^{(n+1)}(\tau)\log(r_1/M) + \sum_{j=2}^{n+1} a_{n,j} f^{(n+2-j)}(\tau)((M/r_1)^{j-1} - 1). \tag{5.5}$$

Here P_n^m is the associated Legendre polynomial, n and m are nonnegative integers, β is an arbitrary constant, $\{a_{n,j}\}$ are certain well-defined constants, and f is an arbitrary smooth function of its argument. This result (especially Equations (5.5)) is a generalization of Whitham's (1974) result. We shall discuss some of its implications in the next section.

As an application of the results of this approach, we consider the pulsating sphere. In this case, the (nondimensional) radial component of the velocity of the surface of the sphere (i.e. on $r = 1$) is given by $V(t) = \sin(\omega t)$. Then we set $n = 0$ and $m = 0$ in the expressions above and find that the leading terms in the solution (5.2) are given by

$$
\begin{aligned}
u^{(0)} &= ((r-1)\omega/((1+\omega^2)r^2))\cos(\omega t) + (1/(1+\omega^2))(\omega^2/r \\
&\quad +1/r^2)\sin(\omega\tau), \\
\rho^{(1)} &= (\omega/((1+\omega^2)r))(\cos(\omega\tau) + \omega\sin(\omega\tau)), \\
u_r^{(1)} &= (b_2/r^2 + 2b_3\omega/r)\cos(2\omega\tau) + (b_3/r^2 - 2b_2\omega/r)\sin(2\omega\tau) \\
&\quad +((29-3\gamma)\omega^2/(16r^3(1+\omega^2)^2) + \omega^2(\omega^2-1)/(r^2(1 \\
&\quad +\omega^2)^2))\cos(2\omega\tau) \\
&\quad -((1+\gamma)\omega^2\log(r))/(4r^2(1+\omega^2)) \\
&\quad +((29-3\gamma)\omega(\omega^2-1)/(32r^3(1+\omega^2)^2) - 2\omega^3/(r^2(1 \\
&\quad +\omega^2)^2))\sin(2\omega\tau) \\
&\quad +0(1/r^4), \\
\rho^{(2)} &= 2b_3\omega\cos(2\omega\tau)/r - 2b_2\omega\sin(2\omega\tau)/r \\
&\quad -1/(4r^4(1+\omega^2)) + (1-\gamma)\omega^2/(4r^2(1+\omega^2)) \\
&\quad +((17+\gamma)\omega^2/(16r^3(1+\omega^2)^2) - (13-3\gamma)(\omega^2 \\
&\quad -1)/(64r^4(1+\omega^2)^2) \\
&\qquad +(5+\gamma)\omega^2(\omega^2-1)/(8r^2(1+\omega^2)^2))\cos(2\omega\tau) \\
&\quad ((13-3\gamma)\omega/(32r^4(1+\omega^2)^2) - (5+\gamma)\omega^3/(4r^2(1+\omega^2)^2) \\
&\qquad +(17+\gamma)\omega(\omega^2-1)/(32r^3(1+\omega^2)^2))\sin(2\omega\tau) \\
&\quad +0(1/r^5),
\end{aligned}
$$

where

$$b_2 = -3\omega^2(14 - 2\gamma + 17\omega^2 + \gamma\omega^2)/(16(1+\omega^2)^2(1+4\omega^2)),$$

$$b_3 = \omega(29 - 3\gamma - 17\omega^2 + 15\gamma\omega^2 - 64\omega^4)/(32(1+\omega^2)^2(1+4\omega^2)).$$

Here τ is defined implicitly by (5.5) with $n = 0$, where

$$f(\tau) = (1+\omega^2)^{-1}(\sin(\omega\tau) - \omega\cos(\omega\tau)).$$

In Figure 2, we have plotted $r(\rho^{(1)} + M\rho^{(2)})(\approx r(\rho - 1)/M)$ as a function of r at a fixed value of t for $\omega = 1.25$ and for two different values of M. We have also plotted Whitham's (1974) first-order approximation to $r(\rho - 1)/M$, as well as an approximation obtained by purely numerical methods. This numerical solution was obtained by first using a special decomposition of the solution to Equations (2.4)-(2.5) and then using a MacCormack predictor-corrector scheme on the resulting equations. As Figure 2 illustrates, there is very good agreement between the perturbation and numerical solutions, even for Mach numbers of 0.2 and 0.3, where a gradual steepening and asymmetry of the wave is apparent.

6. Discussion

We note first that the multiple scales perturbation approach has allowed us to "capture" many of the salient nonlinear features of acoustic wave motion. For example, as the Mach number increases, the nonlinear characteristics of the gradual steepening of the waves, the asymmetries due to "second harmonic" terms, as well as the convergence of the wave to an "N-wave" profile, are all evident in the examples we have considered. These phenomena, of course, are not predicted by linear theory. Thus, it appears that the method has allowed us to obtain approximate analytical solutions which are valid for (much) larger values of M than are the classical linear acoustic solutions. In addition, the solutions assume a rather simple form, being essentially of the same form as the classical linear solutions, but with a different argument. That is, the classical retarded time $t + 1 - r$ in the linear solution has simply been replaced by τ, where τ is defined implicitly. Other perturbation approaches to this problem have resulted in more involved expressions for the quantities of physical interest (cp., e.g., Crow (1970)).

The general form of the equation for τ also provides an insight into the interplay between the effect of the Mach number and the angular variation of the acoustical source on the form of the acoustic wave. For example, if $M = 0$ or if the term $P_n^m(\cos(\theta))\cos(m\psi - \beta)$ is zero (corresponding, say, to a particular value of θ or ψ), then from (5.5) we see that $\tau = t - r + 1$ and hence our solution reduces just to the linear acoustic solution. However, if $M \neq 0$, but the angular term $P_n^m(\cos(\theta))\cos(m\psi - \beta)$ is *zero*, then our solution again reduces to the linear solution and hence the nonzero Mach number has essentially no effect on the form of the solution. These observations motivate us to define an "effective" Mach number $M_{\text{eff.}}$ by

$$M_{\text{eff.}} \equiv M P_n^m(\cos(\theta))\cos(m\psi - \beta).$$

Thus, it is the (angularly dependent) effective Mach number $M_{\text{eff.}}$, and not just M, which determines the magnitude of departure of the wave from its linear form. This phenomena is illustrated in Figure 3 for the example of the oscillating sphere, for which $n = 1$ and $m = 0$ in the formulas above. (The details of this and other three-dimensional examples will be presented elsewhere.) In particular, the linear form of the wave is clearly seen, even for nonzero Mach numbers, when $\cos(\theta)$ assumes a value such that $M_{\text{eff.}}$ is zero.

In comparing our results with Whitham's (1974) results, we note that for spherically symmetric disturbances (for which $n = 0$) our Equations (5.5) reduce to his results. However, in more general problems (for which $n > 0$), our results differ from Whitham's equations, except in the far field, where the log term in the expression for F_n dominates the remaining terms. Thus, in the far field, Equations (5.5) reduce to

$$\tau = t - r + 1 + ((\gamma+1)/2) f^{(n+1)}(\tau)\log(r_1/M) M P_n^m(\cos(\theta))\cos(m\psi - \beta),$$

which is equivalent to Whitham's equation for τ when $n = 0$.

As a final general observation, we note that the analysis presented here may provide some insight into the derivation of appropriate non-reflecting boundary conditions to be used at artificial computational boundaries for a purely numerical simulation of acoustical waves generated by general (arbitrary) sources. For example, the rather simple form of our final results allows us to express the far field behavior of the waves in a concise form. This expression, when used with some of the ideas of Bayliss and Turkel (1980), for example, may allow us to derive the required boundary conditions. Investigations along these lines are continuing.

References

Bayliss, A., and Turkel, E., 1980. "Radiation boundary conditions for wave-like equations," *Comm. Pure Appl. Math.* **33**, pp. 707-726.

Crow, S. C., 1970. "Aerodynamic sound emission as a singular perturbation problem," *Studies in Appl. Math.* **49**, pp. 21-44.

Lighthill, M. J., 1949. "A technique for rendering approximate solutions to physical problems uniformly valid," *Phil. Mag.* **40**, pp. 1179-1201.

Lighthill, M. J., 1961. "A technique for rendering approximate solutions to physical problems uniformly valid," *Z. Flugwiss.* **9**, pp. 267-275.

Nayfeh, A. H., 1973. *Perturbation Methods*, Wiley, New York.

Whitham, G. B., 1956. "On the propagation of weak shock waves," *J. Fluid Mech.* **1**, pp. 290-318.

Whitham, G. B., 1974. *Linear and Nonlinear Waves*, Wiley, New York.

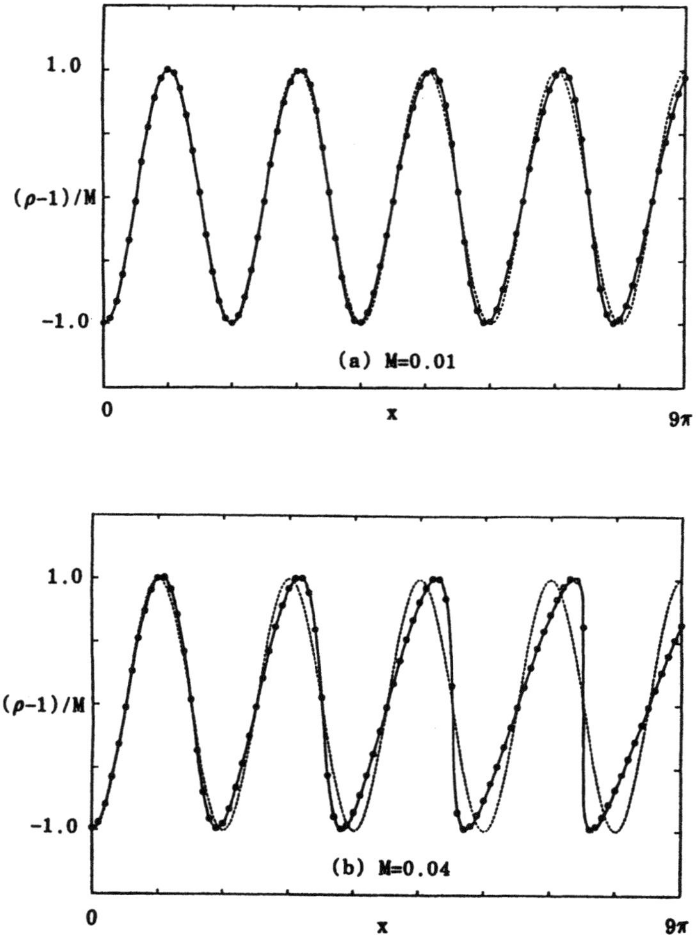

Figure 1: Comparison of $(\rho - 1)/M$ as a function of x at a fixed time for the one dimensional example using the linear solution (short dashed line), the multiple scales solution (solid line), and the numerical solution (circles) for: (a) M=0.01; and (b) M=0.04.

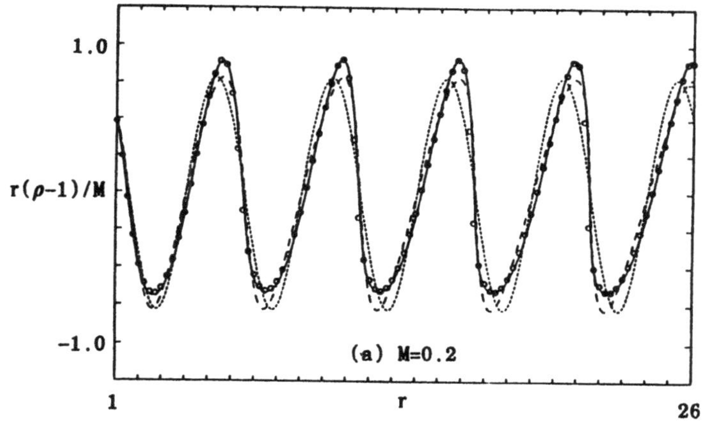

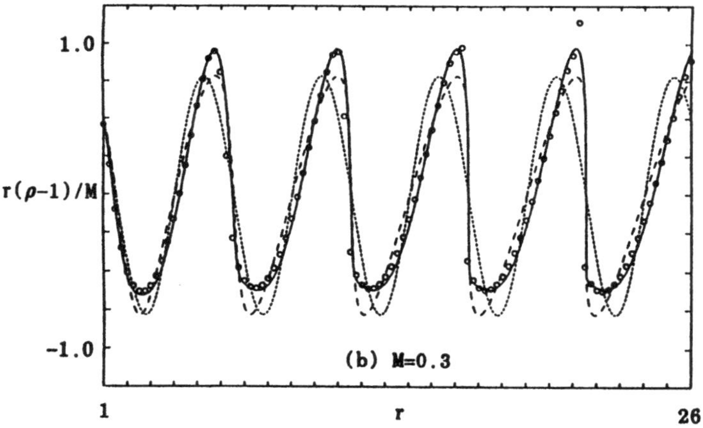

Figure 2: Comparison of $r(\rho - 1)/M$ as a function of r at t=75 for the pulsating sphere example with ω=1.25 using the linear solution (short dashed line), Whitham's solution (long dashed line), the multiple scales solution (solid line), and the numerical solution (circles) for: (a) M=0.2; and (b) M=0.3.

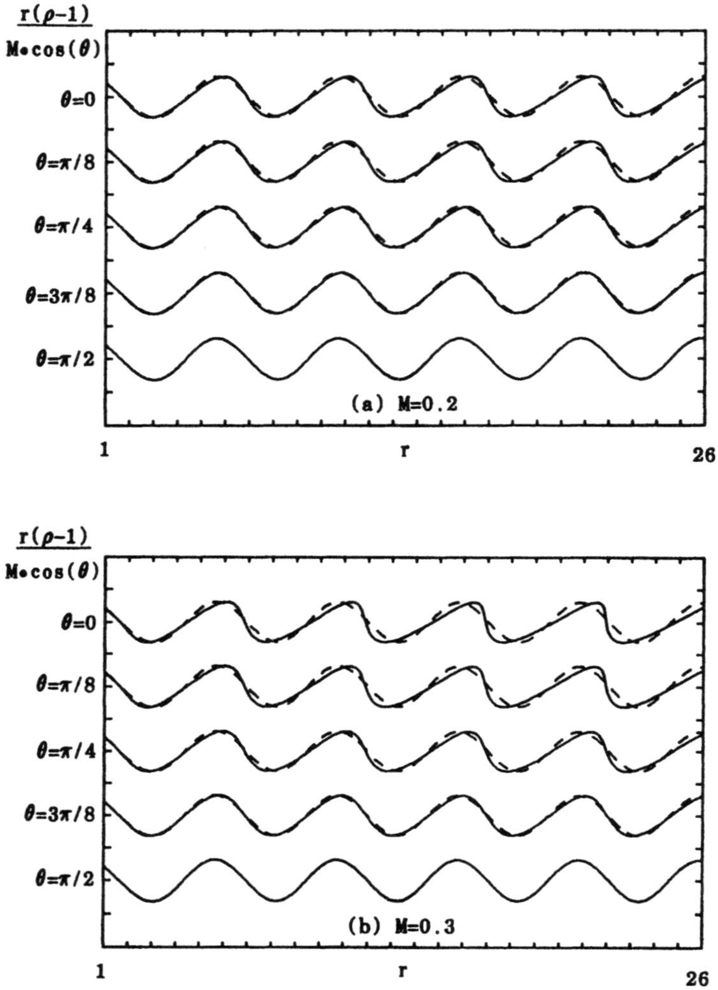

Figure 3: Plot of $r(\rho - 1)/(M\cos(\theta))$ as a function of r at t=75 for the oscillating sphere example with ω=1.25 using the linear solution (short dashed lines) and the multiple scales solution (solid lines) for θ=0, $\pi/8$, $\pi/4$, $3\pi/8$, and $\pi/2$ with: (a) M=0.2; and (b) M=0.3.

ON THE NOISE RADIATED FROM A TURBULENT HIGH SPEED JET

Geoffrey M. Lilley

Department of Aeronautics and Astronautics
University of Southampton United Kingdom

*(with Appendix 1 by Sir James Lighthill F.R.S.
University College London)*

ABSTRACT

The study of noise radiated from aerodynamically unsteady fluid flows was studied by Lighthill, who showed there exists an "acoustic analogy" whereby the exact fluid flow can be replaced by an equivalent distribution of acoustic quadrupole sources in which the sources may move but not the fluid. The problem posed by Lighthill was that although this "acoustic analogy" represented an exact solution of the Navier-Stokes equations for the density or pressure, in applications of engineering interest the flow field is rarely known to the exactness required to provide an accurate prediction of the radiated sound field, its acoustic power output, its intensity distribution with distance and direction, and its spectrum. Approximate models of the flow field for subsonic and supersonic jets at rest or moving at a uniform speed through a stationary atmosphere have been introduced into Lighthill's "acoustic analogy" for the prediction of the radiated sound and to provide a comparison with experimental results. The difficulties in the applications of these models and the limitations in the success achieved with such models has been the subject of continuing study.

This paper gives a brief status report of the current position of the prediction methods used in the noise from jets, and comments on the available experimental data used to verify the accuracy of these models and procedures. Also discussed are the possible computational methods which could exploit current advances made in supercomputer technology in solving the Euler and Navier-Stokes equations for certain classes of flows and which would lead to the evaluation of the radiated sound field from such flows. The problems of computational domains containing embedded flows and their boundary conditions together with the resolution requirements are considered. Also considered are the advantages of the deployment of the "acoustic analogy" over the direct computation of the fluctating pressure and density fields within and external to the flow.

1. Introduction

Lighthill's Theory of Aerodynamic Noise has been the foundation of our present knowledge on the generation of noise from unsteady flows and, in particular, the generation of noise from turbulent high speed jets. Lighthill (1952) made the first major contribution to the study of ACOUSTICS since the pioneering work of Lord Rayleigh in the last century. This theory is exact, even for a turbulent flow, and any lack of agreement in the radiated noise intensity, directivity and spectrum, between experiment and theory is the result of approximations introduced in predicting the Lighthill Integral for a given jet configuration and jet exit Mach Number. In making such approximations we are continually being reminded of Lighthill's warnings that to make unwise approximations at too early a stage in the analysis can lead to gross errors, or even worse, to a non-physical answer for the far-field radiation. Perhaps this is not surprising for the energy radiated away from the flow as noise is only a minute fraction of its turbulent kinetic energy.

The comparisons with experiments on the noise radiated to the far-field from high speed jets show that many of the broad features of Lighthill's Theory (Lighthill, 1954; 1962; 1963) are realized in respect of variations in the total acoustic power output, the intensity and its directivity and spectrum with the jet exit Mach Number and jet diameter. This is a major triumph for the theory since the properties and the characteristics of shear flow turbulence in a turbulent jet are still poorly documented.

The aim, and it remains a challenge, is to find a value to an engineering accuracy for the Lighthill Integral. Up to the present we have had to rely on experiment to confirm the estimates and predictions arising from the theory.

We have so far spoken of the theoretical or computational evaluation of Lighthill's Integral in the prediction of the radiated noise from any unsteady flow. In addition the Lighthill Theory has been used as the basis for setting up empirical theories which use an experimental data base. This is an important aspect of the Lighthill Theory and has led to accurate and rapid noise prediction methods as used by the Aircraft Industry in confirmation of Aircraft Noise Regulations.

Computational aerodynamics has made great advancements in the last 15 years and results for turbulent incompressible and compressible flow have been obtained for a number of turbulent shear flows. These numerical results have complemented the numerous measurements of the components of the Reynolds stress tensor and the integral length scales. In such turbulent shear flows, the structure of the turbulence is complex and measurement and theory act together to provide a complementary picture of the structure of the flow. Measurement has provided both quantitative and graphical information, which in the latter case is in the form of flow visualization. Computation has provided more complete space-time information on flow quantities, such as pressure and vorticity, that cannot be measured using conventional techniques, and at the same time has extended greatly the graphical representation of such flows. This continues to advance and clarify the complex nature of the structure of turbulent shear flows.

However the resolution of fluctuating quantities such as fluid density in space-time calculations continues to present difficulties when conventional computational fluid dynamics (CFD) methods are employed. In CFD a fine grid is required covering the entire flow field, but for acoustic problems must include the boundaries of the inner acoustic field as well. It remains a challenge to introduce the necessary boundary conditions in such numerical flow calculations, particularly for turbulent flow, which reflect the properties of the inner field sound generation, and which at the same time satisfy the wave equation for the radiated sound in the external field.

2. Exact Flow Equations

The exact flow equations for a perfect gas, which is also a viscous and compressible fluid, are presented in the following equations. $p, \rho, h, h_s, \boldsymbol{v}$ are respectively the pressure, density, specific enthalpy, specific stagnation enthalpy, velocity, the viscous stress tensor is $\boldsymbol{\tau}$, and the heat flux vector is $\boldsymbol{q}$, with γ the ratio of the specific heats. Later we introduce the vorticity, $\boldsymbol{\omega} = \text{curl}\boldsymbol{v}$, and the dilatation, $\Theta = \nabla.\boldsymbol{v}$.

Equation of state

$$p = \rho h(\gamma - 1)/\gamma \tag{1}$$

Equation of conservation of mass

$$\partial\rho/\partial t + \nabla.\rho\boldsymbol{v} = 0 \tag{2}$$

Equation of conservation of momentum

$$\partial\rho\boldsymbol{v}/\partial t + \nabla.(\rho\boldsymbol{v}\boldsymbol{v} + p\boldsymbol{I} - \boldsymbol{\tau}) = 0 \tag{3}$$

Equation of conservation of energy

$$\partial\rho h_s/\partial t + \nabla.(\rho\boldsymbol{v}h_s - \boldsymbol{q} - \boldsymbol{\tau}.\boldsymbol{v}) - \partial p/\partial t = 0. \tag{4}$$

Lighthill selected Equations (2) and (3) from which he derived the inhomogeneous wave equation for the fluid density. By taking the time derivative of Equation (2) we find:

$$\nabla.\partial\rho\boldsymbol{v}/\partial t = -\partial^2\rho/\partial t^2 \tag{5}$$

and the divergence of Equation (3) gives:

$$\nabla.\partial\rho\boldsymbol{v}/\partial t = -\nabla.\nabla.(\rho\boldsymbol{v}\boldsymbol{v} + p\boldsymbol{I} - \boldsymbol{\tau}). \tag{6}$$

Now we note $\nabla.\boldsymbol{v}$, and therefore $\nabla.\rho\boldsymbol{v}$, are small quantities, being exactly zero in an incompressible flow. But $\nabla.\partial\rho\boldsymbol{v}/\partial t$ is a measure of the disturbance in a flow and therefore must be known exactly for the correct prediction of the noise generated from an unsteady compressible flow. As Lighthill showed this is best achieved by eliminating it between Equations (5) and (6), so deriving:

$$\partial^2\rho/\partial t^2 - \nabla^2 p = \nabla.\nabla.(\rho\boldsymbol{v}\boldsymbol{v} - \boldsymbol{\tau}). \tag{7}$$

But the wave equation in the external medium is:

$$\partial^2\rho/\partial t^2 - {c_\infty}^2\nabla^2\rho = 0 \tag{8}$$

and hence we derive Lighthill's Equation:

$$\partial^2\rho/\partial t^2 - {c_\infty}^2\nabla^2\rho = \nabla.\nabla.\boldsymbol{T} \tag{9}$$

where the stress tensor $\boldsymbol{T} = \rho\boldsymbol{v}\boldsymbol{v} + (p - \rho c_\infty^2)\boldsymbol{I} - \boldsymbol{\tau}$ and c_∞ is the constant speed of sound in the external medium. Here we assume the fluid external to the flow is essentially at rest, except for the presence of the sound field. In this sound field the disturbances are assumed to be isentropic and we neglect the effects of non-linear steepening of the sound waves due to their finite amplitude near the boundaries of the flow, and their attenuation due to sound absorption. Hence in the external flow,

$$(p - p_\infty) = {c_\infty}^2(\rho - \rho_\infty). \tag{10}$$

The further assumption is made that $\boldsymbol{T}$, which is finite inside the flow, falls sufficiently fast with distance from the flow so that its volume integral over all space is finite.

3. Lighthill's Theory of Aerodynamic Noise

Lighthill gave the solution of Equation (9) for an unbounded flow as:

$$(\rho - \rho_\infty)(\boldsymbol{x}, t) = \frac{1}{4\pi {c_\infty}^2} \frac{\partial^2}{\partial x_i \partial x_j} \int\int\int T_{ij}(\boldsymbol{y}, \tau) \frac{d\boldsymbol{y}}{\mid \boldsymbol{x} - \boldsymbol{y} \mid} \tag{11}$$

where the retarded time is $\tau = t - \mid \boldsymbol{x} - \boldsymbol{y} \mid / c_\infty$. Therefore the density fluctuation at the observer, at $(\boldsymbol{x}, t)$, is generated at a source point $\boldsymbol{y}$ at time τ, where the time difference between t and τ is equal to the time taken for sound to travel between $\boldsymbol{y}$ and $\boldsymbol{x}$. The interpretation of Equation (11) given by Lighthill is that the actual fluid flow satisfying given boundary equations and/or initial conditions is replaced exactly by an equivalent distribution of acoustic sources of strength $\boldsymbol{T}$ per unit volume embedded in a medium at rest having the properties of the outer medium. Lighthill expressed T_{ij} as a distribution of acoustic quadrupoles per unit volume which may move, whereas the background fluid in this "acoustic analogy" must remain at rest. The acoustic source strength, $\boldsymbol{T}$, includes all the physical processes of noise generation plus the interactions and interference of the resulting sound waves with the flow field including the effects of refraction, diffraction, and scattering.

When $\boldsymbol{T}$ is known throughout the flow field Equation (11) gives the density fluctuation at the observer at time t. However $\boldsymbol{T}$ involves the unknown fluid density in all its three terms. Lighthill argued that $\boldsymbol{T}$ could be obtained to a good approximation by calculating its value for an equivalent flow devoid of sound. With this assumption for $\boldsymbol{T}$, Lighthill showed the far-field density could be found to an engineering accuracy when Equation (11) is written in the following form:

$$(\rho - \rho_\infty)(\boldsymbol{x}, t) \approx \frac{1}{4\pi c_\infty^4 x} \iiint \frac{\partial^2 T_{xx}(\boldsymbol{y}, \tau)}{\partial \tau^2} d\boldsymbol{y}. \tag{12}$$

Here T_{xx} denotes the aligned value of the stress tensor in the direction pointing from the source at $\boldsymbol{y}$ to the observer at $\boldsymbol{x}$. The retarded time between source and observer must be included for all source positions in the flow field domain. The integration is over the flow domain only, since as stated previously $\boldsymbol{T}$ is assumed to be zero everywhere outside the flow. It has been shown by Crow (1970) that the assumption that $\boldsymbol{T}$ does not contribute to the integral outside the flow is valid at low flow Mach Numbers but no proof has been given that this is true for all Mach Numbers, even though it appears to represent an acceptable assumption on physical grounds.

Lighthill's Equation for the far-field density was obtained by including in the source term $\boldsymbol{T}$ the quantity $(p - {c_\infty}^2 \rho)$. We can find, alternatively, the corresponding equation for the pressure fluctuations at the far-field observer by making use of the total energy equation. Thus using Equation (4) and noting $\partial\rho/\partial t = -\nabla.\rho\boldsymbol{v}$, we derive:

$$\frac{\partial}{\partial t}\left(\frac{p - \rho c_\infty^2}{c_\infty^2}\right) = -\frac{\gamma - 1}{2c_\infty^2}\frac{\partial \rho v^2}{\partial t} - \nabla.\left(\frac{\rho\boldsymbol{v}(h_s - h_\infty) - \boldsymbol{q} + \boldsymbol{\tau}.\boldsymbol{v}}{h_\infty}\right) \tag{13}$$

and the wave equation for the pressure is:

$$\begin{aligned}\left(\nabla^2 - \frac{1}{{c_\infty}^2}\frac{\partial^2}{\partial t^2}\right)(p - p_\infty) &= \frac{(\gamma - 1)}{2{c_\infty}^2}\frac{\partial^2 \rho v^2}{\partial t^2} \\ -\nabla.\left(\nabla.(\rho\boldsymbol{v}\boldsymbol{v}\boldsymbol{\tau}) - \frac{\partial}{\partial t}\left(\rho\boldsymbol{v}\frac{h_s - h_\infty}{h_\infty} - \frac{\boldsymbol{q} + \boldsymbol{\tau}.\boldsymbol{v}}{h_\infty}\right)\right)\end{aligned} \tag{14}$$

and its solution in the far-field :

$$(p - p_\infty)(\boldsymbol{x}, t) \approx \frac{1}{4\pi {c_\infty}^2 x} \iiint \frac{\partial^2 T_{xx}}{\partial \tau^2} d\boldsymbol{y} \tag{15}$$

where, $T_{xx} = \rho {c_\infty}^2 \left[\frac{v_x}{c_\infty}\left(\frac{v_x}{c_\infty} + \frac{(h - h_\infty)}{h_\infty}\right) + \frac{(\gamma-1)v^2}{2{c_\infty}^2}\left(\frac{v_x}{c_\infty} - 1\right)\right]$ when the diffusive terms are omitted. The additional term to $\rho v_x v_x$, which is $(h - h_\infty)$, or $(h_s - h_\infty)$, comes of importance in flows where the temperature in the jet differs from that in the ambient medium. The usual interpretation is that $\rho v_x v_x$ represents the strength of anisotropic acoustic quadrupoles, $(\gamma - 1)\rho v^2/2$ the strength of isotropic acoustic quadrupoles and $\rho \boldsymbol{v}(h_s - h_\infty)/h_\infty$, at least at low Mach Numbers, the strength of acoustic dipoles involving the temporal enthalpy fluctuations in a heated jet. It follows that in all applications of the Lighthill "acoustic analogy" we must use the exact form of the Lighthill Stress Tensor, $\boldsymbol{T} = \rho \boldsymbol{v}\boldsymbol{v} + (p - c_\infty{}^2)\boldsymbol{I} - \boldsymbol{\tau}$, except possibly at high Reynolds Numbers, where the viscous stress tensor is expected to generate low amplitude noise of high frequency only. This is associated with the small-scale eddies responsible for the bulk of the dissipation of kinetic energy from a turbulent flow. Moreover it is expected that the dominant contribution to the noise radiated from a jet arises from eddies of scale comparable with the main energy bearing eddies as shown by Sir James Lighthill in Appendix 1. All acoustic sources, no matter where their position is within the flow, radiate with the same efficiency, and there is no shielding effect of the 'outer sources' imposed on the 'inner sources'. There is no interaction between the sources which can modify their amplitudes or directivity, for the acoustic sources are totally independent, and their characteristics are those imposed by the real flow field. Acoustic sources radiate to the far-field along straight lines in the direction between the source and the observer.

If the contribution from the sound field is neglected in an approximate evaluation of $\boldsymbol{T}$, small errors only exist in the prediction of the sound power output from the flow and the major part of the noise spectrum. What is missing is the change in the wave-number spectrum of $\boldsymbol{T}$ inside the flow arising from the acoustic-flow interactions in the real flow, which prevent the determination of the true noise directivity and spectrum in the far-field noise radiation.

The Lighthill Integral involves only the fluctuations in $\boldsymbol{T}$, since the mean value of the Integral over all space is zero. In the real flow moving axis formulation these fluctuations in $\boldsymbol{T}$ involve many interactions which contribute to the amplitude and directivity of the 'source function', and include the flow-acoustic interaction. The contribution from the velocity field is shown to be mainly dependent on the local mean shear and the fluctuating Reynolds stresses, where the local mean shear determines the order of magnitude of the dominant frequencies in the turbulence in a moving frame. This is a characteristic found in measurements of jet noise. The velocity components contributing to $\partial^2 \boldsymbol{T}/\partial t^2$ include both the local mean velocity and the turbulent velocity components relative to the local mean velocity, and contribute to its directivity properties. The integration of the 'source function', over the entire volume of the turbulent mixing region, results in an averaging of the contributions from the turbulent flow, and suggests it is sufficient in a simplified model, to assume the turbulence, and consequently the sources of sound, move at some average convection speed, V_c. We also assume, on the basis of the discussion above, that we can put $\boldsymbol{T}$ of order $\rho v_x{}^2$, where v_x is the aligned component of the local turbulent velocity with respect to a far-field observer. Lighthill in his early papers showed how T_{ij} could be replaced by quantities involving the local rate of strain and also showed that sound generated in such a flow would be amplified by the presence of the mean shear. This alternative formulation was used by Lilley (1958) in his estimates of jet noise. Here we prefer to refer the flow field, at low to moderate Mach Numbers, to a mean convection speed of approximately $0.62V_J$, where V_J is the speed of the jet at the nozzle exit. A more complete formulation would allow V_c to change with distance downstream of the jet exit and to vary across the flow field. The main advantage in using a moving axis system to define the acoustic sources, is that in such a moving frame the effect of differences in retarded time is minimized. Lighthill's Integral may be evaluated in a coordinate system at rest provided the value of $\boldsymbol{T}$ is known exactly as a function of space and time and the proper retarded times are incorporated. In the predictions discussed in the next section we use the moving axis system in which the acoustic convection Mach Number $M_c = V_c/c_\infty$ is taken as a constant. (We note, following Lighthill, that the Mach Number in acoustic flow problems is best referenced with respect to the ambient speed of sound, c_∞, rather than the true local speed of sound, c.)

4. The Prediction of Jet Noise in the Far-Field

We will assume the jet mixing regions both in the region near the nozzle exit, and further downstream, are fully turbulent. Transition from laminar to turbulent flow is assumed to take place inside the nozzle and is complete when the flow emerges from the nozzle exit. The noise generated by the turbulence will be an irregular function of the source time, τ, at any location inside the jet, and hence the far-field density at the later time t, will mirror such gross irregularities. To find the characteristics of the far-field noise we introduce the autocorrelation of the far-field Intensity, I($\boldsymbol{x}, t^\star$), defined by:

$$I(\boldsymbol{x}, t^\star) = \frac{{c_\infty}^3}{\rho_\infty} \lim_{T\to\infty} \frac{1}{2T} \int_{-T}^{+T} \rho(\boldsymbol{x}, t)\rho(\boldsymbol{x}, t + t^\star)dt \qquad (16)$$

and its power spectral density,

$$\mathcal{I}(\boldsymbol{x}, \gamma) = \frac{1}{2\pi} \int_{-\infty}^{+\infty} \exp(-i\gamma t^\star) I(\boldsymbol{x}, t^\star)dt^\star \qquad (17)$$

where γ is the frequency at the fixed frame of the observer and $t^\star$ is the time separation between signals arriving at the observer from the adjacent points, $\boldsymbol{y}$ and $\boldsymbol{z}$ in the source region. It is assumed the turbulent field is stationary and hence the statistically averaged properties of the turbulence and the radiated noise are independent of time. If we use Equation (12) above then:

$$\begin{aligned} I(\boldsymbol{x}, t^\star) \quad \approx \quad & \frac{1}{16\pi^2 \rho_\infty c_\infty^5 x^2 \mid 1 - M_c \cos\theta \mid^5} \int\!\!\int\!\!\int d\boldsymbol{y} \\ & \int\!\!\int\!\!\int d\boldsymbol{\delta} \frac{\partial^4 P_\theta(\boldsymbol{y}, \boldsymbol{\delta}, \tau)}{\partial \tau^4} \end{aligned} \qquad (18)$$

given by Ffowcs Williams (1963) and Ribner (1964), where

$$\frac{\partial^4}{\partial \tau^4} P_\theta(\boldsymbol{y}, \boldsymbol{\delta}, \tau) = < \frac{\partial^2 T_{xx}(\boldsymbol{y}, \tau_1)}{\partial {\tau_1}^2} \frac{\partial^2 T_{xx}(\boldsymbol{z}, \tau_2)}{\partial {\tau_2}^2} > . \qquad (19)$$

P_θ is the space-retarded time covariance of the source function, taken at the separated points $\boldsymbol{y}$ and $\boldsymbol{z}$ as measured in a moving frame with the convection Mach Number $\boldsymbol{M}_c$, where the corresponding moving coordinates are $\boldsymbol{\eta}$ and $\boldsymbol{\zeta} = \boldsymbol{\eta} + \boldsymbol{\delta}$, and the respective retarded times are $\tau_1 = t- \mid \boldsymbol{x} - \boldsymbol{y} \mid c_\infty$ and $\tau_2 = t + t^\star - \mid \boldsymbol{x} - \boldsymbol{z} \mid / c_\infty$. Also T_{xx} is its fluctuating part. The relation between the moving axes $\boldsymbol{\eta}$ and $\boldsymbol{\zeta}$ and the fixed axes $\boldsymbol{y}$ and $\boldsymbol{z}$ are given by, $\boldsymbol{\eta} = \boldsymbol{y} - c_\infty \boldsymbol{M}_c \tau_1$, and $\boldsymbol{\zeta} = \boldsymbol{z} - c_\infty \boldsymbol{M}_c \tau_2$, with $\boldsymbol{\delta}$ being the spatial separation of sound sources at $\boldsymbol{\eta}$ and $\boldsymbol{\zeta}$ in the moving frame corresponding to the fixed points $\boldsymbol{y}$ and $\boldsymbol{z}$ respectively. Sound emission takes place as the sources cross $\boldsymbol{y}$ and $\boldsymbol{z}$ respectively. If τ is the retarded time difference between the emissions taking place at $\boldsymbol{y}$ and $\boldsymbol{z}$ then with their reception time difference, $t^\star$, at the far-field observer position, $\boldsymbol{x}$, we find:

$$\tau = (c_\infty t^\star x + \boldsymbol{\delta}.\boldsymbol{x})/(c_\infty x(1 - M_c \cos\theta)) \qquad (20)$$

where in the far-field $|\boldsymbol{x}| \gg |\boldsymbol{y}|$ or $|\boldsymbol{z}|$, the observer is located at (x, θ), and θ is the angle between the direction of V_c and the line joining the source to the observer. We assume as stated above that the flow is statistically stationary and for a circular jet the average noise intensity is independent of the azimuthal angle about the jet axis.

Thus the central problem in Lighthill's Theory of Aerodynamic Noise is the evaluation of $\partial^4 P_\theta(\boldsymbol{y}, \boldsymbol{\delta}, \tau)/\partial\tau^4$ at all points in the flow domain and its sextuple integral with respect to $\boldsymbol{\delta}$ and $\boldsymbol{y}$ over this same flow domain. It is however more convenient to find its four-dimensional Fourier Transform and on defining:

$$\mathcal{P}_\theta(\boldsymbol{y}, \boldsymbol{k}, \omega) = \frac{1}{(2\pi)^4} \int\int\int \exp(-i\boldsymbol{k}.\boldsymbol{\delta}) d\boldsymbol{\delta} \int \exp(-i\omega\tau) P_\theta(\boldsymbol{y}, \boldsymbol{\delta}, \tau) d\tau \tag{21}$$

we find the cross-power spectral density in the far-field is:

$$\mathcal{I}(x, \gamma) \approx \frac{\pi}{2\rho_\infty c_\infty^5 x^2} \int\int\int \gamma^4 \mathcal{P}_\theta(\boldsymbol{y}, \boldsymbol{k}, \omega) d\boldsymbol{y} \tag{22}$$

where the wave-number vector $\boldsymbol{k} = -\gamma\boldsymbol{x}/x c_\infty$ and $\omega = \gamma(1 - M_c \cos\theta)$ is the Doppler shifted frequency, being the frequency of the turbulence in the moving frame, whereas γ is the frequency at the far-field observer. In addition $dt^\star = (1 - M_c \cos\theta) d\tau$.

It is seen that only a small fraction of the source spectrum function is responsible for the noise generation and is just those values of $\mathcal{P}_\theta$ in $(\boldsymbol{k}, \omega)$ space which have a corresponding phase speed equal to the speed of sound.

5. The Simplified Model for Jet Noise Prediction

As stated above, the Lighthill Stress Tensor, $\boldsymbol{T}$, in a turbulent shear flow, is not omnidirectional, but current information does not provide an estimate for its directivity. It is insufficient to take only $\rho v_x v_x$ and to express it in terms of v_1, v_2, and v_3, the components of $\boldsymbol{v}$. All we can say, until more information is available from experiment or numerical studies, is that the amplitude of P_θ is given by $\rho_0^2 u_0^4 F(\theta)$ where $F(\theta)$ is some function of θ and is of order unity. ρ_0 and u_0 are reference values of the local mean density and turbulent velocity and may be taken as functions of the axial distance only. The next requirement is to introduce the space-time properties of P_θ. The space-time covariance of T_{xx}, as shown in Appendix 1, is positive.

We expect it to be influenced by the presence of the sound field and to therefore fall to zero less slowly at large spatial separations than in the case where the sound field is absent. Detailed calculations for the jet show, however, that its local large separation distance behavior has little influence on the characteristics of the far-field radiated sound, which depend more on the envelope through the peaks of the local distributions than the details of the distribution functions themselves. (This would not be true for the noise radiation from an isolated source.) In the case of the high speed jet the mixing region for a turbulent jet grows linearly from almost zero thickness at the jet exit and the dominant eddy sizes at each station along the jet vary in proportion to the width of the jet. Therefore the frequency spectrum of the radiated noise is not wholly dependent on the characteristics of the source function at each station, but more on how those characteristics change with distance along the jet. It is convenient, in this simplified model, to represent P_θ by a Gaussian distribution with respect to each component of the moving frame spatial separation, $\boldsymbol{\delta}$, and the retarded time separation, τ. But other distributions would lead to similar results. An improved non-Gaussian model for the 4th-order covariance, P_θ, is proposed in Appendix 1, by Sir James Lighthill, based on the known properties, calculated or measured, for the known 2nd-order covariances.

We assume a self-preserving formulation for the source function which after integration across the flow degenerates into an axial source distribution. On the assumption that the length scales and the root mean square turbulent velocity components are known throughout the source region, a physically possible form for the effective source function can be proposed which reduces to a function of the axial coordinate y_1 only, where y_1 is measured along the jet axis. We refer to this function as the Axial Source Function, $R(y_1, \gamma)$, where, $R(y_1, \gamma) = \int\int dy_2 dy_3 P_\theta(\boldsymbol{y}, \boldsymbol{k}, \omega)$, and the integration is taken over the cross-sectional area of the jet at any given station y_1. We define $P_\theta(\boldsymbol{y}, \boldsymbol{\delta}, \tau)$ in terms of its moving frame turbulence quantities $\rho_0, u_0, l_1, l_2, l_3$, and ω_0 all of which are functions of y_1. These quantities are respectively characteristic mean densities, root mean square turbulent velocities, turbulent length scales in the directions (y_1, y_2, y_3), and moving frame frequencies. We write, $P_\theta(\boldsymbol{y}, \boldsymbol{\delta}, \omega) = \rho_0^2 u_0^4 f(\boldsymbol{y}, \boldsymbol{\delta} : \delta_i/li : \omega_0 \tau)$, where $f(\boldsymbol{y}, \boldsymbol{\delta}, \tau)$ is assumed equal to a Gaussian distribution for the space and retarded-time separations. We argue that the far field directivity is more a function of flow-acoustic interaction than source anisotropy. It follows that:

$$R(y_1,\gamma) = \int\int dy_2 dy_3 \frac{\rho_0^2 u_0^4 \ell_0^3}{16\pi^2\omega_0}(\ell_\perp/\ell_0)^2 \exp\left(-\frac{\gamma^2\ell_0^2 C_0}{4s_0^2 u_0^2}\right) \quad (23)$$

where the effective Doppler factor is C_0 and where:

$$C_0 = (1 - M_c\cos\theta)^2 + \frac{s_0^2 u_0^2}{{c_\infty}^2}\left(\cos^2\theta + \frac{\ell_\perp^2}{\ell_0^2}\sin^2\theta\right) \quad (24)$$

and is finite when $M_c = 1$. We make the assumption that the transverse scales l_2, l_3 are equal and can be replaced by $\ell_\perp$. The longitudinal length scale l_1 is replaced by ℓ_0. The Strouhal Number of the turbulence $s_0 = \omega_0\ell_0/u_0$ is assumed to be a constant throughout the entire jet flow. The effective cross-sectional areas of the jet, where P_θ is assumed to be constant, are given by:

$$\int\int dy_2 dy_3 = \begin{cases} \pi D_J b & 0 < y_1 < L \\ \pi b^2 & L < y_1 < \infty. \end{cases} \quad (25)$$

The first region covers the initial mixing region to the end of the potential core of length L, and the second region covers the entire jet downstream of the potential core. The mean width of the mixing region at any station y_1 is given by b, and it is assumed that b/ℓ_0 is a constant throughout the entire jet. The convection Mach Number $M_c = V_c/c_\infty$ and the jet Mach Number is $M_J = V_J/c_\infty$. The Strouhal Number of the radiated noise is $S = \gamma D_J/V_J$ where D_J is the jet exit diameter. The Mach numbers M_J and M_c are not true Mach Numbers and are known as the 'acoustic' Mach Numbers, since, as stated above, they are referenced to the constant speed of sound in the external medium. A consequence of these self-similar assumptions and the integrations over the total volume of the jet is that the far-field noise spectrum decreases as $1/S^2$ at high frequencies and is proportional to S^2 at low frequencies. A simple improvement for the far field directivity, to allow for refraction by the basic flow field of the sound field at high frequencies, can be obtained using Snell's Law. Thus the wave number vector of sound in the far-field is modified to give the corresponding source wave number vector at all positions inside the flow, dependent on variations of the local velocity and speed of sound.

These predictions have been compared with measurements using the polar-correlation technique of Fisher et al. (1977) and Tester et al. (1981). The comparison has only been performed at $\theta = 90°$. The agreement at this angle is favorable and suggests that the details of the self-similar source model described above, and the corresponding distribution and strength of the equivalent axial source distribution are qualitatively correct and quantitatively of the correct order of magnitude. Further details of these results are given in Hubbard (1991).

6. Modifications to the Simplified Model

The assumptions used in the Simplified Model above can all be checked, modified or replaced by new data from the computational studies in progress on the structure of turbulent shear flows.

The choice of a moving frame analysis is to some extent arbitrary and was first proposed by Lighthill to minimize retarded time effects in the source terms. However if proper account is taken of retarded time effects then the moving frame analysis for the T_{ij} covariance can be replaced by a stationary frame covariance. As shown above the values of $\boldsymbol{v}, \rho, p$, and h are required as functions of t at all points $\boldsymbol{y}$ and $\boldsymbol{z}$ within the flow region, which becomes the acoustic source region. The corresponding data bank required for the flow variables is enormous for it needs to be stored for detailed interrogation at selected times. It is supposed that a 'map' of $\boldsymbol{v}, \rho, p, h$ can be obtained for each value of t, where the 'map' covers the three-dimensional volume of the flow region. For an exact use of the Lighthill "acoustic analogy" the values of the flow quantities would need to include not only their field values satisfying the appropriate boundary conditions, and the properties of, say, statistical stationary flow, or some necessary initial value conditions, but they should strictly include the contributions from the generated sound field and its interaction and interference with the basic unsteady flow. The size of the flow field precludes the computational domain extending far beyond the flow region. But we note that intermittency of the flow, especially near its boundaries, requires the computational domain to extend to well beyond what is normally specified as the mean flow boundary. At the boundary of the computational domain the vorticity will be zero, but no boundary condition on $\boldsymbol{v}, p, \rho$, and h can be specified except that the flow field beyond this boundary is irrotational. But only part of this irrotational flow is due to the presence of sound waves. The major contribution to the flow, in this 'near field region', is due to the irrotational entrainment flow. If the entrainment flow near the flow boundaries is approximately incompressible then the flow near the boundary both inside and outside the flow is governed by $\nabla^2\Phi = 0$, where Φ is the entrainment velocity potential. The sound field, on the other hand, involves a solution to the wave equation $(\partial^2/\partial t^2 - \nabla^2)\phi = 0$, where $\nabla\phi = v_s$, and v_s is the sound particle velocity. On this computational boundary the total velocity is $\nabla(\Phi+\phi)$ and is continuous across the boundary. The values of $\nabla\Phi$ and $\nabla\phi$ will be of different orders of magnitude and occupy different parts of the wave-number/frequency spectrum. Φ will be a function controlled locally by the internal vortical flow, its amplitude and length scales, whereas ϕ will have larger characteristic length scales typical of the external sound field, and matched wave number and frequency components giving a phase speed equal to the speed of sound. Such a computation, involving the growing mixing region of the turbulent compressible jet presents opportunities and challenges to achieve the necessary resolution for accurate computation of ϕ and Φ.

The importance of retarded time analysis can be seen by noting in a low Mach Number flow that the mean square of the density fluctuations will be proportional to M^4, whereas the mean square of the retarded time noise source strength is proportional to M^8. At high Mach numbers the retarded time noise source strength is reduced to proportionality with M^3. The vortical and entropy or heat modes are so distinct in amplitude and scale from the sound mode, which itself is minute in amplitude and its length scale so long, that it would seem unwise to ignore the Lighthill philosophy, which states the radiated sound may be calculated to a good approximation once the basic flow field is known, even when this has been obtained devoid of the sound field it generates. The Simplified Model described above, together with its comparison with experiment, shows that this procedure gives an acceptable answer for the total acoustic power radiated from a high speed jet over a very wide speed range.

Thus we infer that a more immediate task is to establish the flow field devoid of the sound field that the turbulent flow generates, and to obtain that data bank for the calculation of the space-time covariance of $\boldsymbol{T}$ throughout the flow-field. The calculation of $\partial^2\boldsymbol{T}/\partial t^2$, using this accurate databank, would assist in clarifying the correct procedures for modelling its space-retarded time covariance and finding an improved estimate of the noise characteristics of a jet using the Lighthill "acoustic analogy". Its detailed comparison with experiment would then establish the extent to which the 'no sound' assumption needs correction to allow for flow-acoustic interaction.

7. Noise Prediction Methods Used in the Aircraft Industry

Many of the models currently used for jet noise prediction in the Aircraft Industry include the effects of acoustic-flow interaction and are based on the solutions to the Lilley Equation (Lilley et al., 1975) as derived by Morfey et al. (1976, 1983) for the asymptotic case of high frequencies, corresponding to the case of geometrical acoustics. In that treatment the effects of the source directivity, its spectrum, and convection Mach Number are all established empirically by comparison between the theory and experimental data. That method is being extended to provide predictions for the jet in a moving stream, and the bypass jet. It includes noise radiation from subsonic and supersonic jets and the effects of shock wave-turbulence interaction.

8. Flow-Acoustic Interaction

Noise is generated at a point in a compressible fluid when a time rate of change in volume occurs in the fluid surrounding that point. This equals the divergence of the velocity and when this is non-zero, a disturbance is generated propagating away at the speed of sound. If $\nabla.\boldsymbol{v} = \Theta$, the dilatation, is zero, as in an incompressible fluid, then the rate of change in volume is zero and no noise can be generated. Inside the flow, which we assume to be non-uniform, the dilatation is likely to be small, although it can never be zero in a compressible fluid flow. Both the mean and fluctuating values of Θ need to be determined accurately inside and outside the flow in all problems of aerodynamic noise.

Our flow equations can be written,

$$\frac{D\ln\rho}{Dt} = -\Theta : \frac{D\boldsymbol{v}}{Dt} = -c^2\nabla\Pi + \frac{\nabla.\boldsymbol{\tau}}{\rho} : \frac{D\Pi}{Dt} = -\Theta + \frac{Ds/C_p}{Dt} \quad (26)$$

where $\nabla\Pi = \nabla p/\rho c^2$, $\boldsymbol{\tau}$ is the viscous stress tensor, and s is the specific entropy. The convective wave equation for Θ is found by eliminating $\nabla^2 D\Pi/Dt$ and when all diffusive terms are omitted:

$$\frac{D}{Dt}\left(\frac{1}{c^2}\frac{D\Theta}{Dt}\right) - \nabla^2\Theta = -\frac{D}{Dt}\left(\frac{\nabla \boldsymbol{v} : \boldsymbol{v}\nabla}{c^2} - \frac{\nabla \ln h}{c^2}.\frac{D\boldsymbol{v}}{Dt}\right)$$

$$-\frac{\nabla^2 \boldsymbol{v}}{c^2}.\frac{D\boldsymbol{v}}{Dt} - 2\nabla\boldsymbol{v} : \nabla\left(\frac{1}{c^2}\frac{D\boldsymbol{v}}{Dt}\right). \tag{27}$$

This convected wave equation for the dilatation, Θ, is exact and was first given by Legendre (1981). We note Θ is a small quantity everywhere and $\Theta = -D\ln(\rho/\rho_\infty)/Dt$ with $D\Pi/Dt = D\ln(\rho/\rho_\infty)$ when the flow is isentropic. The similarity with the 'source function' of Lighthill's Theory is easily seen. The 'source function' of Equation (27) is, however, a function of the velocity and enthalpy only. If that part of the flow responsible for the wave motion within it, and which degenerates into the external radiated sound field, is removed from the 'source' function, then it may be regarded as resulting from a flow devoid of sound. The use of this 'source function' for the convected wave equation for Θ, derived for a flow devoid of sound, parallels the determination of $\boldsymbol{T}$ in Lighthill's Equation. However the major difference is that the flow-acoustic interaction which is hidden in Lighthill's "acoustic analogy", can now be investigated in the real flow. This flow-acoustic interaction is just the propagation through the non-uniform flow field of those components of the turbulence having wave numbers and frequencies responsible for the far-field sound radiation. The matching of solutions to Equation (27) inside the flow, with solutions of the wave equation in the external irrotational flow are discussed by Goldstein (1976), who also reviews the work of a number of authors on studies of the convective wave equation. The convected wave equation for Θ may be compared with the corresponding convective equations for Π, and h or h_s.

We described in Section 2 above Lighthill's Theory of Aerodynamic Noise which is based on the exact flow equations. On elimination of the small quantity $\nabla.\partial\rho\boldsymbol{v}/\partial t$, between the equations of conservation of mass and momentum, Lighthill derived the "acoustic analogy" based on the wave equation for the fluctuating density, where the 'source function' is the double divergence of $\boldsymbol{T}$. A later work by Phillips (1960) used the same exact equations of fluid flow but rearranged them to provide a convected wave equation in $\ln(p/p_\infty)$ by the elimination of the small term $\nabla\boldsymbol{v} = \Theta$. Since, $D\ln(\rho/\rho_\infty)/Dt = -\Theta$ we see the similarity with Lighthill's Theory. As shown in Figure 1, the 'source function' for Phillips Equation, when the diffusion terms are omitted, involves the velocity field only, and no longer includes the unknown density, as in Lighthill's Theory. However in order to solve the Phillips Equation for $\ln p$ the velocity and enthalpy in the turbulent flow field must be known everywhere. As in Lighthill's Theory these should include the properties of the sound field generated by the vortical flow.

Lilley et al. (1975), and in a series of earlier papers, in reviewing the work of Lighthill and Phillips, showed, following the work of Landahl (1967) on pressure fluctuations in a turbulent boundary layer, that to find the convected wave equation for the propagation of pressure fluctuations in a compressible shear flow, it was necessary to include in the left-hand-side of Phillips Equation part of the 'so-called source term' from the right-hand-side which contained a linear contribution to $\ln(p/p_\infty)$, and which was amplified by the mean shear. With this term included, the left-hand-side was shown to represent the correct linearized equation for the propagation of acoustic disturbances in a uniform shear flow. This equation had been derived earlier by Pridmore-Brown (1958). An earlier version of this work was due to Ribner (1964) and colleagues, and in unpublished work by Csanady.

In this alternative theory of Aerodynamic Noise the left-hand-side of the resulting equation, as shown in Figure 1, represents the propagation of sound in a non-uniform shear flow while the right-hand-side represents the 'source function'. The success of this alternative theory, and its comparison with Lighthill's "acoustic analogy" has been discussed by Goldstein (1976). This theory, based on the Lilley Equation, shows improvement with experiment in respect of the directivity and the changes in the spectrum with Mach Number, at least for the case of the high speed jet. In the form of the Lilley Equation adopted for the calculation of the noise radiated from a jet the velocity and enthalpy in the left-hand-side are replaced by their mean distributions. Thus the equation becomes linear in Π with the 'source function' containing only non-linear terms in the velocity and enthalpy. This reduced form of the exact equation has been criticized by Crighton (1981), Mohring (1983) and others, on the grounds that no clear distinction can be placed on what constitutes the proper propagation terms and what constitutes the generation terms since both are non-linear. Nevertheless they admit the high and low frequency solutions, as presented by Goldstein (1976), give results in fair agreement with experiment. An empirical theory has already been referred to above for jet noise predictions as used in Industry for Aircraft Noise Predictions, which is based on the results obtained from this theory, and in particular the results obtained by Morfey et al. (1983), Tester et al. (1976) and others, for the asymptotic high frequency approximations.

Once we accept the philosophy introduced by Lighthill that the Theory of Aerodynamic Noise must be based on an exact determination of the dilatation or derivatives of it, we find there are a number of possible forms of the equation of aerodynamic noise, which ultimately depend on the choice of the flow variable used. In all cases it is possible to find a form for the equation which reduces to the homogeneous wave equation outside the flow. Thus from the exact fluid flow equations in the seven flow variables, $\rho, p, h, s, \boldsymbol{v}$, we can derive two equations, in terms of two selected unknowns, followed by the elimination of that variable whose value is known to be negligibly small, and difficult to measure and calculate to a required accuracy. Thus in the convective wave equations for aerodynamic noise discussed above the primary equations involve ($\ln p$ and Θ), and (Π and Θ). The approach used by Howe (1975), following earlier work by Powell, selects (Θ and h_s), and following the elimination of $\partial\Theta/\partial t$, derives the convected wave equation for h_s shown in Figure 1. This equation possesses certain advantages over the other convected wave equations in $\ln p$, Π, or Θ, for certain classes of flows, but in the case of the turbulent jet it appears to offer no such advantages.

All the convected wave equations above can be solved numerically for a given computational domain, provided the flow field is known to a sufficient approximation. These convective wave equations contain all interactions and interference between the basic vortical flow and the disturbance wave motion it generates, and which radiates as sound to the far-field. The major interaction between the flow and the sound generated by it, is one of refraction primarily caused by the presence of the non-uniform mean flow field. Scattering of sound by the turbulence is normally considered of lesser importance, at least in the problem of the radiated noise from high speed jets. The problem of refraction relates to the high frequency solution of the convective wave equation and is analogous to the solution using geometrical acoustics. At lower frequencies the full solution of the convective wave equation is required and diffraction phenomena are encountered showing that the entire flow field cannot contribute to the sound radiation since part of the propagating wave motion remains trapped locally within the flow, although it may escape as sound further downstream due to the growth of the convected flow field with downstream distance. Approximate analytic solutions to the convective wave equation are given by Goldstein (1976), but more general solutions can be found numerically.

9. The Propagation of Sound Waves in an Inhomogeneous Flow

In Section 8 we considered the exact equations of fluid flow in analyzing the acoustic-flow interaction inside the flow field and the sound radiated to a far-field observer. In the present section we consider a hybrid scheme. For convenience we will neglect all diffusive effects. Since the approach used in this section differs from that used above, the notation differs also and is described below. Let us consider we can isolate from a given compressible flow, immersed in a uniform medium at rest, that part of the flow, defined by its velocity $\boldsymbol{w}$, pressure p_0 , and density ρ_0, which excludes the contribution due to the sound waves generated by it. We assume the fluctuating part of this velocity field has respectively a characteristic speed, w_0, and length scale, l_0, with $w_0/c_0 \ll 1$. A particular case is that in which the vortical component of the total velocity is $\boldsymbol{w}$, and in this case $\nabla.\boldsymbol{w} = 0$, and curl$\boldsymbol{w} = \boldsymbol{\omega}$. The equations of motion, continuity and energy for this hypothetical flow are:

$$\frac{D_0\boldsymbol{w}}{Dt} = -\frac{\nabla p_0}{\rho_0} : \frac{D_0\rho_0}{Dt} = -\rho_0\nabla.\boldsymbol{w} : \frac{D_0 \ln \rho_0}{Dt} Dt = \frac{D_0\Pi_0}{Dt} \qquad (28)$$

where $D_0/Dt \equiv \partial/\partial t + \boldsymbol{w}.\nabla$, and $\nabla\Pi_0 = \nabla p_0/\rho_0 {c_0}^2$. The speed of sound is c_0 and ${c_0}^2 = \gamma p_0/\rho_0$. It follows that the convected wave equation for the pressure field results in Phillips Equation:

$$\frac{{D_0}^2}{Dt^2}\Pi_0 - \nabla.({c_0}^2\nabla\Pi_0) = \nabla\boldsymbol{w} : \boldsymbol{w}\nabla \qquad (29)$$

where inside the flow the 'source function', $\nabla\boldsymbol{w} : \boldsymbol{w}\nabla$, has an order of magnitude such that the density fluctuations there are of order $\rho_0 w_0^2/c_0^2$, proportional to the flow kinetic energy. If we follow the arguments above, this equation shows that the fluctuating flow supports a radiated noise field, whose strength is of order $\rho_0 \boldsymbol{w}_0^4/c_0^4$, and thus small compared to the flow kinetic energy. It is further reduced in amplitude with distance from the flow by the inverse square law. These results are equal to those obtained using the Lighthill "acoustic analogy". In this hypothetical flow the acoustic disturbance generated by the flow does not interfere with the flow. Naturally such a flow supports an acoustic field, $\boldsymbol{v}$, where the total fluid velocity is $(\boldsymbol{v} + \boldsymbol{w})$ and the pressure and density are p and ρ respectively. The basic flow distorts the acoustic field and there is flow-acoustic interference. The strength of the acoustic disturbance is:

$$\nabla \Pi^s = \frac{(p - p_0)}{\rho c^2}$$

with $\mid \boldsymbol{v} \mid \ll \mid \boldsymbol{w} \mid$ internal to the flow, and $\Pi^s \ll \Pi_0$.

If we write $D/Dt \equiv \partial/\partial t + (\boldsymbol{v} + \boldsymbol{w})$, then approximately:

$$\frac{D_0^2 \Pi^s}{Dt^2} - \nabla.(c_0^2 \nabla \Pi^s) = \nabla \boldsymbol{v} : \boldsymbol{v} \nabla + 2 \nabla \boldsymbol{w} : \boldsymbol{v}\ \nabla \qquad (30)$$

where $\boldsymbol{v}$ is a function of Π^s. This is the convected wave equation for the acoustic disturbance and its right-side represents respectively the effects of non-linear wave steepening and acoustic-flow interaction. Both are propagation effects inside the flow field. Outside the flow where $\boldsymbol{w} = 0$ and non-linear steepening is absent, this equation reduces to the wave equation for sound propagation to a distant observer. When combined with the equation for Π_0 we recover the third-order convected wave equation for Π as described above. The solution to Equation (30) describes the acoustic field within the flow. This complex disturbance field superimposed on the basic turbulent flow will, in many flows, be insufficiently strong to change that flow or its flow structure, even though the energy of the flow must suffer a reduction (back reaction) since the energy of the disturbance is provided by the flow. This was exposed by Lighthill in his early work on acoustics.

Thus the description of the reduced convected wave equation in Π given in Section 8 above, with its attempt to describe the propagation of acoustic disturbances across the flow field separately from their generation by the flow, is shown to have support from the analysis in the present Section. We draw the conclusion that if the amplitude of the acoustic disturbances is known throughout the source region then the convected wave equation can determine their matched vector wave number components and frequencies corresponding to a given far-field observer and frequency. But the amplitude of the noise generation function in the convected wave equation is similar in amplitude to that in Lighthill's "acoustic analogy". If we find the ensemble averages of the contributions to the noise generation function, as described above, we can find the distribution of P_θ throughout the turbulent flow. What is now required are the changes to its wave number vector and frequency arising from its passage across the flow field in respect of a given observer and frequency. When the flow field is replaced by its mean velocity distribution an approximate determination of this flow-acoustic interaction can be obtained from use of the Whitham-Lighthill non-linear theory of ray acoustics as described by Guiraud (1964) and presented here in Appendix 2.

The high frequency approximation to the solution of the convected wave equation is analogous to ray refraction by the mean flow, and the change in ray direction across the flow generates a near 'zone of silence' in the downstream arc with respect to the jet exit and a refracted zone in the upstream arc. The complete solution to the convected wave equation shows an exponential decay in place of the 'zone of silence', or sharp cut-off. At lower wave numbers lesser changes in the wave number vector take place between the source and observer, although important changes exist in the intensity of the radiated sound and its directivity with jet Mach Number between high and low frequencies. An experimental confirmation of these results is shown in Figure 2.

Turbulent flow-acoustic interaction involves many complex phenomena such as refraction, diffraction, scattering, formation of caustics, as well as shock wave-turbulence interaction, when shock waves are both stationary and moving relative to the flow. In many of these interactions it is not possible to consider the acoustic disturbance so weak that no structural change in the basic flow occurs. The complete convective wave equations described above can still be used in such cases but the approximate and hybrid schemes would then no longer be appropriate.

10. The Computational Problem

The problem of Aerodynamic Noise considered in this review relates to the noise radiated to the far field of a jet at subsonic or supersonic speeds. However our treatment is related more to 'acoustic' jet Mach Numbers less than about 2, since the noise due to the presence of shock waves in the jet, and the formation of weak shock waves or

Mach waves, on and near the jet boundary, and their several interactions with the convecting turbulence is not discussed. The problem discussed here is therefore not one of compressible turbulence since we are considering the generation of noise in a flow in which the structure of the turbulence is only weakly affected by compressibility. The recent computational studies on compressible turbulence, especially in the computation of the fluctuations of the dilatation, contain important new results when the turbulence Mach Number is greater than about 0.4. These results, including the contribution to the dissipation from the dilatation, are important in respect of the turbulent structure of jets when the 'acoustic' jet Mach Number exceeds about 2. Nevertheless although in this paper we are concerned with lower jet Mach Numbers, the knowledge that CFD has made important advances in numerical techniques in recent years, offers the opportunity that progress can be made towards the development of a numerical prediction scheme for the problem of jet noise radiation at these lower Mach Numbers.

The first priority is the numerical calculation of the turbulent flow in the mixing region of a jet, with Large Eddy Simulation (LES) appearing to be the best choice available. Whatever scheme is adopted, however, it is suggested that turbulence should either be initiated at the nozzle lip or alternatively an initial vortex sheet springing from the nozzle lip could generate its own instability followed by the development of a subsequent cascade of subharmonics in a finite amplitude study as presented recently by Morris et al. (1990). In all cases a number of realizations of the flow are required continuing until the turbulence has reached a stationary state. It may be more convenient to consider a jet issuing into a uniform flow rather than into the ambient medium at rest. Differences in jet temperature, density and specific heats from ambient need to be considered. The aim is to provide a complete databank for the space-time properties of the entire flow field along with its space and time derivatives, which can be interrogated to find the properties of the mean velocity field and the several space-retarded time covariances and their related spectrum functions as required in Lighthill's "acoustic analogy" and the convective wave equations.

The direct attack on the calculation of the radiated noise to the far field of a jet presents many difficulties especially in respect of the resolution problem existing between the differences in length scales and amplitudes between the vortical flow and its generated, and accompanying sound field. It is suggested that a first approach would be to find the flow structure in which the noise could be ignored, or, if present, its amplitude would be sufficiently small not to cause any change in the flow structure. Thus a computational domain is required, along with an appropriate grid to provide the necessary resolution of all the flow variables and their derivatives throughout the flow field. Such a computational domain would represent only the flow domain and its near field, but the entrainment flow into it would have to exist. The field of flow would be compressible and viscous, and dissipation would be included. The necessary boundary conditions have been discussed in Section 6 above.

Let us assume the computational domain extends for several jet diameters and the computational boundary is cylindrical with radius R. The developing flow will not be cylindrical, however, and will depend on r, ν, z, t. The mean entrainment velocity $v_E(z)$ can be found from:

$$m(z) = \int\int \overline{\rho v_z} dS : (2\pi R)\rho_\infty v_E = -dm/dz : (\overline{p} - p_\infty) = -\rho_\infty {v_E}^2/2$$

and, $Th = \int\int \overline{(p + {v_z}^2 - p_\infty)} dS = constant$ and independent of z. The $Thrust = Th = \rho_J \pi (D_J^2/4) V_J^2$. The integrals of the rate of mass flux and momentum flux can be taken as ensemble averages over a suitable number of realizations. Upstream of the jet exit, in the absence of an external flow, the entrainment is spherically inwards and its overall entrainment source strength equals half that further downstream. Similarly some adjustment must be made to the entrainment source strength at the downstream end of the computational boundary. Due to the entrainment the pressure at the jet exit will be slightly smaller than p_∞ but its effect on the jet thrust will be negligible. Strictly the entrainment is not cylindrically symmetric and will be a function of the azimuthal angle, ν, and of time. The expression above for v_E gives the mean value for the entrainment velocity and the corresponding pressure at the domain boundary of radius R. But the entrainment velocity at each station along the jet axis must vary with azimuthal angle along with the corresponding pressure variation. Since the entrainment velocity is not known a priori it has to be obtained by iteration but must satisfy the integral conditions expressed above. It would appear the algorithm for this computation needs careful, further study.

From the databank giving the space-time properties of the turbulent shear flow, in the absence of sound waves, but when the turbulence is statistically stationary, estimates can be made of the T_{ij} space-time covariance at all positions inside the flow. If then an observer position is selected at, say, $\boldsymbol{x} = (x, \theta)$, in the far field the retarded times at all positions will be known and so the retarded-time two-point space covariance of T_{xx}, $P_\theta(\boldsymbol{y}, \boldsymbol{\Delta}, \tau)$, could be evaluated for all points $\boldsymbol{y}$ inside the flow, where $\boldsymbol{\Delta}$ is the space separation, and τ the corresponding retarded time difference. Some preliminary trials may show that P_θ possesses certain similarity properties, and if this is so, then this might avoid its detailed estimation everywhere inside the flow field. With that proviso we would require only its peak amplitude and the values of its characteristic length and time scales. In the notation used above these would be ${\rho_0}^2{u_0}^4, l_1, l_2, l_3$, and $1/\omega_0$. These values of P_θ in fixed frame coordinates could be transformed to moving coordinates at what was judged from the data to be the average convection speed of the turbulence. This would then provide a direct comparison with the earlier approximate calculations presented previously. The next step involves the evaluation of the Lighthill Integral in the calculations of the noise intensity and its directivity in the far field. This ignores flow-acoustic interaction. To correct for refraction effects at the higher frequencies due to the interaction between the sound and the mean flow, and other aspects of flow-acoustic interaction, we can use a number of approaches to find the change in the wave number vector for a given frequency of the sound at the specified observer, to find the corresponding wave number at the source as discussed in Section 10 above.

We note that the many turbulent flow models currently being used for the calculation of mainly steady or slowly developing unsteady flows, such as the Reynolds Stress Models and developments of the $k - \epsilon$ Model, cannot be used to model the turbulence in problems of Aerodynamic Noise, where the emphasis is on the properties of space retarded-time covariances of functions such as Lighthill's T_{ij}. These models at best are empirically derived and serve to model turbulent shear flows to a good approximation within the bounds of the databank available from experiment, and supplemented by results from computation. The use of such models for boundary layers, mixing regions, wakes and jets is necessary for the rapid calculation of complex flows such as the external flow around aerofoils, wings, intakes and fuselages, and internal flows in ducts and turbomachinery. But Aerodynamic Noise demands a far more detailed appraisal of the space-time properties of shear flow turbulence. However when the studies as discussed here are completed, we may well find they lead to improved descriptions of models suitable for steady flow analysis.

11. Conclusions

The subject of Aerodynamic Noise is reviewed with reference to the noise generated from the turbulent mixing region of a jet and radiated into free space. The review has centred on two approaches, the first being that of the Lighthill "acoustic analogy", and the second requiring a solution of a convected wave equation for the complete flow field.

It is shown that the triumph of Lighthill's great masterpiece, his "acoustic analogy", lies in its derivation from the exact equations of fluid motion, and presents us with an exact solution of the Navier-Stokes Equations as applied to this problem. As such it was a landmark in the science of fluid dynamics and acoustics and was the foundation stone for the new subject of Aeroacoustics. All work on Aerodynamic Noise has its links with Lighthill's "acoustic analogy". It is shown that although the Lighthill "acoustic analogy" is exact and includes all possible types of fluid flow, and in particular the complex and interacting modes of vorticity, entropy and sound, it presents a difficulty in its application to practical flows since the stress tensor, T_{ij}, which replaces the complete flow field is not known a priori. However, as Lighthill, was able to show at the time he introduced the "acoustic analogy", an estimate of T_{ij} which excluded the presence of all sound from the flow field including the noise generated by the unsteady flow itself, would in many case be a sufficient approximation from which to derive the properties of the radiated noise to the far field. This realization was the major breakthrough in making predictions in numerous problems of Aerodynamic Noise. Before the advances in computer science and the advent of supercomputers, the only possible method to predict the properties of the radiated noise from a high speed air jet was to model T_{ij} based on limited experience from experiment and crude approximations to its space-time properties. The paper discusses one such simplified approach, and its comparison with experimental measurements.

To improve the accuracy of predictions based on Lighthill's "acoustic analogy", and noting the almost impossible task of obtaining accurate information on the properties of T_{ij} from experiment, it is suggested that computational methods, such as LES, could be used to provide the necessary databank for the flow field, which under interrogation would provide the necessary space-retarded-time properties of the T_{ij} covariance in turbulent shear flows, such as the mixing region of a jet, when the flow was statistically stationary.

References

Buckley, R. and Morfey, C. L., 1983. "Effects on jet mixing noise: Scaling laws predicted from flight simulation data," AIAA 83-0748.

Crighton, D. G., 1981. "Acoustics as a branch of fluid mechanics," *J. Fluid Mech.* **106**, p. 261.

Crow, S. C., 1970. "Aerodynamic sound emission as a singular perturbation problem," *Stud. Appl. Math.* **XLIX**, p. 1.

Ffowcs Williams, J. E., 1963. "The noise from turbulence at high speed," *Phil. Trans. Roy. Soc. (A)* **255**, p. 1061.

Fisher, M. J., Harper-Bourne, M., and Glegg, S. A. L., 1977. "Jet noise source location: The polar correlation method," *J. Sound Vibration* **51**, p. 1.

Goldstein, M. E., 1976. *Aeroacoustics*, McGraw-Hill Inc.

Guiraud, J. P., 1964. "Theorie du Bruit Ballistique," ONERA Note Tech., p. 79.

Howe, M. S., 1975. "Contributions to the theory of aerodynamic sound with applications to excess jet noise and the theory of the flute," *J. Fluid Mech.* **71**, p. 4.

Hubbard, H. H., 1991. "Aeroacoustics of flight vehicles: Theory and practice. Vol. 1: Noise sources," NASA Ref. Pub. 1258.

Landahl, M. T., 1967. "A wave-guide model for turbulent shear flow," *J. Fluid Mech.* **29**, p. 3.

Legendre, R., 1981. "Bruits emis par la turbulence," ONERA Publ. 1981-3.

Lighthill, M. J., 1952. "On sound generated aerodynamically: I. General theory," *Proc. Roy. Soc. (A)* **211**, p. 1107.

Lighthill, M. J., 1954. "On sound generated aerodynamically: II. Turbulence as a source sound," *Proc. Roy. Soc. (A)* **222**, p. 1148.

Lighthill, M. J., 1962. "The Bakerian lecture, 1961: Sound generated aerodynamically," *Proc. Roy. Soc. (A)* **267**, p. 1329.

Lighthill, M. J., 1963. "Jet noise," *AIAA J.* **1**(7), pp. 1507-1517.

Lighthill, M. J., 1978. *Waves in Fluids*, C.U.P.

Lilley, G. M., 1958. "On the noise from air jets," ARC 20376 (unpublished).

Lilley, G. M., Morris, P. J., and Tester, B. J., 1975. "On the theory of jet noise and its applications," *AIAA Progress in Astronautics and Aeronautics* **37**.

Mohring, W., 1983. "Problems in flow acoustics," *Reviews in Modern Physics* **55**, p. 3.

Morris, P. J., Giridharan, M. G., and Lilley, G. M., 1990. "On the turbulent mixing of compressible free shear layers," *Proc. Roy. Soc. (A)* **431**, p. 1882.

Phillips, O. M., 1960. "On the generation of sound by supersonic turbulent shear layers," *J. Fluid Mech.* **9**, p. 1.

Pridmore-Brown, D. C., 1958. "Sound propagation in a fluid flowing through an attenuating duct," *J. Fluid Mech.* **4**.

Ribner, H. S., 1964. "The generation of sound by turbulent jets," *Advances in Applied Mechanics* **8**, Academic Press.

Tester, B. J. and Fisher, M. J., 1981. "Engine noise source breakdown: Theory, simulation and results," AIAA-81-2040.

Tester, B. J. and Morfey, C. L., 1976. "Developments in jet noise modelling-theoretical predictions with measured data," *J. Sound Vibration* **46**, p. 1.

Appendix 1

An estimate of the covariance of T_{xx} without using statistical assumptions

by

Sir James Lighthill F.R.S.

(University College, London UK)

We are concerned with mean products like, $< T_{xx}(\boldsymbol{y},t)T_{xx}(\boldsymbol{z},t) >$; or, equivalently (since variations in the factor in T_{xx} are not usually significant), with $< u^2(0)u^2(r) >$. Here, $u(0)$ and $u(r)$ are random variations in the direction of a distant far-field observer x, and r is the vector separation $\boldsymbol{z} - \boldsymbol{y}$.

We first introduce the two-point correlation coefficient,

$$R = \frac{< u(0)u(r) >}{< u^2 >}$$

assuming that both $u(0)$ and $u(r)$ have the same mean square $< u^2 >$. Here, R represents that fraction of the variation of $u(r)$ which follows that of $u(0)$. This suggests that we may use $u(0)$ as one random variable and introduce

$$w = \frac{u(r) - Ru(0)}{(1 - R^2)^{1/2}}$$

as another, but independent, random variable with

$$< u^2(r) >= R^2 < u^2 > +(1 - R^2) < w^2 >,$$

so that, $< u^2 >=< w^2 >$. On these assumptions, we have,

$$\begin{aligned} < u^2(0)u^2(r) > &= < u^2(0) \left(Ru(0) + (1 - R^2)^{1/2} w\right)^2 > \\ &= R^2 < u^4 > +(1 - R^2) < u^2 >< w^2 > \\ &= R^2 < u^4 > +(1 - R^2) < u^2 >^2 \end{aligned}$$

so that the covariance of $u^2(0)$ with $u^2(r)$ is,

$$< \left(u^2(0) - < u^2 >\right) \left(u^2(r) - < u^2 >\right) >=$$

$$< u^2(0)u^2(r) > - < u^2 >^2 = R^2 \left(< u^4 > - < u^2 >^2\right).$$

Thus the correlation coefficient of $u(0)$ with $u(r)$ is

$$\frac{<(u^2(0)-<u^2>)>(u^2(r)-<u^2(r)>)>}{<(u^2(0)-<u^2>)^2} =$$

$$\frac{<u^2(0)u^2(r)> -<u^2>^2}{\left(<u^4> -<u^2>^2\right)} = R^2,$$

a remarkably simple result.

Here there is no assumption about the relative magnitudes of $<u^4>$ and $<u^2>^2$. If Gaussian statistics were employed $<u^4>/<u^2>^2$ would take the value 3 of course. It is interesting that even when at large separations R may become negative before approaching zero as $r \to \infty$, the correlation coefficient R^2 of $u^2(0)$ with $u^2(r)$ remains positive for all values of r. But the main importance of this analysis is that it indicates a correlation length for T_{xx} (distance within which its correlation, here estimated as R, remains significant) comparable with that of the velocity itself. In short, it is eddies of scales only slightly smaller than the main energy bearing eddies that generate most of the acoustic radiation.

Appendix 2

The Whitham-Lighthill non-linear theory of ray acoustics

From the exact equations of fluid flow as presented in Section 10 for a basic flow $(\boldsymbol{w}, p_0, \rho_0, c_0)$ on which is superimposed its acoustic disturbance field, with the total flow $(\boldsymbol{v}+\boldsymbol{w}, p, \rho, c)$, it can be shown following Guiraud (1964) that in a curvilinear system of coordinates with σ measured along a ray tube from source to observer:

$$\frac{\partial v_1}{\partial t} + (c_0 + \boldsymbol{w}.\boldsymbol{e}_1 + (\gamma+1)v_1/2)\,\boldsymbol{e}_1.\nabla v_1 = -v_1 \frac{\partial \ln \text{Æ}}{\partial t} + \frac{\delta}{2}\nabla^2 v_1$$

where δ is the absorption coefficient, and

$$\text{Æ} = \exp\left(\frac{1}{2}\int^t \left(Kc_0 + \frac{\partial \ln \rho_0 c_0}{\partial t'} + \gamma \nabla \boldsymbol{w} + \boldsymbol{e}_1.\boldsymbol{\nabla w}.\boldsymbol{e}_1\right) dt'\right)$$

the ray tube area function, with K equal to twice the curvature of the wave front, contains the properties of the basic flow.

If $e_1.\nabla = \partial/\partial\sigma$ and $x_1 = \sigma - (c_0 + w.e_1)t$ then on transforming the above equation to moving coordinates (x_1, Λ), where $\Lambda = t$, corresponding to the position of the wavefront along the ray tube,

$$\frac{\partial v_1}{\partial t} + \frac{\gamma+1}{2} v_1 \frac{\partial v_1}{\partial x_1} = -v_1 \frac{\partial \ln Æ}{\partial \Lambda} + \frac{\delta}{2}\nabla^2 v_1.$$

Since Æ is independent of x_1, we put $w = v_1$ Æ and introduce the 'age variable' $\lambda = \int^{\Lambda}(\gamma+1)/2Æ d\lambda$, then, with $\nu_E = \delta Æ/(\gamma+1)$

$$\frac{\partial w}{\partial \lambda} + w\frac{\partial w}{\partial x_1} = \nu_E \nabla^2 w$$

is Burgers Equation for the non-linear propagation of sound through a specified flow field. When absorption is neglected then the solution is:

$$\begin{aligned} w &= w(\omega) \\ x_1 &= \omega + \lambda w(\omega), \end{aligned}$$

which is the Whitham-Lighthill non-linear theory of sound propagation (Lighthill, 1978). This solution must be matched to the initial source data.

Figure 1. Equations of Aerodynamic Noise

Name	Variable Eliminated	Dependent Variable	Equation (Diffusion terms omitted)
Lighthill	$\frac{\partial \nabla.\rho \boldsymbol{v}}{\partial t}$	ρ	$\left(\frac{\partial^2}{\partial t^2} - c_\infty{}^2\nabla^2\right)\rho = \nabla.\nabla.\boldsymbol{T}$ $T_{ij} = \rho v_i v_j + (p - \rho c_\infty{}^2)\delta_{ij}$
Phillips	$\frac{D\Theta}{Dt}$	$\ln p$	$\left(\frac{D^2}{Dt^2} - \nabla.c^2\nabla\right)\ln p =$ $\gamma\nabla\boldsymbol{v} : \boldsymbol{v}\nabla$
Lilley	$\frac{D\Theta}{Dt}$	Π $= \ln p^\gamma$	$\frac{D}{Dt}\left(\frac{D^2}{Dt^2} - \nabla.c^2\nabla\right)\Pi$ $+2\nabla\boldsymbol{v} : \nabla(c^2\nabla\Pi) =$ $-2\nabla.\boldsymbol{v}\nabla.\boldsymbol{v} : \boldsymbol{v}\nabla$
Legendre	$\nabla^2\frac{D\Pi}{Dt}$	Θ $= \nabla.\boldsymbol{v}$	$\left(\frac{D}{Dt}\frac{1}{c^2}\frac{D}{Dt} - \nabla^2\right)\Theta =$ $-\frac{D}{Dt}\left(\frac{\nabla\boldsymbol{v} : \boldsymbol{v}\nabla}{c^2} - \frac{\nabla \ln h}{c^2}.\frac{D\boldsymbol{v}}{Dt}\right)$ $-\frac{\nabla^2\boldsymbol{v}}{c^2}.\frac{D\boldsymbol{v}}{Dt} - 2\nabla\boldsymbol{v} : \nabla\left(\frac{1}{c^2}\frac{D\boldsymbol{v}}{Dt}\right)$
Howe	$\frac{D\Theta}{Dt}$	h_s $= h + v^2/2$	$\left(\frac{\partial}{\partial t}\frac{1}{c^2}\frac{D}{Dt} - \nabla^2\right)h_s =$ $\frac{\partial}{\partial t}\left(\frac{1}{c^2}\frac{D}{Dt}\frac{v^2}{2}\right) - \nabla.(\boldsymbol{v}\times\boldsymbol{\omega})$

Figure 2. Directivity of jet noise as found in two sets of experiments at subsonic speeds.

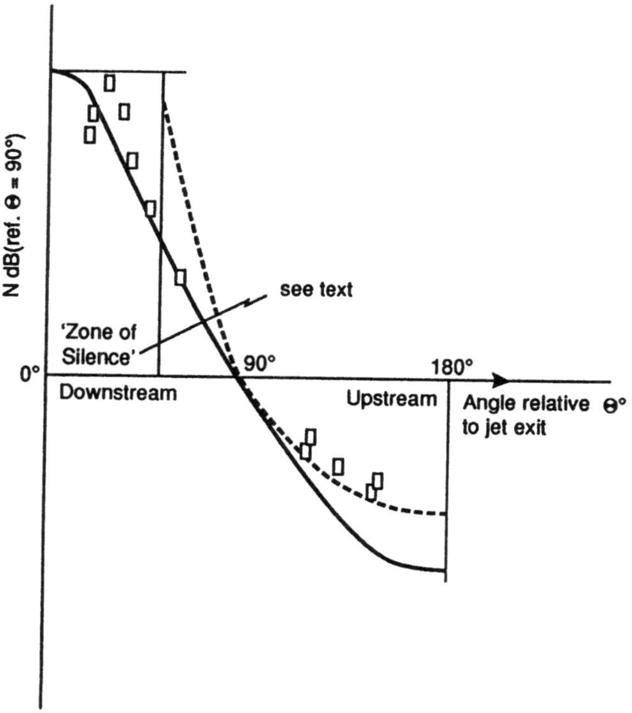

—— $N = 10\log_{10}\left((1 - M_c\cos\theta)^2 + \frac{s_0^2 u_0^2}{c_\infty^2}(\cos^2\theta + \frac{l_1^2}{l_0^2}\sin^2\theta)\right)^{-5/2}$

– – – High frequency approximation (refraction only) using ray acoustics.

A STUDY OF THE SHORT WAVE COMPONENTS IN COMPUTATIONAL ACOUSTICS

Christopher K. W. Tam, Jay C. Webb and Zhong Dong[1]

Department of Mathematics
Florida State University
Tallahassee, Florida 32306-3027

ABSTRACT

The feasibility of performing direct numerical simulations of acoustic wave propagation problems has recently been demonstrated by a number of investigators. It is easy to show that the computed acoustic wave solutions are good approximations of those of the exact solutions of the linearized Euler equations as long as the wavenumbers are in the long wave range. Computed waves with higher wavenumber, or the short waves, generally have totally different propagation characteristics. There are no counterparts of such waves in the exact solutions. The short waves are contaminants of the numerical solutions. The characteristics of these short waves are analyzed here by group velocity consideration. Numerical results of direct simulations of these waves are reported. To purge the short waves so as to improve the quality of the numerical solution, it is suggested that artificial selective damping terms be added to the finite difference scheme. It is shown how the coefficients of such damping terms may be chosen so that damping is confined primarily to the high wavenumber range. This is important for then only the short waves are damped leaving the long waves basically unaffected. The effectiveness of the artificial selective damping terms is demonstrated by direct numerical simulations involving acoustic wave pulses with discontinuous wave fronts.

1. Introduction

Linear acoustic waves are governed by the linearized Euler equations. These waves are nondispersive, nondissipative and isotropic. They propagate at exactly the speed of sound. To calculate these

[1]This work was supported by the NASA Lewis Research Center Grant NAG 3-1267 and in part by the Office of Naval Research Grant No. N00014-89-J-1836 and the Florida State University through time granted on its Cray – YMP Supercomputer.

waves computationally, say, by using a finite difference scheme it is not at all clear that the numerical solutions will retain these characteristics. On the contrary it is well known that finite difference approximation invariably introduces numerical dispersion, dissipation and mesh anisotropy (Vichnevetsky and Bowles, 1982 and Trefethen, 1982). These problems are most severe for short waves or waves with high wavenumbers. For very short waves inevitably there will be numerical dispersion. Further, spurious grid-to-grid oscillations, referred to as parasite waves (Trefethen, 1982), are often found. These numerical contaminants are most undesirable. They degrade the quality of the numerical solutions. If the fully nonlinear Euler equations are used they might even cause extraneous distortion of the waveform and perhaps nonlinear instability.

The purpose of this paper is two fold. The first objective is to provide a mathematical definition of the short wave components from the wavenumber point of view. A concrete example will be provided to illustrate their uncommon and somewhat unexpected wave propagation characteristics. Since these wave components are not part of the physical solution, the second objective is to introduce a method to eliminate these waves. The long waves with small wavenumbers constitute the useful band of waves for numerical simulation. This portion of the wavenumber spectrum must be protected while the short waves are being purged. To meet this constraint, a method of selective damping is proposed. It will be shown that it is possible to add artificial damping terms to a finite difference scheme which remove primarily the short waves while having negligible effects on the long waves. The effectiveness of this method will be demonstrated by examining the solutions of the linearized Euler equations with discontinuous initial data.

2. The Wavenumber of a Finite Difference Scheme

In a recent paper Tam and Webb (1992) introduced a way to find the effective wavenumber and frequency of a finite difference scheme. Consider the approximation of the first derivative $\frac{\partial f}{\partial x}$ at the ℓth node of a uniform grid of spacing Δx. Suppose M values of f to the right and N values of f to the left of this point are used to form the finite difference approximation, i.e.

$$\left(\frac{\partial f}{\partial x}\right)_\ell \simeq \frac{1}{\Delta x} \sum_{j=-N}^{M} a_j f_{\ell+j}. \tag{2.1}$$

Equation (2.1) is a special case of the following finite difference equation in which x is a continuous variable:

$$\frac{\partial f}{\partial x}(x) \simeq \frac{1}{\Delta x} \sum_{j=-N}^{M} a_j f(x + j\Delta x) \tag{2.2}$$

(2.1) can be recovered from (2.2) by setting $x = \ell\Delta x$. The Fourier transform and its inverse of a function are related by:

$$\tilde{f}(\alpha) = \frac{1}{2\pi} \int_{-\infty}^{\infty} f(x) e^{-i\alpha x} dx \tag{2.3a}$$

$$f(x) = \int_{-\infty}^{\infty} \tilde{f}(\alpha) e^{i\alpha x} d\alpha. \tag{2.3b}$$

The Fourier transform of the left and right side of (2.2) are:

$$i\alpha\tilde{f} \simeq \left(\frac{1}{\Delta x} \sum_{j=-N}^{M} a_j e^{i\alpha j \Delta x}\right) \tilde{f}. \tag{2.4}$$

By comparing the two sides of (2.4) it is clear that the quantity

$$\overline{\alpha} = \frac{-i}{\Delta x} \sum_{j=-N}^{M} a_j e^{ij\alpha\Delta x} \tag{2.5}$$

is effectively the wave number of the Fourier transform of the finite difference scheme (2.2) or (2.1). $\overline{\alpha}\Delta x$ is a periodic function of $\alpha\Delta x$ with period 2π. Tam and Webb chose the coefficients a_j so that $\overline{\alpha}\Delta x$ is the best mean least square approximation of $\alpha\Delta x$ over the range $-\pi/2 \leq \alpha\Delta x \leq \pi/2$. For $N = M = 3$ the coefficients are found to be

$$\begin{aligned} a_0 &= 0 \\ a_1 &= -a_{-1} = 0.79926643 \\ a_2 &= -a_{-2} = -0.18941314 \\ a_3 &= -a_{-3} = 0.02651995. \end{aligned}$$

Figure 1 shows the relation $\overline{\alpha}\Delta x$ versus $\alpha\Delta x$ over the interval 0 to π using the above coefficients. For $\alpha\Delta x$ up to 1.45 the curve is nearly the same as the straight line $\overline{\alpha} = \alpha$. Thus the finite difference scheme can provide an adequate approximation to the partial derivative for waves with wave lengths longer than 4.5 mesh spacings. For $\alpha\Delta x$ greater than 1.45 the $\overline{\alpha}(\alpha)$ curve deviates increasingly from the straight line relationship. Because of this the wave propagation characteristics of the short wave components of the finite difference equations would be very different from those of the partial differential equations. We will refer to the wave spectrum for which $\alpha\Delta x > 1.45$ as the short waves and the rest of the spectrum the long waves.

Tam and Webb (1992) also considered a four level explicit time marching scheme. The governing equations, e.g. the Euler equations, provide the time derivative of the dependent variable $\boldsymbol{U}$. Suppose the solution is known up to a time level $t = n\Delta t$. To advance to the next time step a four-level finite difference approximation in the form

$$\boldsymbol{U}^{(n+1)} - \boldsymbol{U}^{(n)} \simeq \Delta t \sum_{j=0}^{3} b_j \left(\frac{d\boldsymbol{U}}{dt}\right)^{(n-j)} \tag{2.6}$$

is used. The finite difference equation (2.6) is a special case of the following difference equation in which t is a continuous variable:

$$\boldsymbol{U}(t+\Delta t) - \boldsymbol{U}(t) \simeq \Delta t \sum_{j=o}^{3} b_j \frac{d}{dt}\boldsymbol{U}(t - j\Delta t) \tag{2.7}$$

(2.7) is identical to (2.6) if t is set equal to $n\Delta t$. By applying the Laplace transform to (2.7) it is easy to obtain

$$-i\frac{i(e^{-i\omega\Delta t} - 1)}{\Delta t \sum\limits_{j=0}^{3} b_j e^{ij\omega\Delta t}}\tilde{\boldsymbol{U}} \simeq \widetilde{\frac{d\boldsymbol{U}}{dt}} \tag{2.8}$$

where $\tilde{}$ represents the Laplace transform and ω is the Laplace transform variable. The Laplace transform of the time derivative of $\boldsymbol{U}$, i.e. the right side of (2.8) is, however, equal to $-i\omega\tilde{\boldsymbol{U}}$. Thus by comparing the two sides of (2.8) the quantity

$$\overline{\omega} = \frac{i(e^{-i\omega\Delta t} - 1)}{\Delta t \sum\limits_{j=0}^{3} b_j e^{ij\omega\Delta t}} \tag{2.9}$$

is the effective angular frequency of the time discretization scheme (2.6). Tam and Webb suggested to use

$$b_0 = 2.30255809 \quad , \quad b_1 = -2.49100760$$
$$b_2 = 1.57434093 \quad , \quad b_3 = -0.38589142.$$

With these coefficients it can be shown that $\overline{\omega}\Delta t$ is almost identical to $\omega\Delta t$ over an extended range of values.

3. The Short Wave Component

In Figure 1 we see that the wavenumber of the optimized 7 point finite difference scheme, $\overline{\alpha}$, is a good approximation of the true wavenumber α when $\alpha\Delta x < 1.45$. But for the short waves ($\alpha\Delta x > 1.45$) the wavenumber of the finite difference scheme deviates more and more from the correct wavenumbers as $\alpha\Delta x$ increases. An important consequence of this discrepancy is numerical dispersion. In this section it will be shown that the propagation characteristics of the short waves are quite different from those of the acoustic waves of the Euler equations and the long waves. These waves are dispersive with propagation speed quite different from the speed of sound.

To fix ideas let us consider the initial value problem associated with the linearized Euler equations in one space dimension in the absence of a mean flow. The dimensionless momentum and energy equations are:

$$\frac{\partial u}{\partial t} + \frac{\partial p}{\partial x} = 0 \tag{3.1}$$

$$\frac{\partial p}{\partial t} + \frac{\partial u}{\partial x} = 0 \tag{3.2}$$

for simplicity we will take the initial conditions to be:

$$t = 0, \qquad u = 0 \tag{3.3}$$

$$p = f(x). \tag{3.4}$$

The above initial value problem has an exact solution for arbitrary $f(x)$ (see Haberman, 1987 or Berg and McGregor, 1966), namely,

$$p(x,t) = \frac{1}{2}[f(x-t) + f(x+t)]. \tag{3.5}$$

This solution implies that half the initial pressure pulse propagates to the left and the other half to the right at the speed of sound (or unity in dimensionless units).

If $f(x)$ is a smooth function such that the dominant part of its Fourier transform consists of wavenumbers lying in the range $\alpha\Delta x <$ 1.45 then the initial value problem can be solved accurately by finite difference method. This is so since only long waves are involved. As an example consider an initial pressure distribution in the form of a Gaussian with a half width of 5.0 (Δx is the length scale) i.e.

$$f(x) = \exp[-\ln\ 2(\frac{x}{5.0})^2]. \tag{3.6}$$

The Fourier transform of (3.6) is also a Gaussian with a half width of 0.4 ln 2 so that almost all the wave components are long waves. On applying the differencing scheme of Section 2 to (3.1) and (3.2) the discretized equations are:

$$u_m^{n+1} = u_m^n - \Delta t \sum_{j=0}^{3} b_j \left[\sum_{k=-3}^{3} a_k p_{m+k}^{n-j} \right] \tag{3.7a}$$

$$p_m^{n+1} = p_m^n - \Delta t \sum_{j=0}^{3} b_j \left[\sum_{k=-3}^{3} a_k u_{m+k}^{n-j} \right] \tag{3.7b}$$

where the superscript is the time index and the subscript the space index. (Note: $\Delta x = 1$ in dimensionless units.) The time evolution of the pressure pulse (3.6) can be easily calculated according to finite difference equations (3.7). Figure 2 gives a comparison between the exact (dotted curve) and the computed (full line) pressure waveform at $t = 200\Delta t$, $2000\Delta t$ and $3800\Delta t$, respectively ($\Delta t = 0.1086$). In this time period the pressure disturbance has propagated over 400 times the characteristic length (half width of the Gaussian) of the problem. As can be seen the speed and wave profile are well maintained. There appears to be minimal numerical dispersion although the entire pressure pulse is resolved only by about 20 mesh points.

3.1. Group velocity, dispersive and parasite waves

A simple way to classify the short waves is to use the concept of group velocity. It can be shown that the nondimensional group velocity of the short waves of the finite difference equation (3.7) is nearly equal to the derivative $\frac{d\bar{\alpha}}{d\alpha}$ of (2.5) or the slope of the curve in Figure 1. This is given in graphical form in Figure 3. The long waves are nearly nondispersive having a dimensionless group velocity essentially equal to unity. The group velocity of the short waves is generally less than unity with the high wavenumber components having negative values. To emphasize this fact, we will call the short waves with positive group velocity "dispersive waves" and those with negative group velocity (group velocity in opposite direction to phase velocity) "parasite waves" (see Trefethen, 1982). The parasite waves have large wavenumbers and hence very short wave lengths. They are responsible for the often observed grid-to-grid oscillations. The grid-to-grid oscillation wave components also have very large propagation (group) velocity; more than twice that of the long waves. Thus once they are excited they can be found at the head of a wave packet.

3.2. Discontinuous initial data

Let us now consider the solution of the discretized equation (3.7) with discontinuous initial data. As a concrete example let $f(x)$ of (3.4) be zero everywhere except for $-M < x < M$, i.e.

$$f(x) = \begin{cases} H(x+M) - H(x-M), & t = 0 \\ 0, & t < 0 \end{cases} \tag{3.8}$$

where H is the unit step function. A solution of the finite difference equation (3.7) and initial conditions (3.3), (3.4) and (3.8) valid for all time can be found computationally. Figure 4 shows the computed pressure waveform of the right propagating waves at 2000 time steps ($\Delta t = 0.1086, M = 50$). As can be seen, overall, the solution consists of a rectangular pulse moving at the speed of sound. However, the quality of the solution is degraded by the presence of numerous fine scale oscillations. Far ahead of the main pulse moving at more than twice the speed of sound are the parasite waves. One characteristic of these waves is their rapid spatial oscillations (grid-to-grid). Superimposed on top of the main wave pulse are the dispersive waves. They have longer wave lengths than the parasite waves. The parasite and the dispersive waves are undesirable elements of a numerical solution (discretization errors). They could obscure small amplitude phenomena of the solution. They could provide a false impression of the speed of propagation of the medium. In short, they are pollutants of computational acoustics.

4. Artificial Selective Damping

The short waves are numerical contaminants of computational acoustics. To improve the quality of the computed solution it is imperative that they be automatically removed from the computation as soon as they are generated. One way to implement this strategy is to add artificial damping to the finite difference scheme. The idea of adding artificial damping terms is not new. For example, earlier Jameson, Schmidt and Turkel (1981) proposed the addition of a second order and or a fourth order damping term to their finite volume scheme with the purpose of suppressing wiggles in regions with severe pressure gradients. Lele (1990), on the other hand, suggested periodically filtering the computed solution to eliminate the high frequency components. Here a procedure for designing artificial damping terms which selectively damp out the short waves but have minimal effect on the long waves will be discussed. The effectiveness of adding such selective damping will be demonstrated by numerical simulation.

4.1. The basic concept

Suppose artificial damping terms are added to the right side of the linearized momentum equation (3.1).

$$\frac{\partial u}{\partial t} + \frac{\partial p}{\partial x} = D. \tag{4.1}$$

Upon applying spatial discretization the equation for the mth mesh point is

$$\frac{du_m}{dt} + \ldots = -\mu \sum_{j=-3}^{3} c_j u_{m+j}. \tag{4.2}$$

Here a linear damping term proportional to u_{m+j} has been assumed. μ is a constant. For convenience, the same size stencil as for the spatial discretization is used. On generalizing (4.2) to a difference equation with continuous variable in x and then taking the Fourier transform of the equation, it is easy to find,

$$\frac{d\tilde{u}}{dt} + \ldots = -\mu \left[\sum_{j=-3}^{3} c_j e^{ij\alpha\Delta x} \right] \tilde{u} \tag{4.3}$$

where $\tilde{u}$ is the Fourier transform of u. (Note: $\Delta x = 1$ in dimensionless unit. It is kept in (4.3) merely for clarity.) Clearly the quantity

$$\widetilde{D}(\alpha\Delta x) = \sum_{j=-3}^{3} c_j e^{ij\alpha\Delta x}$$

is the (relative) damping constant for the wavenumber α. Since the coefficients c_j are arbitrary they will be chosen so that $\widetilde{D}$ is practically zero for $\alpha\Delta x$ in the long wave range but is finite and positive in the short wave number range.

To ensure that $\widetilde{D}$ is an even function of $\alpha\Delta x$ we set $c_{-j} = c_j$. Thus

$$\widetilde{D} = c_0 + 2\sum_{j=1}^{3} c_j \cos(j\alpha\Delta x). \tag{4.4}$$

The right side of (4.4) is a truncated Fourier cosine series. The coefficients can be adjusted so that $\widetilde{D}$ has the desirable properties. As a template for the choice of c_j the function $\widetilde{D}$ may be made to approximate the Gaussian function

$$\exp\left[-\ln 2\left(\frac{\alpha\Delta x - \pi}{\sigma}\right)^2\right]. \tag{4.5}$$

This function peaks at $\alpha\Delta x = \pi$ and decays quickly towards the long wave number range. These are the desirable properties of a selective damping function. It is not necessary to make $\widetilde{D}$ the best approximation of (4.5) over the whole range $0 \leq \alpha\Delta x < \pi$. It is sufficient to use the range $0 \leq \alpha\Delta x < \pi/2$. Therefore, the c_j's are determined by minimizing the integral

$$\int_0^{\pi/2} \left\{\widetilde{D}(\alpha\Delta x) - \exp\left[-\ln 2\left(\frac{\alpha\Delta x - \pi}{\sigma}\right)^2\right]\right\}^2 d(\alpha\Delta x).$$

To standardize the damping curve the function $\widetilde{D}(\alpha\Delta x)$ is normalized so that $\widetilde{D}(\pi) = 1.0$. As an example, with σ (the half-width of the Gaussian) taken to be 0.3π the values of the c_j's are found to be

$$c_0 = 0.351061040 \quad , \quad c_1 = -0.242824317$$

$$c_2 = 0.074469480 \quad , \quad c_3 = -0.007175683. \tag{4.6}$$

A graph of the damping function $\widetilde{D}(\alpha\Delta x)$ is given in Figure 5.

4.2. Numerical results

On adding the artificial damping terms to the discretized Euler equations (3.7), the difference equations of the explicit time marching scheme become

$$u_m^{n+1} = u_m^n - \Delta t \sum_{j=0}^{3} b_j \left[\sum_{k=-3}^{3} (a_k p_{m+k}^{n-j} + \mu c_k u_{m+k}^{n-j}) \right]$$

$$p_m^{n+1} = p_m^n - \Delta t \sum_{j=0}^{3} b_j \left[\sum_{k=-3}^{3} (a_k u_{m+k}^{n-j} + \mu c_k p_{m+k}^{n-j}) \right]. \qquad (4.7)$$

To test the effectiveness of the selective damping terms (4.7) has been computed up to $n = 4000$ with discontinuous function (3.8) as initial data. The constant μ is taken to be 0.3. In computing the numerical solution it is noticed that the parasite waves which are readily seen in Figure 5 are damped out almost immediately as they are generated. The damping coefficient for the dispersive waves is smaller. Their presence in the numerical solution can still be seen after 200 time steps. However, at $t = 500\Delta t$ they are almost completely eliminated. Figure 6 shows the calculated pressure waveform at 2000 time steps. By comparing with the waveform of Figure 5 it is easy to see that the artificial selective damping is extremely effective. Essentially all the spurious waves have been damped out. The remaining solution consists of two rectangular pulses propagating with the speed of sound, one to the left and the other to the right consistent with the exact solution.

5. Summary and Discussion

In this paper the question of whether direct computation methods can be used to calculate acoustic wave propagation problems accurately is addressed. Results of numerical simulations strongly suggest that such methods can, indeed, provide high quality numerical solutions to this type of problem. This, however, is true only for long waves. The reason for this is that the wavenumber of a finite difference scheme invariably differs from the true wavenumber at the high end of the wavenumber spectrum. Waves in this range are called short waves whereas those in the low wavenumber range are referred to as long waves. The dispersion relations of the long waves are essentially the same as those of the original partial differential equations so that the long waves of the computation scheme and the waves of the governing partial differential equations automatically have the same propagation characteristics. On the other hand the dispersion relations of the short waves are different. Hence they are not capable of simulating the corresponding waves of the original equations.

For the linearized Euler equations the characteristics of the short waves can be analyzed and understood through group velocity consideration and standard dispersive wave theory (see Whitham, 1974 and Tam, Webb and Dong, 1993). It is shown that the finite difference scheme can support short waves which cause grid-to-grid oscillations. These waves propagate with ultra-fast speeds and can be easily found ahead of the main waves.

Short waves are contaminants of a computation scheme. They should be purged as soon as they are generated. Here the idea of adding artificial selective damping terms to the finite difference scheme to suppress the short waves without materially affecting the long waves is discussed. It is shown how the damping coefficients can be chosen so that artificial damping is confined primarily to the high wave number range. The effectiveness of selective damping is considered and demonstrated by a concrete numerical example.

References

Berg, P. W. and McGregor J. L., 1966. *Elementary Partial Differential Equations*, Holden Day, San Francisco.

Haberman, R., 1987. *Elementary Applied Partial Differential Equations*, 2nd Edition, Prentice Hall.

Jameson, A., Schmidt, W. and Turkel, E., 1981. "Numerical solutions of the Euler equations by finite volume methods using Runga-Kutta time-stepping schemes," AIAA Paper 81-1259.

Lele, S. K., 1990. "Compact finite difference schemes with spectral-like resolution," Center for Turbulence Research Manuscript 107, NASA Ames Research Center/Stanford University.

Tam, C. K. W. and Webb, J. C., 1992. "Dispersion-relation-preserving schemes for computational aeroacoustics," AIAA Paper 92-02-033.

Tam, C. K. W., Webb, J. C. and Dong, Z., 1993. "A study of the short wave components in computational acoustics," *J. Computational Acoustics* (to appear).

Trefethen, L. N., 1982. "Group velocity in finite difference schemes," *SIAM Review* **24**, 113.

Vichnevetsky, R. and Bowles, J. B., 1982. *Fourier Analysis of Numerical Approximations of Hyperbolic Equations*, SIAM, Philadelphia.

Whitham, G. B., 1974. *Linear and Nonlinear Waves*, Wiley-Interscience.

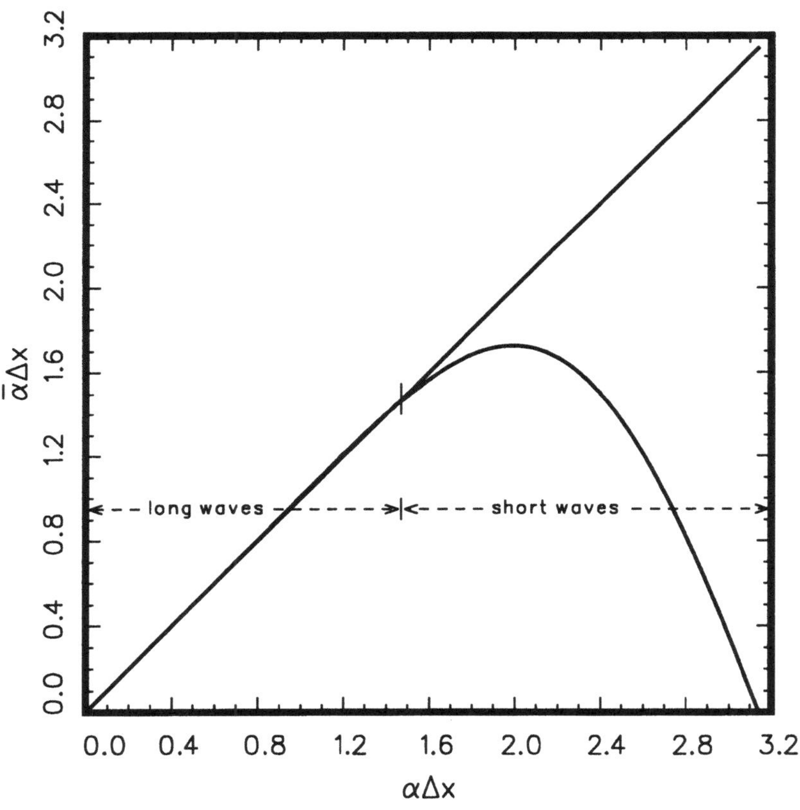

Figure 1. $\overline{\alpha}\Delta x$ versus $\alpha\Delta x$ for the optimized 4th order central finite difference scheme (N=M=3)

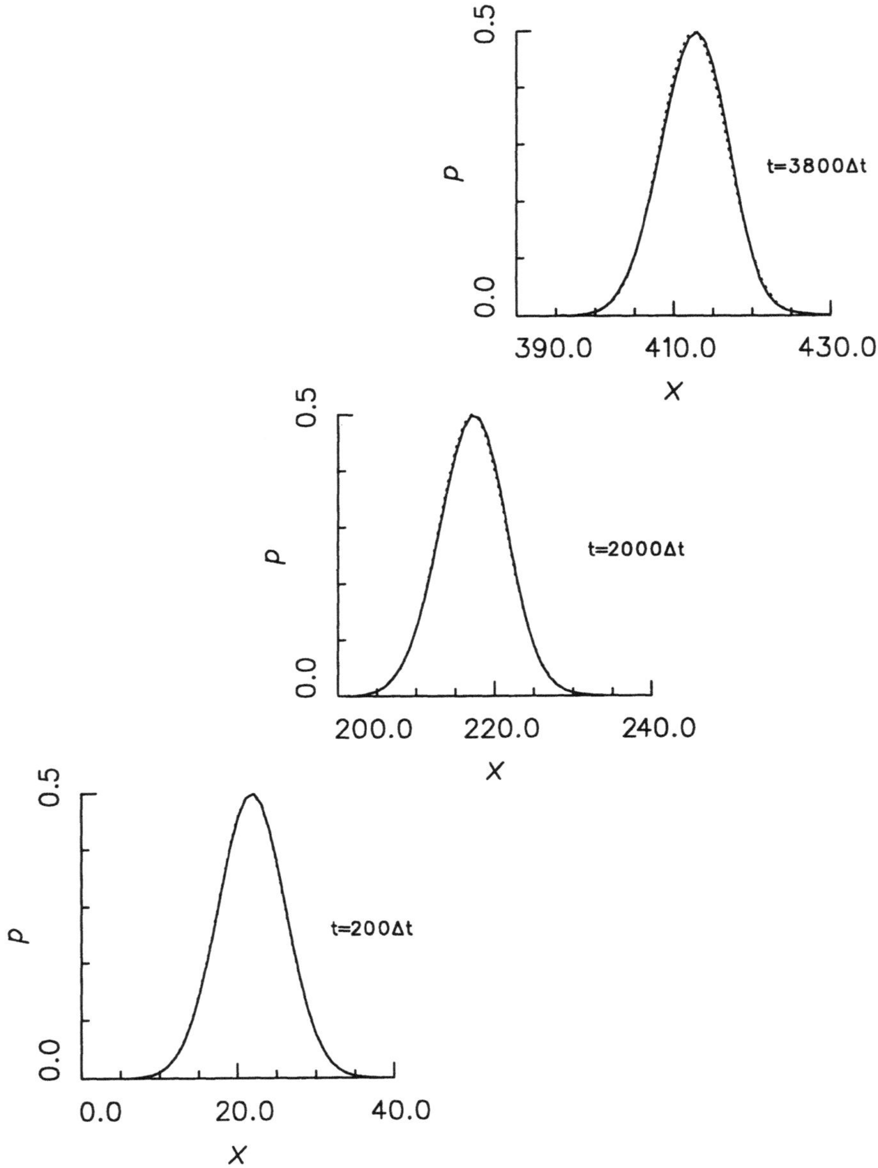

Figure 2. Pressure waveform for the pulse propagating to the right. Initial disturbance has Gaussian profile.

··· exact solution — computed solution

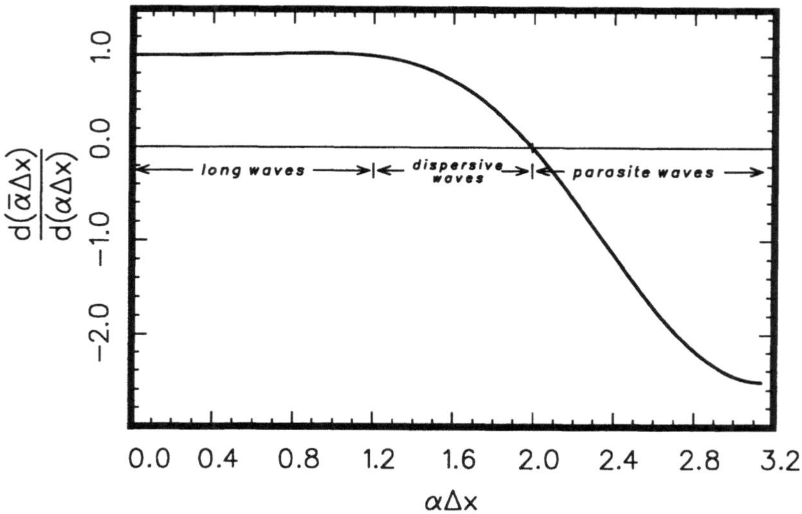

Figure 3. Slope of $\overline{\alpha}\Delta x$ versus $\alpha\Delta x$ curve (N=M=3)

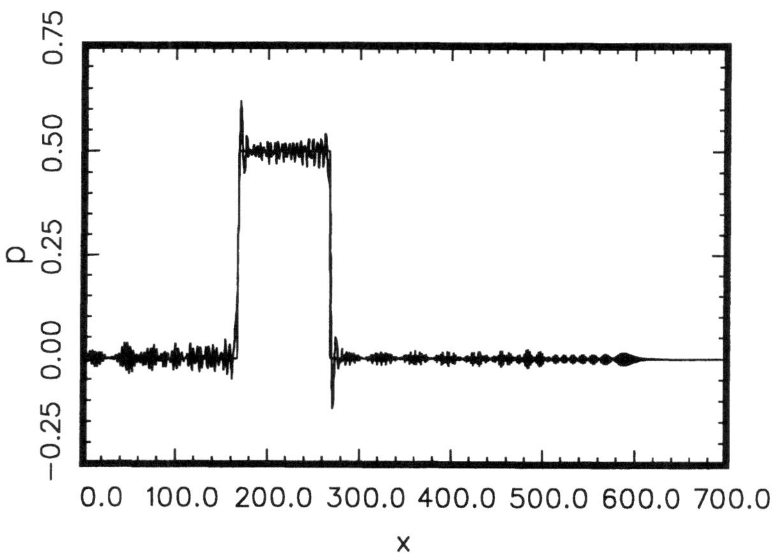

Figure 4. Pressure waveform initiated by a disturbance with rectangular profile showing dispersive and parasite waves. 2000 time steps

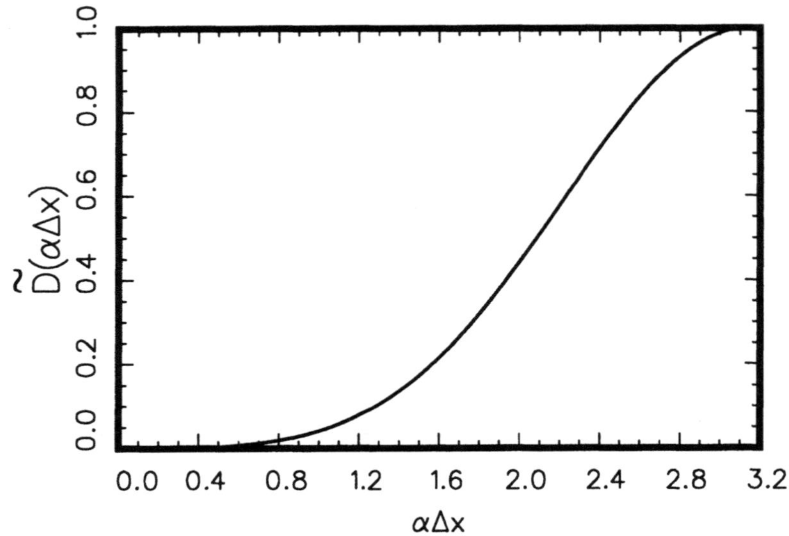

Figure 5. Fourier Transform of the artificial selective damping terms. Gaussian template with 0.3π half-width

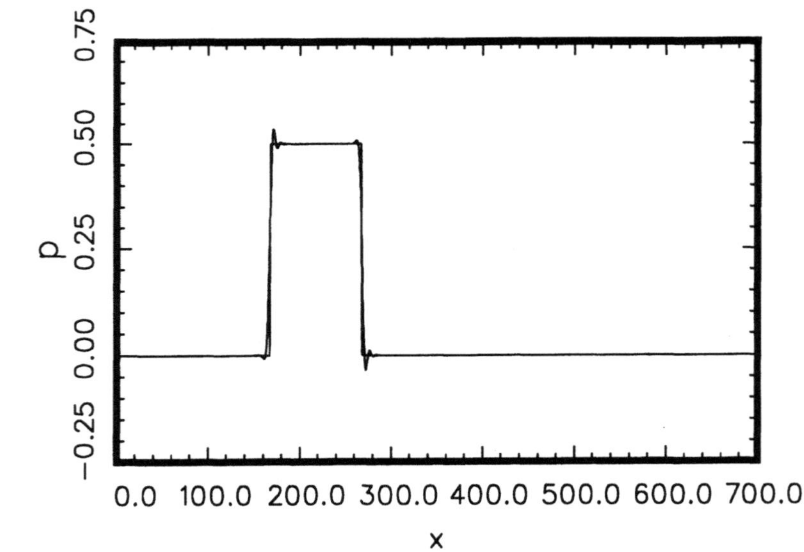

Figure 6. Pressure waveform initiated by a disturbance with rectangular profile at 2000 time steps computed with artificial selective damping terms, $\mu = 0.3$, half-width of Gaussian $= 0.3\pi$

MATHEMATICAL ASPECTS OF ACOUSTICS

THE ACOUSTIC ANALOGY AS A TOOL OF COMPUTATIONAL AEROACOUSTICS

F. Farassat

NASA Langley Research Center
Hampton, Virginia 23681-0001

ABSTRACT

The acoustic analogy proposed by Lighthill in 1952 in his jet noise theory provides an exact governing equation of noise generation away from the flow region. The generalization of Lighthill's equation in the presence of moving surfaces by Ffowcs Williams and Hawkings in 1969 was a major advance. The primary impact of the governing equation of noise generation known as the Ffowcs Williams-Hawkings (FW-H) equation has been in the prediction of the noise of rotating blades such as high speed propellers and helicopter rotors. Since the mid-seventies many researchers have derived solutions to the FW-H equation in both the time and frequency domains. Sophisticated computer codes for noise prediction based upon the acoustic analogy have been developed with highly realistic blade geometry, kinematics and aerodynamic input. As a result of the demand by aeroacousticians, much effort has gone into the development of high resolution aerodynamics for use in acoustic codes. The acoustic analogy is now in a mature stage. The aircraft industry is beginning to rely on codes based on the acoustic analogy to control aircraft propulsion system noise. In this paper, we argue that, because of the exact nature of the FW-H equation, the availability of high quality aerodynamic data, sophisticated acoustic codes and high performance computers, the acoustic analogy is an important and useful tool of computational aeroacoustics. Some examples for advanced propellers and rotors are presented.

1. Introduction

Aircraft engine and propulsion system noise has become an important public issue today. Increased helicopter utility because of its vertical take-off and landing ability in civilian and military environments has focused attention on the noise issue. Fixed wing aircraft using conventional and advanced propellers and turbofan engines can generate considerable noise during take-off and landing. Airline passengers have increased substantially in recent years. Current projections show further growth in the future. This growth in passengers must be met by a combination of an increase in aircraft utility, the use of larger aircraft and increasing the number of airliners. The last remedy may require the construction of more airports. To increase aircraft utility, one clear option is night flights which are now prevented for most large aircraft because of the airport curfew due to the community complaint against noise. Larger aircraft make more noise and airport construction is almost always opposed by surrounding communities because of the noise issue. Therefore, all current remedies to meet the increase in passengers are governed by the issue of aircraft noise. The angry public is forcing their legislature to enact strict noise regulations. Future noise standards (Stage 4) are bound to be lower than present (Stage 3) standards by perhaps more than 6 PNdB.

Current noisy aircraft in crowded communities reduce the noise impact by engine cut-backs and by selecting a flight speed, altitude, and path to minimize the noise footprint. This is not a completely satisfactory solution since the engine noise of many aircraft is so high that this procedure may not be sufficient to meet noise standards. Also the issue of aircraft safety during such maneuvers precludes extreme departures from conventional take-off and landing procedures. Aircraft propulsion noise reduction must, therefore, be included in the design stage. The engine and aircraft industries are responding to this demand. Thus far, semi-empirical noise prediction approaches have sufficed. But the need for accurate predictions in the design stage to meet future noise standards, as well as the increased commercial advantage of low noise aircraft necessitates sophisticated theories and computer codes. Progress in this direction has been made steadily in the last few decades and the pace is increasing because of the increased government funded research and other factors such as the development of supercomputers. In this paper, we briefly describe the successes of one of the most useful computational aeroacoustic approaches based on the acoustic analogy. We will concentrate on rotating blade noise prediction.

Since the early days of flight, many researchers have worked on the understanding and calculation of the noise of rotating blades. One remembers the names of Hilton, Gutin, Deming, Ernsthausen, Garrick and Watkins, Hubbard, Lordi, Lowson, Leverton, and many others. The concept of acoustic analogy was introduced in the study of another source of aircraft noise, namely the turbulence of a jet engine which was utilized in airliners shortly after World War II (Lighthill, 1952). The acoustic analogy was used to explain qualitatively many of the measured characteristics of noise from jets. Curle extended Lighthill's theory to the case of turbulent flow in the presence of stationary reflecting surfaces (Curle, 1955). One of the most significant theoretical advances has been the generalization of Lighthill's theory to the case of noise generation by turbulence in the presence of moving surfaces (Ffowcs Williams and Hawkings, 1969). The governing equation of noise generation for moving surfaces is now known as the Ffowcs Williams-Hawkings (FW-H) equation. The current capability in propeller and helicopter rotor noise prediction is largely based on the use of this equation.

Since the mid-seventies when the first applications of the FW-H equation appeared, many technical papers on the subject of noise prediction have been published. In some areas of propeller and rotor noise analysis, particularly for deterministic sources, significant advances have been made over a long period of time. One of the aims of this paper is to show that the acoustic analogy is capable of supplying excellent quantitative, as well as qualitative results such as trends and rules for design. Many researchers in the U.S. and abroad have been involved in the development of noise prediction tools using the acoustic analogy. One is now able to look back and evaluate the usefulness of the acoustic analogy in practical applications. The present workshop on computational aeroacoustics (CAA) provides the opportunity for this evaluation.

In the next section, the concept of the acoustic analogy is described first and the FW-H equation is presented. Some topics related to noise prediction are discussed in the subsections. In the final section, several examples of applications of the FW-H equation will be presented for propellers and helicopter rotors. The possibility of application of the acoustic analogy to ducted fans will also be discussed.

2. The Acoustic Analogy

2.1. Complications associated with noise prediction of rotating machinery

Before we discuss the acoustic analogy, it is appropriate to present the complicating features associated with prediction of the noise of rotating machinery. Since a designer requires accurate prediction of the noise for optimal design, a prediction model should incorporate the physics, the geometry, and the kinematics of the blades as accurately as possible given the available computers and computation time. This means that one can make a range of approximations from very simple to very realistic and complex models. It is safe to say that for very accurate models, one needs to use the modern supercomputers.

Helicopter rotors have a complex blade motion such as coning, flapping, cyclic and collective pitch changes. Aerodynamics and structural dynamics of the blades are coupled. Accurate prediction of even the blade motion is a difficult task at present. This means that the unsteady aerodynamics of helicopter rotors is inherently a very difficult problem. There are many rotor noise generation mechanisms such as blade-vortex, rotor-turbulence and rotor-wake interactions. The planform of rotor blades is usually rectangular except possibly at the tips. The large blade thickness produces shock waves on blades requiring inclusion of nonlinearities in acoustic analysis.

Advanced technology propellers use thin blades that are highly loaded and operate at supersonic tip speeds. In addition, the blades are highly twisted and swept. The large disc loading causes blade deformation which affects both aerodynamics and noise. Therefore, aerodynamics and structural dynamics of the blades are coupled just like helicopter rotor blades. Because of axial motion of propellers, this problem is much easier to handle for propellers than helicopter rotors.

A problem of considerable interest is the prediction of the noise of large modern turbofans known also as ducted fans. The noise generated in the external duct system of these propulsors dominates the core noise. The unsteady aerodynamics of the external duct system is complex because of the interaction of fan flow with the guide vanes and the pylon. The full aerodynamic problem has not yet been solved satisfactorily. Liner modeling of the duct surface is an additional problem which requires further research.

2.2. The hierarchy of noise prediction techniques

What are the avenues open to us for noise prediction of rotating machinery? Roughly speaking, we have three approaches in the order of chronological development and perhaps accuracy:

1. high resolution aerodynamics in the near field and acoustic analogy for radiation to far field,
2. high resolution aerodynamics in the near field and Kirchhoff's formula for radiation to the far field,
3. fully computational aerodynamics and acoustics.

The first approach based on the acoustic analogy is by far the most mature of the three approaches. It is the subject of the present paper. The second approach based on the Kirchhoff formula was proposed by Hawkings to incorporate the effect of nonlinearities which appear in the acoustic analogy approach as quadrupole sources. It has been somewhat successfully used in the prediction of the blade-vortex interaction noise of helicopter rotors. In connection with this approach, we must mention that the correct Kirchhoff formula for moving surfaces derived by Morgans (1930) was rederived by a modern approach (Farassat and Myers, 1988). It is not at present obvious that this approach is superior to the acoustic analogy since it has not yet been applied to the full rotor or propeller noise prediction. One reason is that unsteady aerodynamic calculations usually deteriorate in the region where the Kirchhoff boundary surface is located. This shortcoming of data specification leads to serious errors in noise prediction. Finally, the fully computational technique requires new algorithms to control numerical dispersion and dissipation. The range of frequencies of interest in computational acoustics dictates very small grid sizes in space and time. This necessitates the use of supercomputers which are not yet available. Some efforts in use of this approach have been made in calculation of high speed impulsive noise of helicopter rotors (Baeder, McCroskey, and Srinivasan, 1986) and in direct simulation of noise from turbulence (Lele, 1992).

2.3. What is the acoustic analogy?

Lighthill in his now classical jet noise theory paper (Lighthill, 1952) reasoned that since the region of agitated fluid is confined to the jet flow itself and the density (or pressure) perturbation satisfies the wave equation away from the jet, one may derive a noise generation formula as follows (see Figure 1). Manipulate the conservation laws in such a way that the left side is the wave operator acting on perturbation density. In practice, only a part of the wave operator involving the time derivatives can be obtained in this way. The Laplacian part is then subtracted from both sides of the equation to get the wave operator on the left side. The inhomogeneous term on the right side is called the source term and in Lighthill's theory is a quadrupole source with the Reynolds stress as the dominant component. The wave equation governing the jet noise generation is

$$\Box^2 p' = \frac{\partial^2 T_{ij}}{\partial x_i \partial x_j} \tag{1}$$

where $p' = c^2 \rho'$, ρ' is the perturbed density, c is the speed of sound in the undisturbed medium and T_{ij} is the Lighthill stress tensor which is nonzero in the flow region only.

Note that the effect of flow nonlinearities has been mathematically simulated in Equation 1 by fictitious sources in an undisturbed medium. This means that somehow we must have the complete information in the near field flow region. In the case of jet noise where the turbulence generates the noise, this is a tall order. Until recently, computer simulation of T_{ij} was impractical but this situation is changing. In the case of rotating blades, the deterministic sources are the most important contributors to the noise and the specification of source strengths is much more manageable. Below we will address the problem of rotating blades.

2.4. The Ffowcs Williams-Hawkings (FW-H) equation

This equation is derived by manipulation of conservation of mass and momentum equations similar to the derivation of Lighthill's jet noise formula. The difference is that now a moving surface described by $f(\vec{x}, t) = 0$ is present. The most direct approach is to set the problem in the generalized function space (Gel'fand and Shilov, 1964; Jones, 1982; Kanwal, 1983). In this space, conservation laws are valid for discontinuous functions provided that all derivatives are viewed as generalized derivatives. To bring out the contributions of the moving surface to the source terms of the wave equation, each parameter of flow is multiplied by $H(f)$, where $H(\cdot)$ is the Heaviside function which is unity outside the moving surface. Lighthill's procedure then gives the FW-H equation (Ffowcs Williams and Hawkings, 1969)

$$\Box^2 p' = \frac{\partial}{\partial t}\left[\rho_0 v_n \delta(f)\right] - \frac{\partial}{\partial x_i}\left[l_i \delta(f)\right] + \frac{\bar{\partial}^2}{\partial x_i \partial x_j}\left[T_{ij} H(f)\right] \quad (2)$$

where ρ_0 is the density of undisturbed medium, v_n is the local normal velocity of the surface, l_i is the force intensity (force/unit area) acting on the medium and T_{ij} is the Lighthill stress tensor. We have assumed that $|\nabla f| = 1$ over the surface $f = 0$. The Dirac delta function is denoted by $\delta(\cdot)$ and the bar on the derivative of the last term indicates generalized differentiation. The three inhomogeneous source terms on the right of Equation 2 are known as the thickness, loading and quadrupole sources, respectively. In practice, the force intensity l_i is usually approximated accurately by $p\ n_i$ where p is the local surface pressure and n_i is the local unit outward normal to the surface. Other derivations of the FW-H equation are available in the literature (Farassat, 1977; Kanwal, 1983). We emphasize that the source terms on the right side of Equation 2 are assumed known either experimentally or from aerodynamic calculations.

We see that one of the main advantages of the acoustic analogy here is the separation of the aerodynamic and acoustic problems. Much effort has gone into computational fluid dynamics (CFD) in recent years. The quality of both steady and unsteady aerodynamic calculations is improving continually. The fully computational acoustic approach is not feasible today and the acoustic analogy which utilizes CFD results is the ideal alternative. Equation 2 is, in principle, exact. Powerful methods of linear analysis can be used to find the solution. In particular, the Green's function approach is commonly used. By bringing the observer onto the surface and treating the formal solution of Equation 2 as an integral equation in the unknown surface pressure p, even the aerodynamic data can be calculated for use in noise prediction (Hanson, 1991; Farassat and Myers, 1986; Dunn, 1991).

2.5. Research on Ffowcs Williams-Hawkings equation

Research on the FW-H equation has progressed in two levels since its derivation in 1969. At the basic research level, the acoustic analogy was used by many researchers, as well as Ffowcs Williams and associates to understand fundamental acoustic problems. These problems are usually difficult and are highly idealized to yield qualitative and scaling results. Some of these problems are discussed by Ffowcs Williams (1984). At the applied research level which has been highly successful for rotating blades, exact analytic solutions in the time or frequency domains are used to develop computer codes for noise prediction. Attempts are made to model the blade geometry and kinematics as accurately as possible. Examples of researchers in this group are Hanson, Farassat, and Crighton and his associates. It is interesting to note that based on an asymptotic analysis of a result of Hanson, Crighton and associates have obtained formulas which give both qualitative and quantitative information on rotating blade noise (Crighton and Parry, 1992).

3. Some Applications

In this section we present some examples of noise prediction based on the acoustic analogy for propellers and helicopter rotors. Most of these examples are by researchers at NASA Langley Research Center. Many other examples have been published in the literature by researchers in the helicopter and propeller industry. As seen from these examples, the inclusion of deterministic sources has resulted in the most spectacular successes. At present, the specification of source strengths for nondeterministic sources (caused by turbulent effects) is difficult. It is hoped that the current efforts in direct simulation of turbulence will also have an impact on the prediction of the noise from nondeterministic sources. Recently, Langley researchers have looked into the possibility of application of the acoustic analogy to the prediction of the noise of ducted fans. The presence of the duct which filters propagating spinning models complicates the problem. However, preliminary calculations appear promising and a new application of the acoustic analogy may be in the horizon.

Two major research programs in the U.S. have produced a wealth of theoretical and experimental results published mostly in the open literature. First was the propfan or advanced technology propeller research program from the mid-seventies until 1990. The second was the NASA-Industry Helicopter Program $(NR)^2$ in the 1980's. One current U.S. government noise research program is on large ducted fans.

3.1. Propeller noise

Propellers have been in use since the early days of aviation. Modern high speed advanced technology propellers (propfan) may still come into use because of their high efficiency. Most commuter aircraft use propellers for propulsion. Currently, there is a strong need for quiet and efficient propellers. Much can be done to meet these needs.

We briefly present the basis for two theoretical formulations in the time domain used extensively in computer codes developed at NASA Langley. Readers should consult the excellent article on propeller and propfan noise in a recent aeroacoustics publication (Magliozzi, Hanson, and Amiet, 1991) in which frequency domain methods are emphasized. Most prediction methods for propellers include thickness and loading noise only. Quadruple sources, because of the low thickness of blade sections, do not appear to be important. We use two formulations known as formulations 1A (low speed) and 3 (high speed). Formulation 1A has surface integrals over the actual blade surface S and has the Doppler factor $1 - M_r$ in the denominator of the integrads (Farassat and Succi, 1983; Brentner, 1986). We note that the solution for the thickness noise equation

$$\Box^2 p'_T = \frac{\partial}{\partial t}\left[\rho_0 v_n \ \delta(f)\right] \tag{3}$$

is

$$\begin{aligned} 4\pi p'_T(\vec{x}, t) &= \frac{\partial}{\partial t}\int_{f=0}\left[\frac{\rho_0 \ v_n}{r(1-M_r)}\right]_{\text{ret}} dS \\ &= \int_{f=0}\left[\frac{1}{1-M_r}\frac{\partial}{\partial \tau}\frac{\rho_0 \ v_n}{r(1-M_r)}\right]_{\text{ret}} dS \end{aligned} \tag{4}$$

where the subscript "ret " stands for retarded time. The solution for the loading noise equation

$$\Box^2 p'_L = -\frac{\partial}{\partial x_i}\left[l_i \delta(f)\right] \tag{5}$$

is

$$\begin{aligned} 4\pi p'_L(\vec{x}, t) &= \frac{1}{c}\frac{\partial}{\partial t}\int_{f=0}\left[\frac{l_r}{r(1-M_r)}\right]_{\text{ret}} dS \\ &+ \int_{f=0}\left[\frac{l_r}{r^2(1-M_r)}\right]_{\text{ret}} dS \end{aligned} \tag{6}$$

where $l_r = l_i \hat{r}_i$ and $\hat{r}_i$ is the component of the unit radiation vector $\vec{r}$. For the first term on the right of Equation (6), bring the time derivative inside the integral using $\frac{\partial}{\partial t} = \frac{1}{1-M_r}\frac{\partial}{\partial \tau}$. Formulation 1A is then the sum of Equations (4) and (6) after time derivatives with respect to the source time are taken.

The derivation of the high speed formulation is very lengthy. Farassat and Myers have recently published a more direct derivation using some advanced concepts from generalized function theory and differential geometry (Farassat and Myers, 1991). We give a glimpse of this derivation for thickness noise only. Let an open surface (e.g. a panel) on the blade be described by $f = 0$ and $\tilde{f} > 0$. The edge of this surface is described by $f = \tilde{f} = 0$. We assume that $\nabla \tilde{f} = \vec{\nu}$, the geodesic normal of the edge. Then, the contribution of this open surface to thickness noise is

$$\begin{aligned}\frac{\partial}{\partial t}\left[\rho_0 v_n \; H(\tilde{f})\delta(f)\right] &= \rho_0 \dot{v}_n \; H(\tilde{f})\delta(f) \\ &- \rho_0 \hat{v}_n^2 \; H(\tilde{f})\delta'(f) \\ &- \rho_0 v_n v_\nu \; \delta(f)\delta(\tilde{f}) \qquad (7)\end{aligned}$$

where $H(\cdot)$ is the Heaviside function and $v_\nu = \vec{v} \cdot \vec{\nu}$. The first term gives a surface integral. The second term gives both a surface integral and a line integral over the edge. The cap ^ on $\hat{v}_n$ denotes the restriction of v_n to the surface $f = 0$. This restriction is needed in the interpretation of $\delta'(f)$. The last term is a line integral over $f = \tilde{f} = 0$. Similar manipulations are applied to the loading term of the FW-H. eq. The solution of the wave-equation with line and surface sources is published elsewhere (Farassat, Padula, and Dunn, 1987; Farassat and Myers, 1991).

The computer code ASSPIN (formerly DFP-ATP) developed by Dunn, Farassat and Padula at NASA incorporates these two theoretical formulations with automatic switch between formulations as needed (Farassat, Padula, and Dunn, 1987; Dunn and Farassat, 1992). Propeller noise prediction is incorporated in the comprehensive aircraft noise prediction code of NASA known as ANOPP. The high speed formulation of this code is a predecessor of formulation 3 used in ASSPIN. Figure 2 shows the measured and predicted noise of a general aviation propeller operating at an angle of attack. Classical quasi-steady aerodynamic calculations were used for this figure in ANOPP. The agreement of measured and predicted data is very good for all harmonics of the blade passage frequency (BPF). Figure 3 shows the measured and predicted waveforms of a small supersonic propfan (SR-3). The large positive peaks of the predicted waveforms are integrated in time by the microphone surface. Again it is seen that the agreement of measured and predicted data is good. Figure 4 shows the comparison of the measured and predicted noise spectra for a contra-rotating unducted fan using ASSPIN. The averaged blade surface pressures were used in these calculations. It is seen that the predicted data agree well with measured data in and near the rotor planes where blade unsteadiness is not important. Recent developments in unsteady aerodynamics make it possible to do a much better prediction than in Figure 4.

Current advanced technology propellers are highly loaded and use thin blades made of composite materials. Blade deflection becomes important in aerodynamic and acoustic calculations. Interactive methods between aerodynamic and aeroelastic codes must be used to find blade shape and load. A comprehensive system of codes involving aerodynamics (Adamczyk), aeroelastic (NASTRAN) and acoustics (ASSPIN and MRS-BLP fuselage boundary layer propagation) codes was developed at Langley. Figure 5 shows how these codes are linked. A recent article presents the development of this system and results of the analysis for a large eight bladed (SR-7L) propfan (Dunn and Farassat, 1992). Figure 6 presents some results from this analysis which show the power of the current noise prediction codes. It was found that aeroelastic deformation of the blades has a significant effect on the aerodynamic and noise of propfans.

3.2. Helicopter rotor noise

As mentioned earlier, a helicopter rotor generates noise by many different mechanisms. In contrast to propellers, only recently have sufficiently good unsteady aerodynamic calculations become available for noise prediction. As expected, deterministic source specification has been much more successful than nondeterministic sources due to turbulence effects. It must be mentioned that at present, further theoretical and experimental research is needed in unsteady aerodynamics for noise prediction because of the complexity of the flow around rotors. Nevertheless, considerable success has been achieved since the mid-70s in prediction of rotor noise based exclusively on the acoustic analogy. Some comprehensive reviews of the state-of-the-art of helicopter rotor noise prediction have been published recently (George, 1978; JanakiRam, 1990; Brentner and Farassat, 1992; Lowson, 1992; Schmitz, 1991; Burton, Schlinker, and Shenoy, 1985). The reviews of JanakiRam and Brenter and Farassat are particularly relevant to this paper because of their emphasis on modern computational aeroacoustics. Lowson's review article addresses all aspects of helicopter noise generation and prediction. His treatment of turbulence induced noise is interesting and fairly complete.

The three areas in which the acoustic analogy has been very successful are: (i) thickness and steady or periodic loading noise; (ii) blade-vortex interaction noise; and (iii) high-speed impulsive noise. Time domain methods are more popular in helicopter rotor noise prediction. The two noise prediction codes used by NASA and the U.S. helicopter industry are WOPWOP (Brentner, 1986) and ROTONET. Both these codes use formulation 1A of the author. As an example of thickness and loading noise calculation, Figure 7 shows a comparison of measured and predicted rotor noise of a two bladed helicopter rotor. The disagreement of the data at the first harmonic is due to the neglect of quadrupole sources. However, the predicted spectrum is in reasonably good agreement with the measured spectrum over a large range of frequencies.

Figure 8 shows a comparison of the measured and predicted noise by blade-vortex interaction of a four bladed model rotor (Brentner and Farassat, 1992). Both the acoustic waveforms and spectra are shown. The agreement between the two data sets is excellent. Here, the measured blade surface pressure is used in acoustic calculation. Other blade-vortex interaction predictions based on theoretical aerodynamic calculations have also achieved considerable success (Brentner and Farassat, 1992).

Quadrupole sources are known to generate noise with the directivity and intensity of thickness noise for thick blades moving at high speeds (Hanson and Fink, 1979; Schmitz and Yu, 1979). In general, the inclusion of quadrupole sources around rotors in forward flight involves large amounts of aerodynamic data. It is becoming feasible to handle such data on recent computers with large memories. The solution of the quadrupole noise equation

$$\Box^2 p'_Q = \frac{\partial^2}{\partial x_i \partial x_j} [T_{ij} H(f)] \tag{8}$$

can be written as

$$4\pi p'_Q(\vec{x},t) = \frac{\partial^2}{\partial x_i \partial x_j} \int_{F>0} \frac{T_{ij}}{r} \, c \, d\Omega \, d\tau \tag{9}$$

where $F(\vec{y};\vec{x},t) = f(\vec{y}, t - r/c) = [f(\vec{y},t)]_{\text{ret}}$ and Ω is the surface of the sphere $r = c(t-\tau)$. The center of this sphere is at $\vec{x}$ and the integration is over all source times $\tau \leq t$. In the far field, $\frac{\partial^2}{\partial x_i \partial x_j} \int [\cdot]$ in Equation 8 can be approximated by $\frac{\partial^2}{c^2 \partial t^2} \int \hat{r}_i \hat{r}_j [\cdot]$ (Schmitz and Yu, 1979). This reduces in the calculation time of the quadrupole noise. The exact conversion of the space derivatives to time derivative was found later (Farassat and Brentner, 1988). This result has not yet been used for quadrupole noise prediction.

Experimental evidence indicates that quadrupole sources become important when shock waves appear around the blades. This noise is particularly dominant when the phenomenon of delocalization (Schmitz, 1991) occurs. For this reason, the present author proposed that the shock surface noise is the major component of the quadrupole sources. Using the quadrupole source term of the FW-H equation, the shock surface sources were identified as monopole and dipole terms. The governing equation of shock noise generation is

$$\Box^2 p'_S = \Delta \left(\frac{\partial T_{ij}}{\partial x_i} \right) n'_j \delta(k) + \frac{\partial}{\partial x_i} \left[\Delta T_{ij} n'_j \delta(k) \right] \tag{10}$$

where $k(\vec{x},t)$ is the shock surface, n'_j is the component of the unit normal to this surface and Δ is the jump in a quantity across the shock surface (Farassat and Brentner, 1988; Farassat and Myers, 1991). It can be shown that the volume quadrupole sources cancel some contribution of the dipole and monopole sources on the shock. However, a physical reasoning can be given to show that the shock surface sources should be the dominant cause of the high-speed impulsive noise of helicopter rotors (Farassat, Lee, Tadghighi, and Holz, 1991). Some numerical results from Equation 10 support this conclusion (Tadghighi, Holz, Farassat, and Lee, 1991). More research is needed in this area. Recently, Equation 10 has been used by Glegg to study shock-turbulence interaction noise. The preliminary unpublished results explain some experimentally observed trends. Once again, the acoustic analogy is providing the quantitative analysis of an important noise generation mechanism.

4. Concluding Remarks

It is hoped that readers have become convinced that, at least for the case of deterministic sources, the acoustic analogy is an important and useful tool for computational acoustics of rotating blade machinery. Many factors have contributed to the present popularity of the acoustic analogy:

1. The development of high speed digital computers with large memories has made it possible to use very realistic physical and geometrical models in computer codes,

2. High resolution unsteady aerodynamic data are becoming available from advanced CFD codes,

3. The difficult problem of aerodynamic-aeroelastic coupling of blades can be solved independent of the acoustic problem to get the blade loads and the deformed blade surface shape for acoustic calculations,

4. The aircraft industry has built up experience with current sophisticated computer codes based on the acoustic analogy.

It is clear that the future trend will be toward the increased use of the acoustic analogy for both deterministic and non-deterministic sources. The aerodynamicist will be involved in solving complicated unsteady flow problems with high resolution while the acoustician will be involved in developing both analytic formulations and prediction codes which utilize aerodynamic data. These acoustic codes model the blade geometry, the kinematics and the various noise generation mechanisms as accurately and efficiently as possible. Smart preprocessors for sorting and scaling of the aerodynamic data will be used to decide about the various time scales of fluctuations. This helps in the automatic selection of the space and time grid sizes in the acoustic calculations.

Acknowledgement

The author is grateful to Professor William R. Sears for making him interested in the aeroacoustics and aerodynamics of rotating blades. This paper is affectionately dedicated to Professor Sears on the occasion of his 80th birthday.

References

Baeder, J. D., McCroskey, W. T., and Srinivasan, G. R., 1986. "Acoustic propagation using computational fluid dynamics," In *Proc. of the 42nd AHS Annual Forum & Tech. Display.*

Brentner, Kenneth B., 1986. "Prediction of helicopter rotor discrete frequency noise–a computer program incorporating realistic blade motion and advanced acoustic formulation," NASA TM-87721.

Brentner, Kenneth S. and Farassat, F., 1992. "Helicopter noise prediction: The current status and future direction," In *Presented at the DGLR-AIAA 14th Aeroacoustics Conf.* (to appear in *J. Sound Vibration*, 1993), Aachen, Germany, May 11-14, 1992.

Burton, D. E., Schlinker, R. H., and Shenoy, R., 1985. "The status of analytical helicopter noise prediction methods," NASA CR-172606.

Crighton, D. G. and Parry, A. B., 1992. "Higher approximation in the asymptotic theory of propeller noise," *AIAA Journal* **30**(1), pp. 23–28.

Curle, N., 1955. "The influence of solid boundaries upon aerodynamic sound," *Proc. Roy. Soc. A* **231**, pp. 505–514.

Dunn, M. H. and Farassat, F., 1992. "High-speed propeller noise prediction – a multidisciplinary approach," *AIAA Journal* **30**(7), pp. 1716–1723.

Dunn, Mark H., 1991. *The Solution of a Singular Integral Equation Arising from a Lifting Surface Theory for Rotating Blades*, Ph.D. Thesis, Old Dominion University.

Farassat, F., 1977. "Discontinuities in aerodynamics and aeroacoustics: The concept and applications of generalized derivatives," *J. Sound Vibration* **55**(2), pp. 165–193.

Farassat, F. and Brentner, K. S., 1988. "The uses and abuses of the acoustic analogy in helicopter rotor noise prediction," *Journal of the AHS* **33**, pp. 29–36.

Farassat, F., Lee, Yung-Jang, Tadghighi, H., and Holz, R., 1991. "High-speed helicopter rotor noise – shock waves as a potent source of sound," In *Proc. of the AHS/RAeS International Technical Specialists' Meeting on Rotorcraft Acoustics and Rotor Fluid Dynamics*, Philadelphia, PA.

Farassat, F. and Myers, M. K., 1986. "Aerodynamics via acoustics: Application of acoustic forumulas for aerodynamic calculations," AIAA-86-1877.

Farassat, F. and Myers, M. K., 1988. "Extension of Kirchhoff formula to radiation from moving surfaces," *J. Sound Vibration* **123**(3), pp. 451–460.

Farassat, F. and Myers, M. K., 1991. "Aeroacoustics of high speed rotating blades: The mathematical aspect," *To appear in the Proceedings of 3rd IMACS Symposium on Computational Acoustics*, Cambridge, MA, June, Harvard University.

Farassat, F., Padula, S. L., and Dunn, M. H., 1987. "Advanced turboprop noise prediction based on recent theoretical results," *J. Sound Vibration* **119**(1), pp. 53–79.

Farassat, F. and Succi, G. P., 1983. "The prediction of helicopter rotor discrete frequency noise," *Vertica* **7**(4), pp. 309–320.

Ffowcs Williams, J. E., 1984. "The acoustic analogy – thirty years on," *IMA Journal* **32**, pp. 113–124.

Ffowcs Williams, J. E. and Hawkings, D. L., 1969. "Sound generated by turbulence and surfaces in arbitrary motion," *Phil. Trans. Roy. Soc. A* **264**, pp. 321–342.

George, A. R., 1978. "Helicopter noise – state of the art," *Journal of the Aircraft* **15**(11), pp. 707–715.

Gel'fand, I. M. and Shilov, G. E., 1964. *Generalized Functions*, Volume 1 of *Properties and Operations*, Academic Press, New York.

Hanson, D. B., 1991. "Unified aeroacoustic analysis for high speed turboprop aerodynamics and noise (Vol. 1), Development of theory for blade loading, wakes and noise," NASA CR-4329.

Hanson, D. B. and Fink, M. R., 1979. "The importance of quadrupole sources in prediction of transonic tip speed propeller noise," *J. Sound Vibration* **62**(1), pp. 19–38.

JanakiRam, R. D., 1990. "Aeroacoustics of rotorcraft," AGARD-R-781.

Jones, D. S., 1982. *Generalized Functions*, McGraw-Hill Publishing Company, London, Second Edition.

Kanwal, R. P., 1983. *Generalized Functions– Theory and Technique*, Academic Press, New York.

Lele, S. K., 1992. "Direct computation of aerodynamic noise," *In these Proceedings.*

Lighthill, M. J., 1952. "On sound generated aerodynamically – I. General theory," *Proc. Roy. Soc. A* **211**, pp. 564–587.

Lowson, M. V., 1992. "Progress towards quieter civil helicopters," *Aeronautical Journal* **96**(956), pp. 209–223.

Magliozzi, B., Hanson, D. B., and Amiet, R. K., 1991. "Propeller and propfan noise. Aeroacoustics of flight vehicles: Theory and practice, Vol. 1: Noise sources," NASA Ref. Publ. 1258.

Morgans, W. R., 1930. "The Kirchhoff formula extended to a moving surface," *Phil. Mag.* **9**, pp. 141-161.

Schmitz, F. H., 1991. "Rotor noise," NASA RP-1258.

Schmitz, F. H. and Yu, Y. H., 1979. "Theoretical modelling of high speed impulsive noise," *Journal of the AHS* **24**(1), pp. 10–19.

Tadghighi, H., Holz, R., Farassat, F., and Lee, Yung-Jang, 1991. "Development of shock noise prediction code for high-speed helicopters – The subsonically moving shock," In *Proc. of the 47th Annual Forum & AHS Technology Display*, Phoenix, AZ.

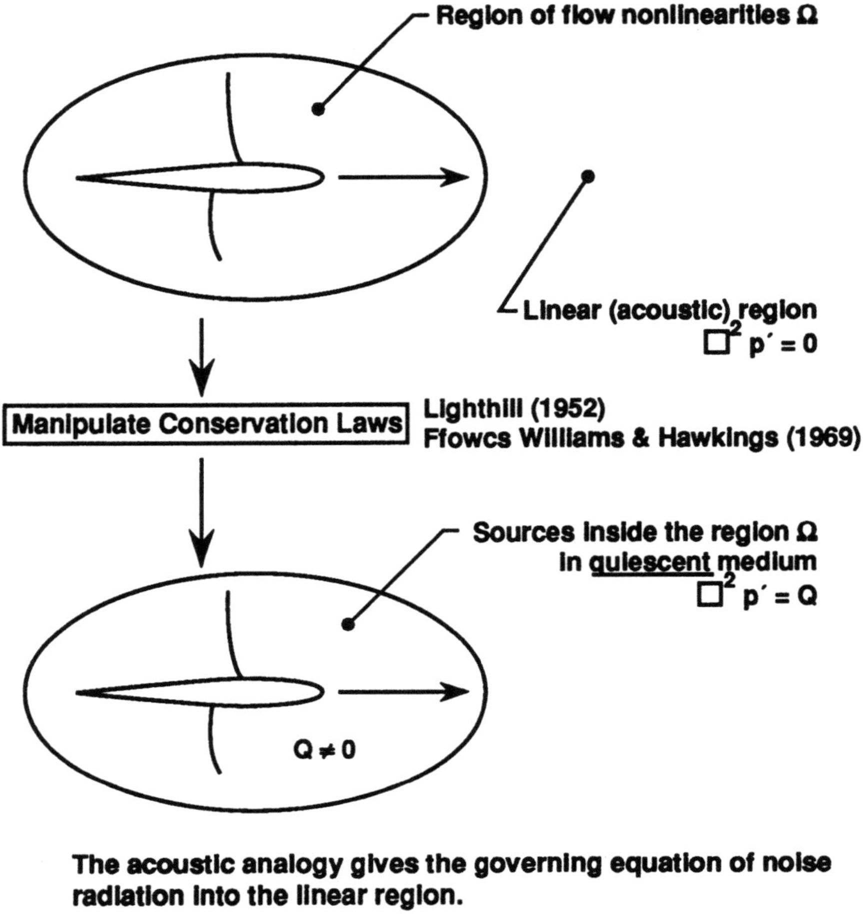

Figure 1: The concept of the acoustic analogy.

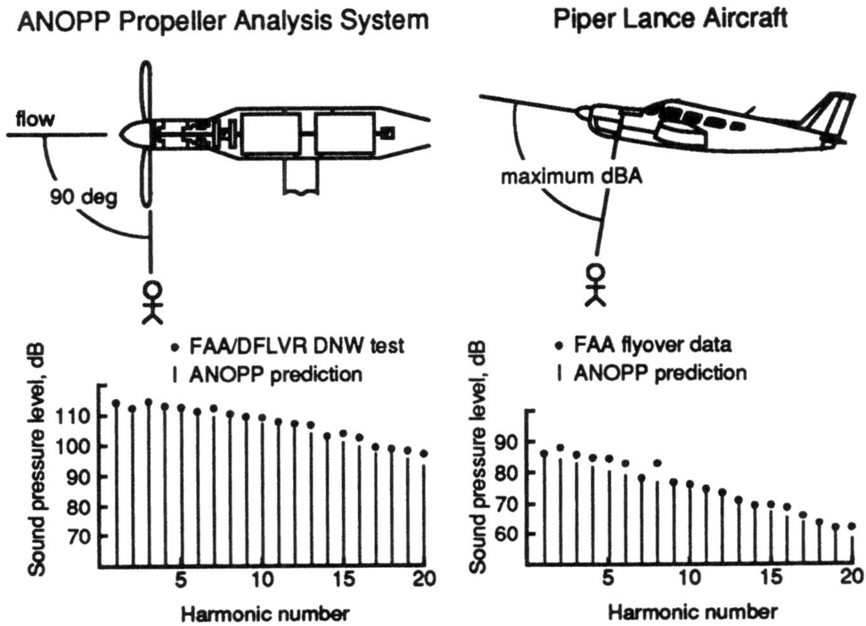

Figure 2: Examples of general aviation propeller noise prediction and comparison with measured data. Aerodynamic input into the acoustic code is from classical propeller theory.

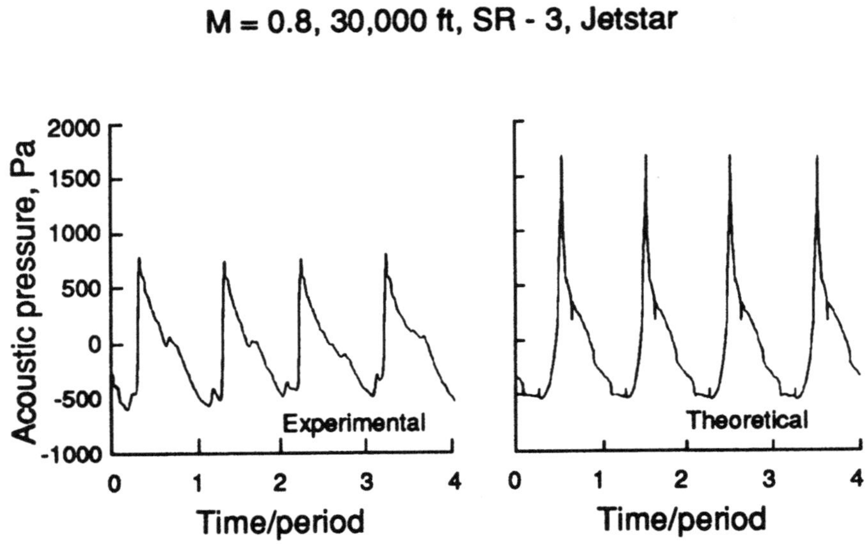

Figure 3: Example of advanced propeller noise prediction and comparison with measured acoustic signature. (Farassat, Padula, and Dunn, 1987, Fig. 16). Aerodynamic input to the acoustic code is from Adamczyk's code.

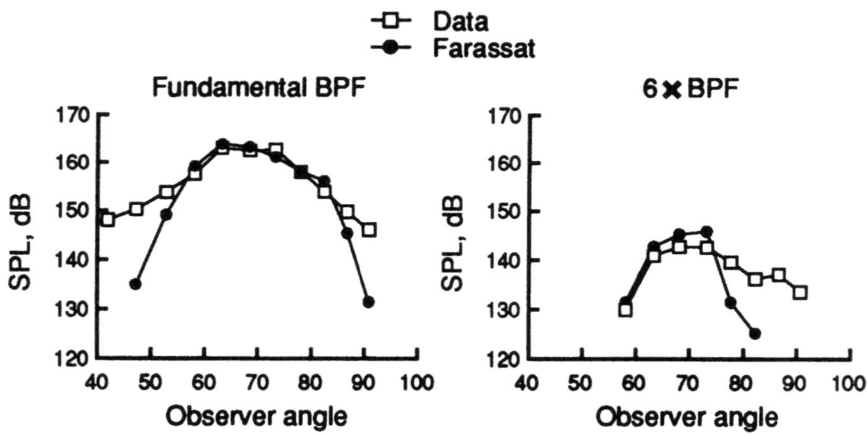

Figure 4: Example of advanced contrarotating propeller noise prediction. Observer angle is the visual angle of a ray intersecting the fan axis midway between rotors. Averaged aerodynamic blade surface pressure was obtained from Adamczyk's code for noise prediction.

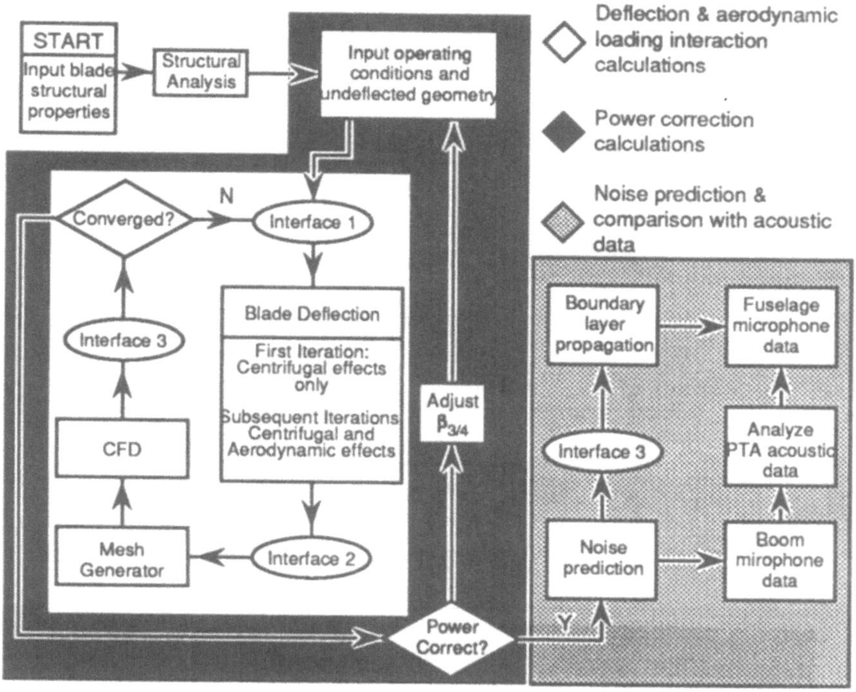

Figure 5: The comphrehensive advanced propeller noise prediction procedure. Four state-of-the-art of aeroelastic, aerodynamic and acoustic codes are used in the system. See Dunn and Farassat (1992) for details.

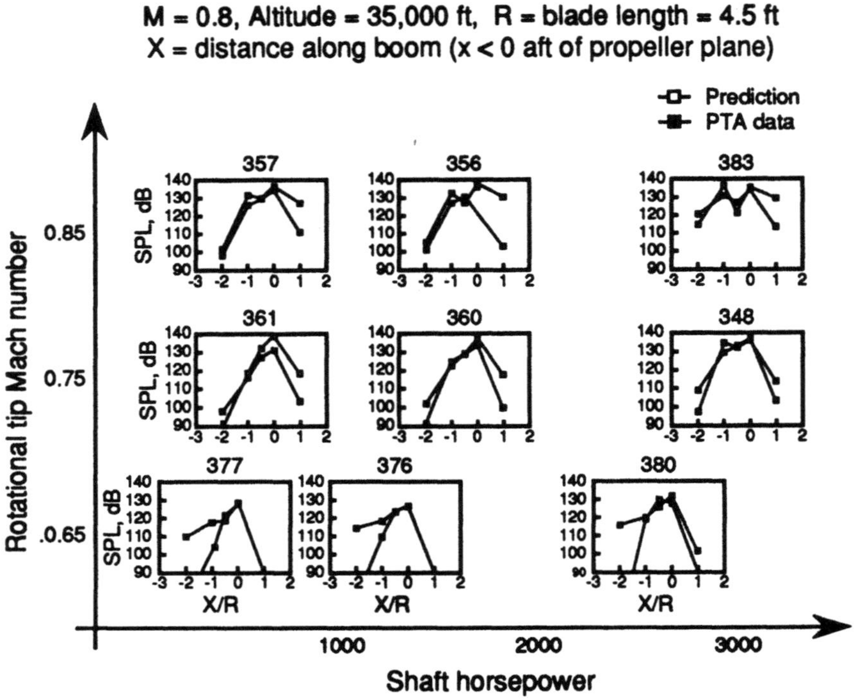

Figure 6: Comparison of measured and predicted axial directivity for a full scale advanced propeller SR-7L. See Dunn and Farassat (1992) for more details.

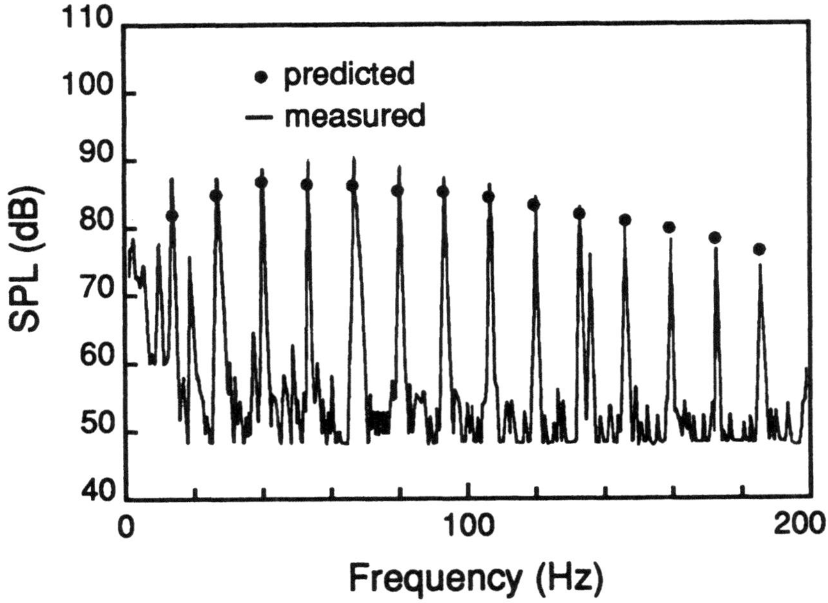

Figure 7: Comparison of measured and predicted rotor noise. Observer is 13.3° below rotor plane at emission time. From Farassat and Succi (1983).

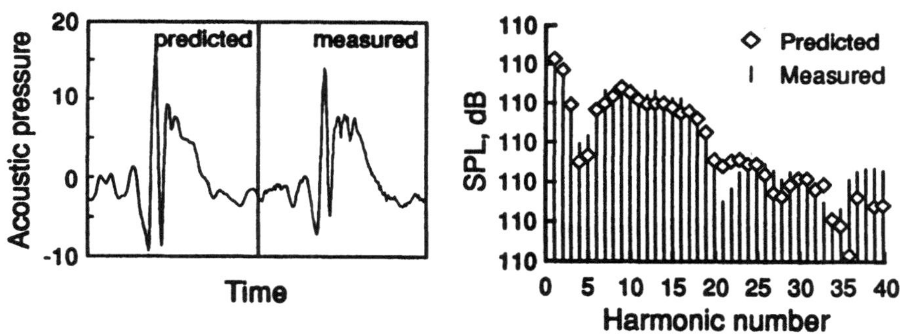

Figure 8: Comparison of measured and predicted blade-vortex intersection noise. The microphone is 30° below the rotor plane and 150° azimuthal angle. From K. S. Brentner, M. M. Marcolini, and C. L. Burley, AHS/RAeS Technical Specialists' Meeting on Rotorcraft Acoustics and Rotor Fluid Mechanics, Philadelphia, PA, October 1991.

PROBLEMS WITH DIFFERENT TIME SCALES AND ACOUSTICS

Heinz-Otto Kreiss

Department of Mathematics
University of California
Los Angeles, California 90024

1. Introduction

In applications, the initial value problem for systems of partial differential equations which allow solutions on different time scales typically has the form

$$\begin{aligned} \partial u/\partial t &= \frac{1}{\varepsilon}P_0(\partial/\partial x)u + P_1(u,\partial/\partial x)u + \nu P_2(\partial/\partial x)u + f(x,t), \\ & \quad t \geq 0, \\ u(x,0) &= f(x), \quad x = (x_1,\ldots,x_s) \in \boldsymbol{R}^s. \end{aligned} \tag{1.1}$$

Here $u = (u^{(1)},\ldots,u^{(n)})'$ is a real vector function with n components. The operators P_0 and P_1 are first order differential operators of the form

$$P_0(\partial/\partial x) = \sum_{j=1}^{s} A_j \partial/\partial x_j, \quad A_j = A_j^* \in \boldsymbol{R}^{n\times n},$$

$$P_1(u,\partial/\partial x) = \sum_{j=1}^{s} B_j(u) \partial/\partial x_j, \quad B_j = B_j^* \in \boldsymbol{R}^{n\times n},$$

i.e., the coefficients are real $n \times n$ Hermitean matrices. The A_j are constant matrices and the B_j are polynomials in the components of u. The operator $P_2(\partial/\partial x)$ is a second order differential operator with constant real coefficients. In applications, νP_2 represents the dissipation present in the system. The parameters $\varepsilon > 0$ and $\nu > 0$ are small constants which measure the difference in time scales and the level of dissipation, respectively. (The coefficients of P_0, P_1, P_2 have been normalized to be of order $O(1)$).

An example of a problem with different time scales is given by low-Mach-number flow. In a slightly simplified form it is given by

$$\begin{aligned} \frac{d\underline{u}}{dt} + \text{grad } p &= \nu \nabla \underline{u} \\ M^2 \frac{dp}{dt} + \text{div } \underline{u} &= 0, \ M \ll 1. \end{aligned} \tag{1.2}$$

Here $\underline{u}$ denotes the velocity vector and p represents the pressure. The problem (1.2) has the form (1.1) if we symmetrize the system by introducing $Mp = \tilde{p}$ as a new variable. In particular

$$P_0 = \begin{pmatrix} 0 & 0 & 0 & 1 \\ 0 & 0 & 0 & 0 \\ 0 & 0 & 0 & 0 \\ 1 & 0 & 0 & 0 \end{pmatrix} \frac{\partial}{\partial x} + \begin{pmatrix} 0 & 0 & 0 & 0 \\ 0 & 0 & 0 & 1 \\ 0 & 0 & 0 & 0 \\ 0 & 1 & 0 & 0 \end{pmatrix} \frac{\partial}{\partial y} + \begin{pmatrix} 0 & 0 & 0 & 0 \\ 0 & 0 & 0 & 0 \\ 0 & 0 & 0 & 1 \\ 0 & 0 & 1 & 0 \end{pmatrix} \frac{\partial}{\partial z}.$$

A discussion of the system from the point of view of different time scales has recently been presented by Kreiss, Lorenz, and Naughton (1991). See also Klainerman and Majda (1982).

Typical questions for problems with different time scales are

1) Can one choose the initial data such that the fast scale is not excited?

2) What is the interaction between the fast and the slow scale?

3) Can one derive asymptotic expressions?

2. Systems with Constant Coefficients

In this section we assume that the coefficients B_j of the operator P_1 are constant matrices. In this case, we can use Fourier expansion to transform the system (1.1) and obtain

$$\begin{aligned} \hat{u}_t(\omega, t) &= |\omega| \hat{P}(\omega) \hat{u}(\omega, t) + \hat{F}(\omega, t), \\ \hat{u}(\omega, 0) &= \hat{f}(\omega). \end{aligned} \tag{2.1}$$

Here $\omega = (\omega_1, \ldots, \omega_s)$ denotes the (real) dual variable to the space variable $x \in \mathbb{R}^s$ and

$$\hat{P}(\omega) = \frac{1}{\varepsilon} P_0(i\omega') + P_1(i\omega') + \nu|\omega| P_2(i\omega'), \tag{2.2}$$

where $\omega' = \omega/|\omega|, |\omega|^2 = \sum \omega_j^2$. (For simplicity of presentation, we will always assume that $\hat{f} = \hat{F} = 0$ for $\omega = 0$.)

A suitable assumption for the dissipative term P_2 is the following eigenvalue condition

Assumption 2.1. *The eigenvalues of $P_2(i\omega')$ are non-positive.*

This assumption ensures that the initial value problem is well posed. In our example (1.2) we have

$$P_2(i\omega') = \begin{pmatrix} -|\omega'| & 0 & 0 \\ 0 & -|\omega'| & 0 \\ 0 & 0 & 0 \end{pmatrix}.$$

Thus the above eigenvalue condition is satisfied. The next assumption concerns the large part $\varepsilon^{-1}P_0$ in our system.

Assumption 2.2. *For each w', the eigenvalues $\kappa = \kappa(\omega')$ of $P_0(i\omega')$ split into two groups $M_{1,2} = M_{1,2}(\omega')$ in the following way:*

$$\textit{If } \kappa \in M_1, \textit{ then } |\kappa| \geq 1. \ \textit{If } \kappa \in M_2, \textit{ then } \kappa = 0.$$

For our (symmetrized) example (1.2) we obtain

$$P_0(i\omega') = \begin{pmatrix} 0 & 0 & 0 & i\omega_1' \\ 0 & 0 & 0 & i\omega_2' \\ 0 & 0 & 0 & i\omega_3' \\ i\omega_1' & i\omega_2' & i\omega_3' & 0 \end{pmatrix} \tag{2.3}$$

i.e.,

$$\kappa_{1,2} = \pm i|\omega|', \ \kappa_{3,4} = 0.$$

Thus both M_1, M_2 consist of two eigenvalues.

In the general case, Assumption 2.2 ensures existence of a unitary transformation $U_0(\omega')$ such that

$$U_0^*(\omega')P_0(i\omega')U_0(\omega') = \begin{pmatrix} R_0(\omega') & 0 \\ 0 & 0 \end{pmatrix}, \tag{2.4}$$

where the eigenvalues of $R_0(\omega')$ are exactly the values in M_1. Therefore,

$$|R_0^{-1}(\omega')| \leq 1. \tag{2.5}$$

Now consider the matrix $\hat{P}(\omega)$ given in (2.2). If $\varepsilon\nu|\omega| \ll 1$ we can determine a transformation

$$S = U_0(\omega')(I + \varepsilon U_1(\omega'))(I + \nu\varepsilon|\omega|T(\omega', |\omega|))$$

which transforms $\hat{P}(\omega)$ to blockdiagonal form, i.e.,

$$S^{-1}\hat{P}(\omega)S = \begin{pmatrix} \frac{1}{\varepsilon}R_0 + R_{11} + \nu|\omega|Q_{11} & 0 \\ 0 & R_{22} + \nu|\omega|Q_{22} \end{pmatrix}. \qquad (2.6)$$

(Note that $0 < \varepsilon \ll 1$ is always assumed.) Thus, for $\varepsilon\nu|\omega| \ll 1$, we can introduce new variables $\hat{v}(\omega, t)$ into (2.1) by

$$\hat{u}(\omega, t) = S\hat{v}(\omega, t)$$

and obtain

$$\begin{aligned} \hat{v}_t^I &= |\omega|\left(\frac{1}{\varepsilon}R_0 + R_{11} + \nu|\omega|Q_{11}\right)\hat{v}^I + (S^{-1}\hat{F})^I, \\ \hat{v}_t^{II} &= |\omega|(R_{22} + \nu|\omega|Q_{22})\hat{v}^{II} + (S^{-1}\hat{F})^{II}, \qquad (2.7) \\ \hat{v}(\omega, 0) &= S^{-1}\hat{f}. \end{aligned}$$

In this case, $\hat{v}^I$ is highly oscillatory whereas $\hat{v}^{II}$ varies slowly. We can decompose the whole solution u correspondingly if we make

Assumption 2.3. *The initial data f and the forcing function F satisfy*

$$\hat{f}(\omega) = \hat{F}(\omega, t) \equiv 0 \text{ for } |\omega| \geq \delta(\nu\varepsilon)^{-1}, \quad 0 < \delta = \text{const.} \ll 1. \qquad (2.8)$$

Under these assumptions the solution can be decomposed as

$$u = u^I + u^{II}, \qquad (2.9)$$

where u^I varies on the "fast" and u^{II} on the "slow" scale.

The assumption (2.8) is not a strong restriction in applications. For example, in low-Mach-number flows one has

$$|\hat{u}(\omega, t)| \approx e^{-\gamma\sqrt{\nu}|\omega|}, \quad \gamma > 0.$$

Therefore, if $|\omega| \geq \delta(\nu\varepsilon)^{-1}$, then $\hat{u}(\omega, t)$ is exponentially small.

3. The Bounded Derivation Principle

We consider now the full nonlinear problem (1.1). If one is not interested in the fast time scale then one can prepare the initial data such that the fast scale is not excited. This is done by the

Bounded derivative principle: *Choose the initial data such that $p \geq 1$ time derivatives are bounded independently of ε at $t = 0$.*

This is justified since one can show that if p time derivatives are bounded independently of ε at $t = 0$, then the same is true at later times. To make the results more precise, we define

Definition 3.1. Let $T > 0$ and let $w(x, t, \varepsilon)$ denote a vector function defined for $x \in \boldsymbol{R}^s, 0 \leq t \leq T, 0 < \varepsilon \leq \varepsilon_0$. We assume that w is 2π-periodic in each x_j. We say that w is slow to order p in $0 \leq t \leq T$ if all space derivatives $D^k w$ have p continuous time derivatives in $0 \leq t \leq T$ and

$$\sup_{0<\varepsilon\leq\varepsilon_0} \max_{0\leq t\leq T} \|\frac{\partial^j D^k w}{\partial t^j}(\cdot, t, \varepsilon)\| < \infty \tag{3.1}$$

for $j = 1, 2, \ldots, p$ and all space derivatives $D^k w$.

One can prove (see Browning and Kreiss (1982)).

Theorem 3.1. *If the initial data of (1.1) are chosen such that p time derivatives are bounded independently of ε at $t = 0$, then the solution is slow to order p in some time interval $0 \leq t \leq T$. Here T does not depend on ε.*

As an example we consider the low Mach number flow (1.2). One time derivative is bounded at $t = 0$ if and only if the initial data are such that

$$\text{div } \underline{u} = M^2 d, \quad d = 0(1). \tag{3.2}$$

Two time derivatives are bounded at $t = 0$ if and only if

$$\text{div } \underline{u}_t = M^2 q \quad q = 0(1) \tag{3.3}$$

i.e.

$$\Delta p + \sum_{i,j} D_i D_j (u_i u_j) = M^2 q. \tag{3.4}$$

By Theorem 3.1 the relations (3.2), (3.4) hold also for later time, i.e., if we start t close to incompressible flow we will stay close to such a flow also at later time.

If we choose the initial data such that also higher time derivatives are bounded independently of ε then d and q are determined as solutions of elliptic equations. Therefore the flow on the slow scale is determined by the initial vorticity.

4. Asymptotic Expansions

Now we are interested in both the fast and the slow time scale and their interaction. We use asymptotic expansions (see Erlebacher et al. (1990); Klainerman and Majda (1982); and Kreiss et al. (1991)) to describe the solution of our example. The first step is to construct the slow part of the solution. By the previous section it is to first approximation given as the solution of the incompressible equations

$$\frac{d\underline{U}_0}{dt} + \text{grad} P_0 = \nu \Delta \underline{U}_0 \qquad (4.1a)$$

$$\text{div} \underline{U}_0 = 0$$

with initial data

$$\text{vorticity}(\underline{U}_0) = \text{ vorticity } \underline{u}. \qquad (4.1b)$$

We change the dependent variables in (1.2) by

$$\underline{u} = \underline{U}_0 + \underline{u}'$$

and obtain

$$\begin{aligned} &\underline{u}'_t + L_1 \underline{u}' + Q_1 + \text{grad } p' = \nu \Delta \underline{u}' \\ &M^2(p'_t + L_{21} p' + L_{22} \underline{u}' + Q_2) + \text{div } \underline{u}' = M^2 G \end{aligned} \qquad (4.2a)$$

with initial data

$$\text{vorticity}(\underline{u}') = 0, \ \text{div} \underline{u}' = \text{div } \underline{u}, \ p' = p - P_0 \qquad (4.2b)$$

where

$$\begin{aligned} G &= -(P_{0t} + (\underline{U}_0 \cdot \nabla) P_0), \\ L_1 \underline{u}' &= (\underline{U}_0 \cdot \nabla) \underline{u}' + (\underline{u}' \cdot \nabla) \underline{U}_0, \\ L_{21} p' &= (\underline{U}_0 \cdot \nabla) p', \ L_{22} \underline{u}' = (\underline{u}' \cdot \nabla) P_0 \end{aligned}$$

and Q_1, Q_2 stand for the nonlinear terms.

The second term in the asymptotic expansion of the slow part of the solution is given by

$$\begin{aligned} \underline{U}_{1t} + L_1\underline{U}_1 + \text{grad}P_1 &= \nu\Delta\underline{U}_1 \\ \text{div}\underline{U}_1 &= M^2 G \end{aligned} \tag{4.3a}$$

with initial data

$$\text{vorticity}(U_1) = 0. \tag{4.3b}$$

We make now the substitution

$$\underline{u}' = U_1 + \underline{u}'', \; p' = P_1 + p''$$

and obtain equations for $\underline{u}'', p''$ which are of the form (4.2). The forcing is of the order $O(M^4)$. This process can be continued. We obtain the asymptotic expansion of the slow part of the solution, i.e.,

$$\underline{u} = \sum \underline{U}_j + \tilde{\underline{u}}, \; p = \sum P_j + \tilde{p}.$$

$\tilde{u}, \tilde{p}$ satisfy equations of type (4.2) where the forcing M^2G is replaced by zero. They vary on the fast (acoustic) scale. Therefore we introduce a new variable $t' = Mt$ and obtain

$$\begin{aligned} \tilde{\underline{u}}_{t'} + M(L_1\tilde{\underline{u}} + Q_1) + M \text{ grad } \tilde{p} &= \nu M \Delta \tilde{u} \\ M(\tilde{p}_{t'} + M(L_{21}\tilde{p} + L_{22}\tilde{\underline{u}} + Q)) + \text{ div } \tilde{\underline{u}} &= 0. \end{aligned} \tag{4.4}$$

On the acoustic scale we can replace (4.4) to first approximation by

$$\begin{aligned} \tilde{\underline{u}}_{t'} + M \text{ grad } \tilde{p} &= 0 \\ M\tilde{p}_{t'} + \text{ div } \tilde{\underline{u}} &= 0 \end{aligned} \tag{4.5a}$$

with initial data

$$\begin{aligned} \text{vorticity } (\tilde{\underline{u}}) &= 0, \text{ div } \tilde{\underline{u}} = \text{div}(\underline{u} - \textstyle\sum U_j) \\ \tilde{p} &= p - \textstyle\sum P_j. \end{aligned} \tag{4.5b}$$

Thus the fast part of the solution is described by wave equations.

The interaction between the fast and the slow part of the solution is small because in Kreiss, Lorenz, and Naughton (1991) we have proved that the vorticity of $\underline{\tilde{u}}$ stays small even on the slow scale. We have

Theorem 3.1. *If at* $t = 0$

$$\text{div } \underline{\tilde{u}} = O(M), \ \tilde{p} = O(1), \ \text{vorticity } (\underline{\tilde{u}}) = 0$$

then

$$\text{vorticity } (\underline{\tilde{u}}) = O(M^2)$$

in any finite time interval $0 \leq t \leq T$ *(slow scale).*

Remark. The assumption that div $\underline{\tilde{u}} = O(M)$ at time $t = 0$ is natural. One can prove (see Erlebacher et al. (1990) and Kreiss et al. (1991)): If div $\underline{\tilde{u}} = O(1)$ at $t = 0$ then $\tilde{p} = O(1/M)$ for $t > 0$. Thus large pressure waves are generated.

We summarize our results. The solution of (1.2) can approximately be represented by

1) The solution of the incompressible equations (4.1)

2) The compressible adjustment (4.3)

3) The fast part (4.5).

5. Acoustics

There are a very large number of different situations and therefore we restrict ourselves to two extreme cases.

1) At $t = 0$ we give a velocity distribution in a box. In this

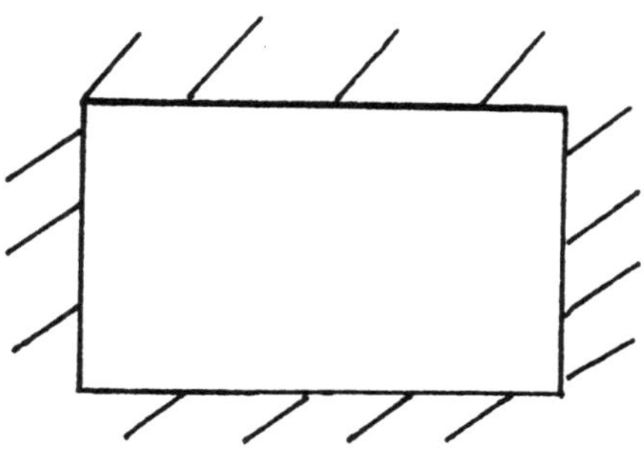

situation we have to calculate all three parts of the solution.

To compute the slow part we use a numerical Navier-Stokes solver. The fast part can be obtained by analytic techniques.

2) Now we consider the whole space with initial data consisting of a velocity patch.

In this case the fast part will be radiated away and therefore we need only to consider the slow part. The main problem is to estimate the decay rate of $\underline{u}$ away from the patch because it will determine when we can replace the Navier-Stokes equation by simpler equations. We need the following well-known estimate.

Consider the Laplace equation

$$\Delta w = D^p F$$

in the whole space R^3 where F is a rapidly decaying function and D^p stands for a p:th derivative. Then

$$w(x) = \frac{1}{4\pi} \int_{R^3} \frac{D_y^p F(y)}{|x-y|} dy.$$

Integration by parts shows that for large $|x|$

$$D_x^j w \sim |x|^{-(p+j+1)}. \tag{5.1}$$

We can use (5.1) to estimate the decay rate of $\underline{U}_0, \underline{U}_1, P_0$ and P_1. We know that the vorticity is a rapidly decaying function and therefore the relation

$$\Delta \underline{U}_0 = D(\text{vorticity})$$

shows that

$$|\underline{U}_0| \sim |x|^{-2}. \tag{5.2}$$

Introducing (5.2) into

$$\Delta P_0 + \sum D_i D_j (U_{0i} U_{0j}) = 0$$

gives for large $|x|$

$$|P_0| \sim |x|^{-3}, \text{ grad } P_0 \sim |x|^{-4}. \tag{5.3}$$

Therefore by (4.1)

$$\frac{\partial \underline{U}_0}{\partial t} + O(|x|^{-5}) = O(|x|^{-4}).$$

Thus in any finite time interval

$$|\underline{U}_0| = O(|x|^{-4}). \tag{5.4}$$

P_1 satisfies also a pressure equation. In this case the leading term is

$$\Delta P_1 = M^2 P_{0tt}$$

i.e. for large $|x|$

$$|P_1| \sim M^2 |x|^{-1}. \tag{5.5}$$

Therefore

$$|\underline{U}_1| \sim M^2 |x|^{-2}. \tag{5.6}$$

The above results are essentially contained in the Lighthill theory of sound.

The above results are only valid as long as $|x| M \ll 1$ because the asymptotic expansion is based on the assumption that we can replace the finite sound speed by an infinite sound speed. If we want to calculate U, P for other values of $|x|$ we cannot make this assumption. However, in view of the decay rate of $\underline{u}$ we can neglect for large $|x|$, the convection terms in (1.2) and need only to solve the wave equation. Thus the numerical task is to construct a scheme which consist of a Navier-Stokes solver locally and a wave-equation solver in the far field.

References

Browning, G. and Kreiss, H.-O., 1982. "Problems with different time scales for nonlinear partial differential equations," *SIAM J. Appl. Math.* **42**, pp. 704-718.

Erlebacher, G., Hussaini, M. Y., Kreiss, H.-O., and Sarkar, S., 1990. *Theoretical and Computational Fluid Dynamics* **2**, p. 73.

Klainerman, S. and Majda, A., 1982. "Compressible and incompressible fluids," *Comm. Pure Appl. Math.* **35**, pp. 629-641.

Kreiss, H.-O., Lorenz, J., and Naughton, M. J., 1991. "Convergence of the solutions of the compressible to the solutions of the incompressible Navier-Stokes equations," *Advances Appl. Math.* **12**, pp. 187-214.

VALIDATION METHODOLOGY

VALIDATION METHODOLOGY: REVIEW AND COMMENTS

Allan D. Pierce

Graduate Program in Acoustics
The Pennsylvania State University
157 Hammond Building
University Park, Pennsylvania 16802

ABSTRACT

A definition is suggested for computational aeroacoustics, and the role of validation methodology is briefly discussed. Some brief remarks are made concerning the subject session. The question is raised as to why it is especially appropriate in 1992 to contemplate the initiation of major research efforts in computational acoustics, and a list of possible reasons is given.

1. Introduction

Each session chairman participated in a panel discussion at the end of the Workshop and was asked to summarize his or her session and to provide his or her own critical views on the general subject of computational aeroacoustics. The present author chaired the session on Validation Methodology, which included seven papers.

2. A Definition of Computational Aeroacoustics

The advance announcements sent out by the conveners of this workshop, along with their selections of invited speakers and the manner in which they organized the program, suggest the following definition:

Computational aeroacoustics is the direct simulation of acoustic fields generated by flow and of the interaction of acoustic fields with flows. The phrase "direct simulation" is understood to imply that results are obtained by computation and that the methodology proceeds directly from fundamental physical principles without reliance on empirical results or heuristic conjectures. The scientific discipline of computational aeroacoustics includes the interpretation and application of such simulations.

(This definition was formulated several weeks after the Workshop during conversations with Guatam SenGupta, James D. Baeder, and Alan Powell.)

The term "aeroacoustics" has been in widespread use for at least three decades and is used to refer to: (i) those aspects of acoustics associated with aerodynamics, (ii) the generation of sound generated by flow (alternately referred to as aerodynamic sound), or (iii) all acoustical topics (including those sometimes referred to as hydroacoustics) that simultaneously involve both flow and sound. In recent years, there has been a trend to prefix the adjective "computational" to any scientific discipline (e. g., computational fluid dynamics, computational mechanics, computational physics, computational acoustics) to distinguish any aspect of the discipline that either pertains or relies substantially on the use of computers. Within the context of the workshop, however, a more restricted definition, along the lines of that articulated above, seems appropriate. The reason for this is that a substantially new approach to aeroacoustics research is now becoming possible and we need to take stock of its prospects and its critical issues, and we need a short phase to distinguish this approach from other approaches. This new approach is that which conforms to the definition given above.

3. The Validation Methodology Session

A general concern is that, should direct simulation of aeracoustic phenomena become feasible, one be able to assess the extent to which such simulations are correct and realistic. Thus one needs a validation methodology. Because computational aeroacoustics, in the sense as defined above, is an embryonic discipline, the development of appropriate validation methodologies has not yet been given much attention. The seven papers in the subject session were presumably solicited with the intent that they collectively give some insight into the possible current options for validating direct simulation algorithms and of the possible directions for parallel research on validation methodology.

The diverse scope of the session is indicated by the following abbreviated mnemonic phrases which came to the chairman's mind after a cursory hearing of each of the seven talks: (i) three line vortices near a solid surface (*Comte-Bellot*), (ii) sound transmission out a duct through an elastic window (*Courtier-Arnoux*), (iii) noise from high speed jets, a description of the present state of the predictive art (*Seiner*), (iv) finite difference approaches to computational aeroacoustics (*Hardin*), (v) sound radiation by vortex-structure interaction and by vortex-vortex interaction (*Kambe*), (vi-A) computational power requirements of computational aeroacoustics (*Morris*), (vi-B) solvable two-dimensional problems involving slowly-decaying vortex fields (*Morris*), and (vii) a perspective overview of significant new ideas and current work pertaining to computational aeroacoustics (*Goldstein*). (The paper by Morris had two clearly distinct parts, and these are distinguished in the above list.)

The present writer did not attempt a detailed summary of the individual papers in the panel discussion, and has not seen their written versions as of the time of this writing. The reader, with the collected volume of written papers in hand, is presumably in a position to make a more detailed and accurate assessment, so it seems prolix to attempt such in the present report. However, it is perhaps fair to state some ephemeral impressions.

- There is much semi-empirical information on aeroacoustic phenomena that can be used to validate results of direct simulation.
- There are a variety of relatively simple aeroacoustic phenomena involving line and ring vortices for which theoretical approaches have had comparatively good success and which can be simulated with modest computational requirements.
- Computational methods are now being extensively applied to acoustics within the context of what may be termed "classical theoretical approaches," such as boundary element formulations and the acoustic analogy approach.
- Direct simulation of realistic aeroacoustic phenomena of engineering interest is not quite within our grasp at present, but one can expect that it will be, within a few years, given a concerted research effort to make an intelligent use of the next generation of supercomputers.

4. Why 1992?

The present writer's general impression is that aeroacoustics as a scientific discipline has been relatively stagnant for at least the last ten years, and even perhaps the last twenty years.

The publication in 1952 of Lighthill's first paper on sound generated aerodynamically must in retrospect be regarded as an epochal event. No other paper in the history of acoustics has been cited so extensively. It initiated a period of creative efforts which involved many researchers of outstanding intellect. However, it is now 40 years since the beginning of that period, and one might legitimately speculate that the train of thought initiated by Lighthill's 1952 paper has more-or-less run its course. A lot of first-rate scientific talent has been attracted to the subject over the past four decades and their efforts have yielded a rich and prolific literature.

During this past four decades there certainly has been no lack of competent people willing to attempt numerical calculations, so why should there be any hope that an enhanced involvement of computers can yield any dramatic new progress to the field of aeroacoustics? If arguments to this effect can be made, then why couldn't they have been made, say, ten years ago?

The actual occurrence of this Workshop in the Spring of 1992 and the fact that it attracted a diverse group of eminent researchers from related disciplines suggests that there may well be some perception that a window of opportunity for substantial new developments in aeroacoustics is just now starting to open. Why wasn't this the case ten years ago?

5. Possible Answers

The writer offers the following list of answers to the "why 1992?" query raised in the preceding section. To a major extent these answers are based on the overall impression gleaned from the participation in the workshop.

- *A confluence of three research communities.* People from three distinct communities (*aeroacoustics*, *computational fluid dynamics*, and *turbulence* participated in the workshop. Although there has been some limited interaction between the groups in the past, it has not been extensive. The author's impression is that, to some extent, research in each of these areas is on a plateau and that reasonable numbers of talented workers in each discipline are looking for a new thrust area. Computational aeroacoustics needs to draw upon the past accomplishments of each community, and the time is now ripe for the necessary interactions to take place.

- *Dramatic increases in computer power.* That such has occurred relatively recently and that we are now approaching the threshold of capabilities for realistic simulations is evident from some of the papers presented in this paper. The writer is thinking in particular of the paper by Morris.

- *Capability to simulate turbulent flows.* New techniques for large eddy simulation have led to dramatic and impressive progress in turbulence. That this is the case was supported by several papers at the workshop, especially that by Steven Orszag.

- *Maturer understanding of numerical methods.* The numerical solution of partial differential equations dates back at least as far as the 1940's, and has yielded a large literature. The problems and potential pitfalls, however, are exceedingly formidable, and we are still progressing toward a mature understanding of this topic. The level of maturity, particularly in those aspects that relate to computational fluid dynamics, is much higher now than it was a decade and a half ago. It is perhaps fair to state that, even apart from considerations of lesser computational power in the past, anyone attempting realistic aeroacoustics in the present era can in principle do so with fewer frustrations.

- *Instability mechanisms much better understood.* There is some perception that the actual physics of the generation of sound by flow is intrinsically tied to the triggering and growth of instabilities in the flow field. If this is the case then one might anticipate that the algorithms that relate to aeroacoustics *must* model the physics of instabilities with some degree of realism. The instability mechanisms of potential interest extend far beyond the classical ones based on linearized perturbations of parallel flows. That the fluid mechanics community in recent years has made substantial new progress in understanding stability is commonly known among aeroacoustics research workers, and the paper at the workshop by Andrew J. Majda is exemplary of the increasingly sophisticated understanding in this area.

The above proffered answers are of course tentative and the reader should feel free to rebut any of them or to supply alternate answers.

WALL PRESSURE AND RADIATED NOISE GENERATED BY 3 LINE VORTICES IN CHAOTIC MOTION CLOSE TO A SOLID SURFACE

Luc Collorec, Daniel Juve, and Geneviève Comte-Bellot

Laboratoire de Mécanique des Fluides et d'Acoustique
URA CNRS 263
Ecole Centrale de Lyon, 36 Avenue Guy de Collongue
69131 Ecully, France

ABSTRACT

Starting from a numerical simulation of the behavior of three inviscid point vortices near a flat wall, we computed the wall pressure evolution and the far field acoustic directivity and intensity. Depending on their initial positions, the trajectories of the vortices exhibit patterns ranging from regular to chaotic. The chaotic patterns affect the wall pressure, create broadband noise and are found to increase the far field acoustic intensity.

1. Introduction

Vorticity has always been a major variable of interest in fluid dynamics (Lamb, 1945; Yates, 1978; Mitchell, Lele and Moin, 1992). Fluid flows can often be easily investigated, described and explained in terms of the vorticity distribution. In turbulence, the energy spectrum of a flow evolves because of energy transfer among "eddies." More recently, people have tried to understand turbulence not only statistically but also through "events," or coherent regions with strong vorticity localization (Brasseur and Lin, 1991). In the domain of acoustics, the direct link between the repartition of vorticity and the radiated noise is well known although there are different theories to model this link (Powell, 1963; Möhring, 1978). Questions that arise naturally are what is the noise created by interacting coherent vortical structures and what are the wall pressure fluctuations?

As these structures have strong vorticity localization, it may be possible to describe the structure of the vorticity field through simple models such as arrays of point vortices, permitting the use of invariants (Aref, Rott and Thomann, 1992; Gröbli, 1877).

In this description another issue arises that is of major interest in fluid dynamics: the chaotic behavior that exists in many flows. Chaos may influence the evolution of coherent structures leading to another question: what are the effects of chaotic trajectories of coherent regions with strong vorticity on noise emission and wall pressure?

2. Dynamics of Motion

Many aspects of chaotic phenomena have been studied in the field of fluid mechanics both on theoretical and experimental grounds. We considered an inviscid flow consisting of the velocity field induced by point vortices which evolve according to the Euler equations. Chaos originates in the strong non-linear interactions among the point vortices. This formulation is only an approximation to the full Navier Stokes equations but its simplicity brings some advantages. First, this is a potential flow formulation. Second, this is a non-dissipative system so that we can apply the properties of Hamiltonian mechanics, including its invariants. Finally, the phenomena arising in this system are easier to understand, providing insights into more complicated flows governed by the full Navier-Stokes equations.

2.1. Formulation of the model

Aref (1989) has shown that in infinite space at least four vortices are required to produce chaos. However, with the introduction of a simple boundary, he demonstrated that the number dropped to three. In the present context we considered this latter system and generated chaotic flows based on three vortices of equal strength and sign with an initial geometry reported by Conlisk, Guezennec and Elliott (1989). Here we consider not only the trajectories of the vortices but also the wall pressure fluctuations and farfield acoustics arising from their motion. Using the complex potential W, we define our initial field as:

$$W = \sum_{j=1}^{N} \frac{i\Gamma_j}{2\pi}[ln(z - \overline{z}_j) - ln(z - z_j)] \tag{1}$$

where z_j is the position of the jth vortex on the complex plane.

From expression (1) we can derive the velocity of each vortex resulting from the velocity field induced by all the other point vortices. This leads to a set of non-linear but ordinary differential equations.

Hence we need only the initial position of each vortex in order to solve the system of equations given by:

$$u_n - iv_n = \frac{1}{2\pi}\left(\sum_{j=1}^{N}\frac{i\Gamma_j}{z_n - \overline{z}_j} - \sum_{\substack{j=1\\ j\neq n}}^{N}\frac{i\Gamma_j}{z_n - z_j}\right). \tag{2}$$

The initial position of the three vortices is defined with a position parameter a as shown in Fig. 1. In the calculation, a is varied from $a = 0.433$ (equilateral triangle) to $a = 0.0$ (a degenerate triangle in which the vortices are aligned perpendicular to the wall). Note that all variables are expressed using System International. A fourth order Runge Kutta method was used to solve the system of equations.

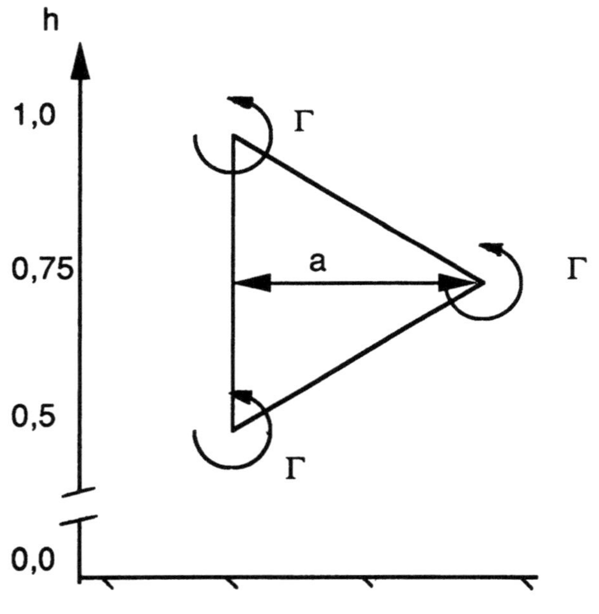

Figure 1. Starting position of the vortices.

2.2. Behavior of the vortices's trajectories

We show two cases in Figs. 2 and 3 with $\Gamma = 2$. Figure 2, corresponding to $a = 0.433$, shows regular motion of the vortices. Figure 3, corresponding to $a = 0.00$, demonstrates chaotic motion. These two figures show the most extreme behaviors from the calculations.

When $a = 0.433$ a regular epicycloidal-like movement is observed. As a decreases, we observe progressively less regular motion, ending at $a = 0.00$ with what appears to be random motion.

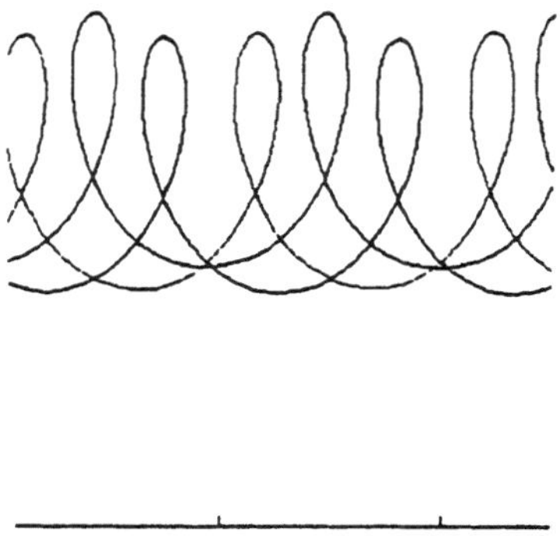

Figure 2. Trajectories of the vortices for $a = 0.433$.

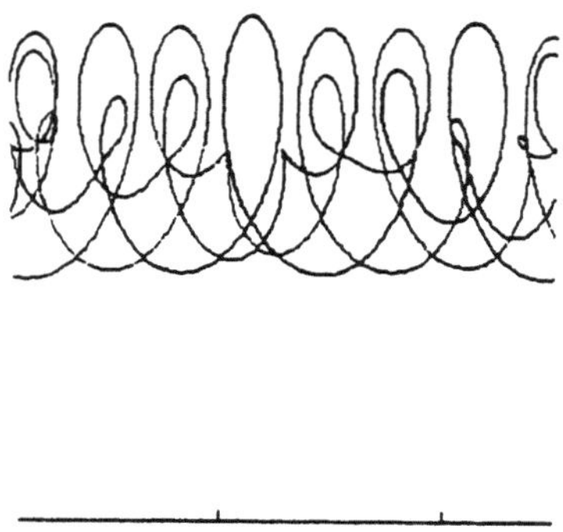

Figure 3. Trajectories of the vortices for $a = 0.00$.

2.3. Centered coordinates

It is possible to define a center of vorticity as:

$$\begin{cases} x_c = \sum_{n=1}^{3} \frac{\Gamma_n x_n}{\sum \Gamma_n} \\ y_c = \sum_{n=1}^{3} \frac{\Gamma_n y_n}{\sum \Gamma_n}. \end{cases} \tag{3}$$

Using the symmetries of our system, it can be demonstrated that the center of vorticity has a trajectory parallel to the wall. The coordinates of the vortices relative to the center of vorticity are called centered coordinates. It is possible to split the movement of each vortex into a translation and a rotation about the center of vorticity.

The speed of the center of vorticity results from the speed induced by the image vortices across the wall. By replacing the group of image vortices by a single equivalent vortex placed at its center of vorticity, we can estimate analytically the translational speed of the center of vorticity of the "real" vortices. This estimation was found to be very accurate in our simulations, although some dependence on distance from the wall may, in general, be expected.

Figs. 4 and 5 show the movement of individual vortices relative to their center of vorticity, that is, with the translational component removed. For $a = 0.433$ the remaining rotational motion exhibits mostly circular trajectories with slight osculation. When $a = 0.00$, however, the osculation is greatly increased. One interpretation for this difference in behavior is that as the parameter a decreases, vortices can approach the center of vorticity and are allowed more complex movement. It has been conjectured by Conlisk, Guezennec and Elliot that the center of vorticity is nearly a stagnation point, thus permitting vortices to easily shift their orbits at this location. The closer a vortex moves to the center of vorticity, the more dramatic the change in its trajectory. The starting position of the vortices is thus very important to the trajectories which follow.

Figure 4. Centered coordinates for $a = 0.433$.

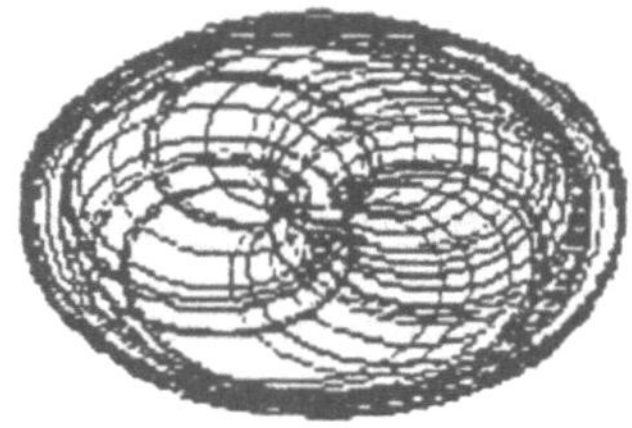

Figure 5. Centered coordinates for $a = 0.00$.

2.4. Leading Lyapunov exponent

A primary feature of chaos is sensitivity to initial conditions. In a "classical" system if we start with two close initial conditions, the distance between the two trajectories remains finite as the system evolves. However, in a chaotic system the distance between two initially close trajectories will increase exponentially. This increase may not be the same in each direction. In fact, for a conservative system, increases in some directions must be balanced by decreases in other directions. The coefficient in the exponential separation of two trajectories is the Lyapunov exponent. A Lyapunov exponent is associated to each coördinate of each vortex. The greatest exponent (called the leading exponent) determines whether the system is chaotic or not. A high Lyapunov exponent indicates that the system

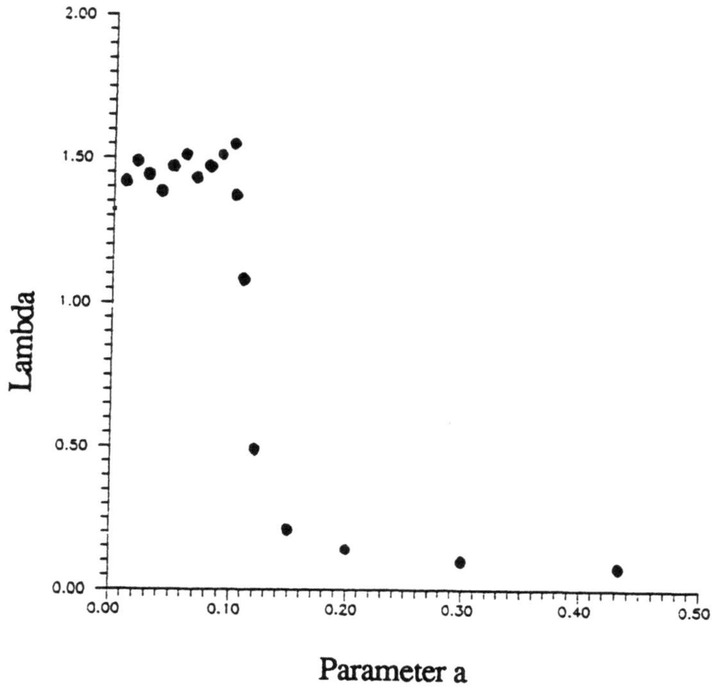

Figure 6. Leading Lyapunov exponent.

has a very chaotic behavior.

We used the method of Wolf et al. (1985) to obtain the evolution of the leading Lyapunov exponent as a function of a . The results are shown in Fig. 6. As a varies from 0.433 to 0.00 we observe three regimes:

1. For a between 0.433 and 0.150 the leading Lyapunov exponent is small and positive and its variation is smooth. The trajectories of the vortices are quasi-periodic with regular features.

2. For a between 0.150 and 0.100 there is a sharp rise in the Lyapunov exponent, corresponding to a transition to chaos. A maximum value is achieved near $a = 0.100$. At this value, the corresponding trajectories are chaotic and the spectrum of centered coordinates have strong broadband features.

3. For a between 0.100 and 0.00 the exponent is high in general, implying chaotic behavior. But there are some variations of the

exponent (e.g., $a = 0.070$) due to varying mixtures of chaotic and regular trajectories. Since there could be any number of additional dips between the plotted points, we do not show the Lyapunov exponent as a curve; there may exist some very narrow parameter bands where the vortices exhibit a quasi-periodic behavior.

3. Wall Pressure

We computed the hydrodynamic pressure at a point on the wall as the vortex system passed over. The computation was implemented using the unsteady Bernoulli expression:

$$P = -\rho_0 \frac{\partial \phi}{\partial t} - \rho_0 \frac{V^2}{2} \tag{4}$$

where ϕ is the velocity potential and ρ_0 is the density of the fluid (in the following $\rho_0 = 1.3$).

Figs. 7 and 8 show the evolution of the wall pressure as the three-vortex system passes over the prescribed point. Fig. 7 represents a case where the motion of the vortices is regular and Fig. 8 represents a case where the trajectories of the vortices are chaotic. The pressure as a function of time is similar in the two cases. A characteristic shape similar to Figs. 7 and 8 was also obtained by a single equivalent vortex placed at the center of vorticity of the vortex system. The oscillations in Figs. 7 and 8 come from the motion of the group of vortices around the center of vorticity. The more chaotic is the motion of the vortices, the more intense and the higher is the frequency of the wall pressure fluctuations.

4. Far Field Acoustics

4.1. Powell-Hardin formulation

The vorticity distribution $\vec{w}$ permits us to calculate the acoustic farfield (Hardin, 1977). To simplify the calculation we assume a 3D pressure field and limit the active emission length of the vortices to a finite length l. Hence, we have:

$$p = \frac{-\rho_0}{4\pi c_0^2} \frac{r_i r_j}{r^3} \left[\frac{d^2}{dt^2} \int y_i L_j d\vec{y} \right]^* \tag{5}$$

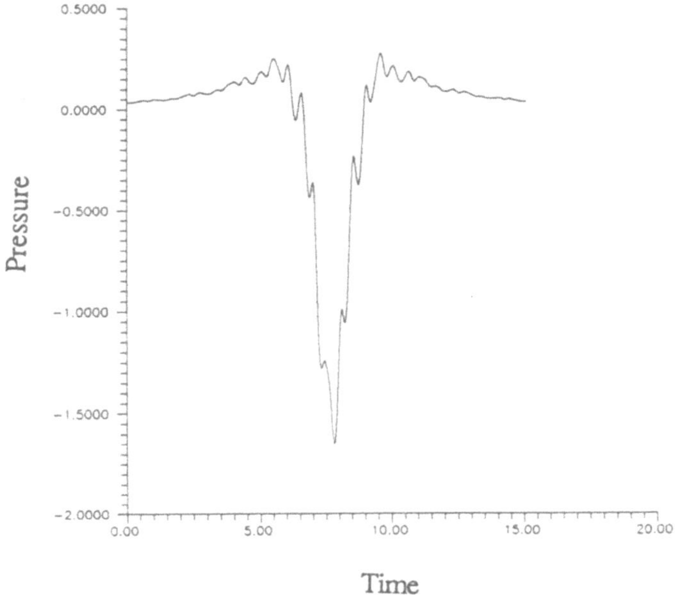

Figure 7. Wall pressure for $a = 0.433$.

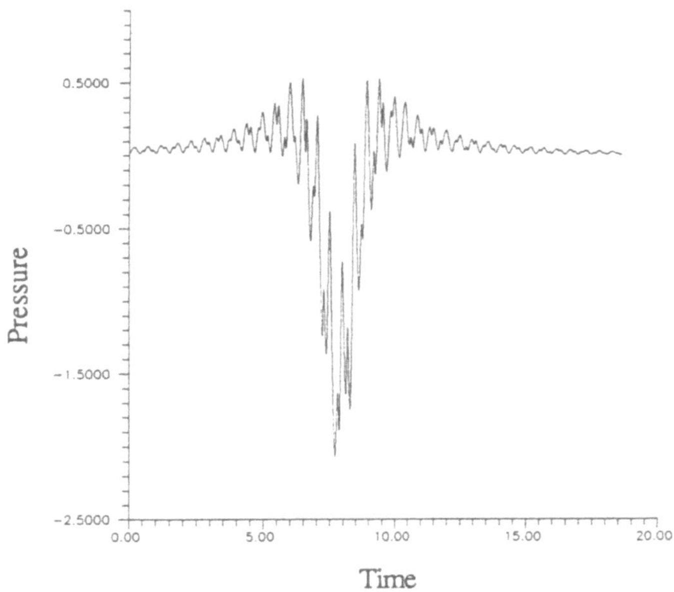

Figure 8. Wall pressure for $a = 0.0$.

where $\vec{r}$ is the position of the "listener," $\vec{L} = -\vec{u}\times\vec{\omega}$, $\vec{u}$ is the velocity of the fluid and c_0 is the speed of sound. The asterisk indicates that expression in square brackets must be taken at the delayed time $t^* = t - \frac{r}{c_0}$. We assume that the source region is compact, thus allowing the same delayed time for each vortex. Taking the vorticity distribution to be a sum of six delta functions, including three image vortices modelling the wall, we obtain the following expression:

$$p = \frac{-\rho_0 l}{4\pi c_0^2 r^3}\frac{r_i r_j}{}\frac{d^2}{dt^2}[P_{ij}]^* \tag{6}$$

with

$$P_{ij} = \sum_{n=1}^{2N} \begin{bmatrix} -\Gamma_n x_n v_n & x_n u_n \Gamma_n & 0 \\ -y_n v_n \Gamma_n & \Gamma_n u_n y_n & 0 \\ 0 & 0 & 0 \end{bmatrix} \tag{7}$$

with (x_n, y_n) being the position of the nth vortex, (u_n, v_n) its speed, l the length of the vortex line and Γ_n the strength of the vortices. The symmetries between a vortex and its image ($\Gamma' = -\Gamma,\ x' = x,\ y' = -y,\ u' = u,\ v' = -v$) lead to $P_{12} = P_{21} = 0$.

The quantity $\sum_{n=1}^{2N} \Gamma_n(x_n v_n - y_n u_n)$ is an invariant of our system which is similar to a kinetical moment in point mechanics (Lamb, 1945). Thus, $\frac{dP_{11}}{dt} = -\frac{dP_{22}}{dt}$ and

$$p = \frac{\rho_0 l}{4\pi c_0^2 r}\frac{r_1^2 - r_2^2}{r^2}\frac{d^2}{dt^2}\left[\sum_{n=1}^{2N} \Gamma_n x_n v_n\right]^* . \tag{8}$$

The sound directivity is therefore quadrupolar as expected and is time independent, given by the factor $\frac{r_1^2 - r_2^2}{r^2}$.

4.2. Use of Möhring's equation

Möhring proposed an expression linking sound pressure and vorticity (Möhring, 1978) in a fluid system having the following requirements:

- low Mach number
- compact flow
- compact source.

Under these conditions he deduced the following expression for farfield acoustic pressure fluctuations:

$$p(\vec{r},t) = \frac{\rho_0}{c_0^2}\frac{r_i r_j}{r^3}\frac{d^3}{dt^3}Q_{ij}(t - r/c_0) \tag{9}$$

where Q_{ij} is composed by a second order moment of vorticity:

$$Q_{ij}(t) = \frac{1}{12\pi}\int y_i(\vec{y}\times\vec{w})_j d\vec{y}. \tag{10}$$

In our case:

$$Q_{ij} = \frac{1}{12\pi}\sum_{n=1}^{2N}\begin{bmatrix} \Gamma_n x_n y_n & -x_n^2\Gamma_n & 0 \\ y_n^2\Gamma_n & -\Gamma_n x_n y_n & 0 \\ 0 & 0 & 0 \end{bmatrix} \tag{11}$$

Due to the symmetries of our system we have $Q_{12} = Q_{21} = 0$. Here the vorticity is not confined to a finite region because our system consists of line vortices. However we computed the pressure according to Möhring's expression to compare with our first expression (8) for the acoustic far field. We found the same results. This equality can be explained by rewriting $\frac{d^3Q_{11}}{dt^3}$. Using the fact that $\sum_{n=1}^{2N}\Gamma_n(x_n v_n - y_n u_n)$ is invariant, we have:

$$\frac{d}{dt}\left(\sum_{n=1}^{2N}\Gamma_n x_n v_n\right) = \frac{d}{dt}\left(\sum_{n=1}^{2N}\Gamma_n y_n u_n\right) \tag{12}$$

and

$$\begin{aligned} \frac{d^3}{dt^3}[Q_{11}] &= \frac{d^2}{dt^2}\sum_{n=1}^{2N}\Gamma_n(x_n v_n + y_n u_n) \\ &= 2\frac{d^2}{dt^2}\left[\sum_{n=1}^{2N}x_n v_n\Gamma_n\right] \\ &= -2\frac{d^2}{dt^2}[P_{11}]. \end{aligned} \tag{13}$$

Thus Möhring's expression leads to the same directivity and time evolution as the Powell-Hardin equation.

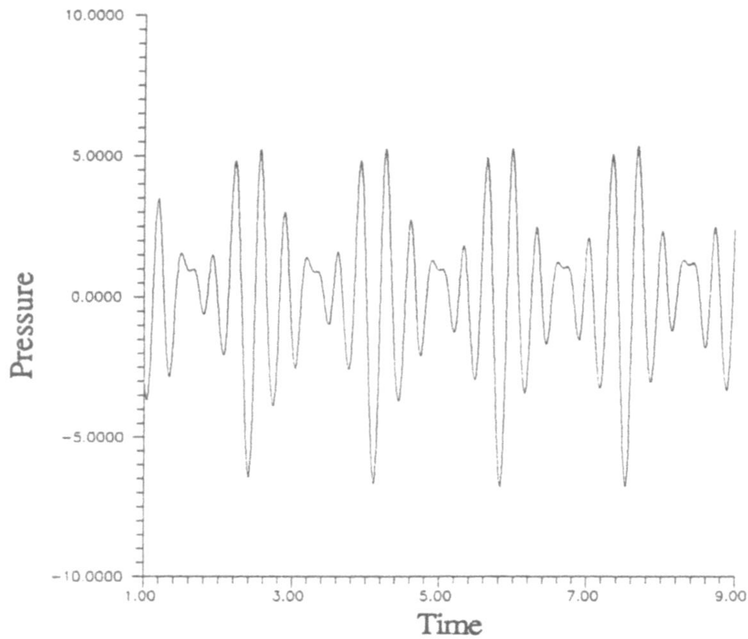

Figure 9. Time evolution of the acoustic farfield for $a = 0.433$.

4.3. Results

As we have shown above, the directivity of the farfield is quadrupolar with maxima both parallel and perpendicular to the wall.

Figs. 9 and 10 show the time evolution of the farfield acoustics, $\frac{d^2}{dt^2}[P_{11}]$, for $a = 0.433$ and $a = 0.00$. In the regular case ($a = 0.433$), the evolution of the pressure is periodic but non-sinusoidal. This is linked to the quasi-circular, centered trajectories of the vortices. As the parameter a decreases, more freedom is allowed in the movement of the vortices, so the quantity P_{11} evolves more and more rapidly, increasing the frequency and amplitude of the acoustic far field. This result is apparent from the corresponding spectra in Fig. 11 for $a = 0.433$ and Fig. 12 for $a = 0.00$ (the reference level is 1). In Fig. 11 we observe a few peaks whose frequencies match those of the Fourier transform of a centered coordinate of a vortex. The time derivative tends to sharpen the harmonics. In Fig. 12, only one peak exists and the spectrum exhibits mostly a broadband shape due to the erratic trajectories of the vortices.

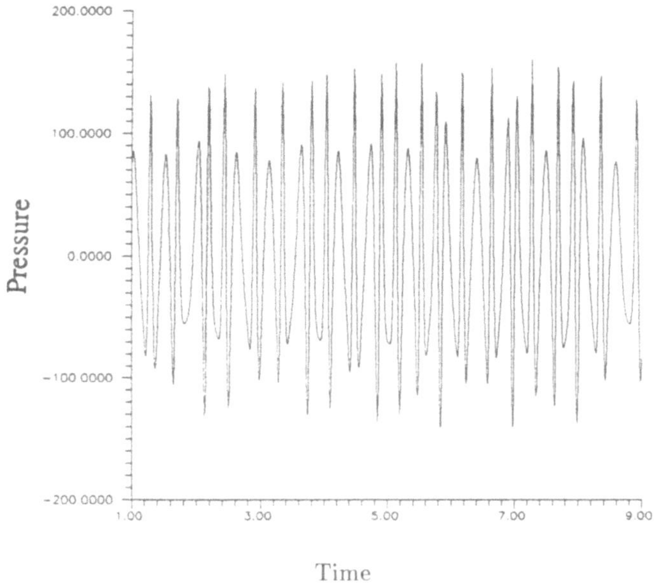

Figure 10. Time evolution of the acoustic farfield for $a = 0.00$.

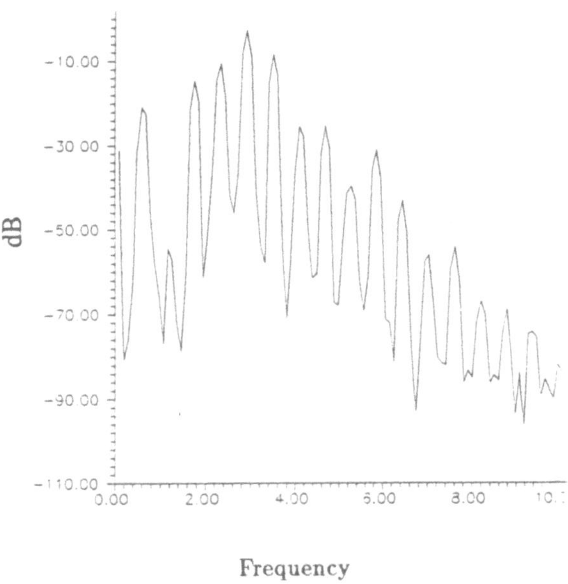

Figure 11. Spectrum of the acoustic farfield for $a = 0.433$.

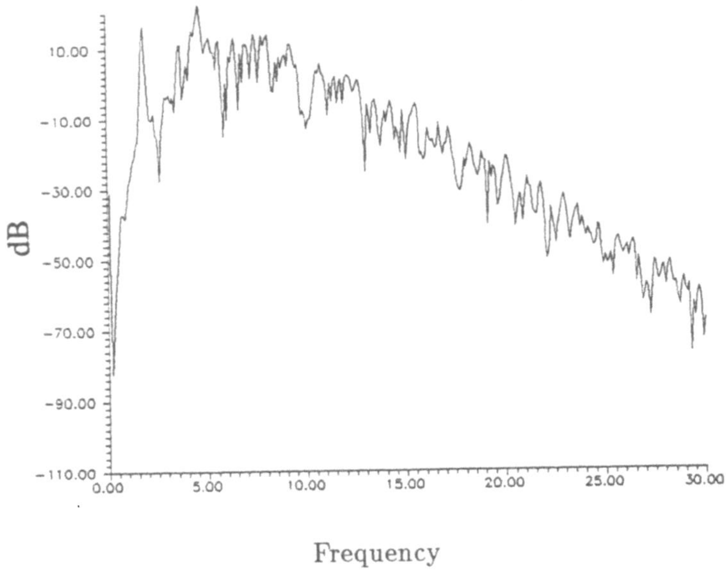

Figure 12. Spectrum of the acoustic farfield for $a = 0.00$.

Cases with greater values of the vortex strength were also calculated, leading to an increase of the level of the acoustic farfield. This increase is non-linearly related to the strength of the vortices because increasing the strength of the vortices increases the non-linear interactions among the vortices.

5. Influence of a Convective Flow

In this section, we summarize some effects created by the introduction of a convective flow.

5.1. Uniform convective flow

In this simplest case, we add only an uniform velocity field parallel to the wall. The addition of this uniform velocity does not change the chaotic or non-chaotic behavior of the vortices. The trajectories are stretched but the centered coordinates are the same.

As the translational speed of the vortices along the wall increases, the vortices affect the pressure at a fixed point of the wall over less time. During this reduced period, the vortices have less time to rotate, so that they create fewer oscillations in the pressure on the wall.

5.2. Shear flow

We initiated a study of the effects of a uniform shear on vortex trajectories, as occurs in wall-bounded shear flow. This is done by simply adding the velocity field of a parallel shear flow to the velocity field created by the point vortices.

We find that if the shear is too strong, the group of vortices "explode." The mechanism of this "explosion" is always the same: two vortices create a vortex pair that comes spinning very close to the wall, and the third vortex rapidly moves away from the wall to an asymptotic distance from the wall. We attempted to quantify the critical shear when "explosion" occurs using a dimensionless number based upon the distance between the vortices, the strength of the vortices and the shear. The difficulty, however, is that the "explosion" occurs suddenly without warnings and may happen after a long period of stability. It is possible, therefore, that "explosion" eventually occurs at all values of shear.

6. Conclusion

In this paper we have studied the wall pressure and the acoustic far field created by three point vortices moving near a flat wall. Depending on the initial position of the vortices, chaos may occur in their trajectories. The chaotic motion of the vortices has minor effects on the wall pressure - primarily in the production of higher frequency pressure fluctuations. Concerning the acoustic far field however, there are more dramatic effects. Chaotic vortex trajectories enhance high frequencies, lead to broadband noise and increase the acoustic intensity. In comparison with the acoustic farfield levels for a non-chaotic behavior of the vortices, we found that the noise created by vortices in chaotic motion is as much as 26 dB above the noise radiated by vortices in regular motion.

This study will be extended to vortices with finite cores.

References

Aref, H. and Pomphrey, N., 1982. "Integrable and chaotic motions of four vortices. I. The case of identical vortices," *Proc. R. Soc. London Ser. A* **380**, p. 359.

Aref, H., Rott, N., and Thomann, H., 1992. "Gröbli's solution of the three-vortex problem," *Ann. Rev. Fluid Mech.*, p. 1.

Brasseur, J. G. and Lin, W. Q., 1991. "Structure and statistics of intermittency in homogeneous turbulent shear flow," *Advances in Turbulence* **3**, Springer-Verlag, p. 3.

Conlisk, A. T., Guezennec, Y. G., and Elliott, G. S., 1989. "Chaotic motion of an array of vortices above a flat wall," *Phys. Fluids A* **1**(4), p. 704.

Gröbli, W., 1877. "Spezielle probleme über die Bewegung geradliniger paralleler Wirbelfäden," Zürcher und Furrer, Zürich.

Hardin, J. C., 1977. "Noise calculation on the basis of vortex flow models," ASME Symposium on Noise and Fluids Engineering, Atlanta, GA.

Lamb, H., 1945. *Hydrodynamics*, Dover.

Mitchell, B. E., Lele, S. K., and Moin, P., 1992. "Direct computation of the sound from a compressible co-rotating vortex pair," AIAA 92-0374.

Möhring, W., 1978. "On vortex sound at low Mach number," *J. Fluid Mech.* **85**(4), p. 685.

Powell, A., 1964. "Theory of vortex sound," *JASA* **36**(1), p. 177.

Wolf, A., Swift, J. B., Swinney, H. J., and Vastano, J. A., 1985. "Determining Lyapunov exponents from a time series," *Physica 16D*, p. 285.

Yates, J. E., 1978. "A study of scattering, production, and stimulated emission of sound by vortex flows," ARAP Report No. 363.

ACOUSTIC SCATTERING BY AN ELASTIC WINDOW IN A RIGID DUCT: NUMERICAL APPROACHES

S. Courtier-Arnoux

Chef du Groupe "Modèles de Vibration
et de Rayonnement des Structures"
Département "Acoustique et
Mécanique Vibratoire"
EDF/Direction des Etudes et Recherches
1, Av. du Gl de Gaulle
F - 92141 Clamart Cedex
FRANCE

1. Introduction

A coupled problem is considered, where acoustic and mechanic phenomena interact (Courtier-Arnoux et al., 1991).

A given geometry tends to simulate a two dimensional pipe whose mechanical properties differ over some finite portion of its length. Sound waves propagate through the pipe and radiate out of the flexible section of the pipe.

Three numerical methods have been compared on the given configuration, in order to model the coupled problem :

direct coupling,

external coupling

and interface coupling.

Besides, an analytical approach has been developed by J. Scott et al. (1991).

The direct coupling method is issued from the mechanic theory using modal approach and finite element method, and leading to computation of both acoustic and mechanic fields by minimization of an energy function. Some special finite element has been introduced in order to represent the coupling between the response of the structure and the response of the fluid: finite element (p, u, ϕ) where p and ϕ characterize the acoustic properties of the fluid (p is the acoustic pressure and ϕ is the displacement potential function), whereas u is the displacement of the structure.

The external coupling method uses a temporal approach and starts from an acoustic code using a finite difference method and solving the linearized Euler equations in the fluid. A special boundary condition has been introduced on the fluid boundary corresponding to the elastic part of the pipe, where the problem is coupled. The vibration of the structure is decomposed upon the first dynamic modes of the elastic window (beam model) and the transfer function gives the relevant value of the boundary condition in the temporal space.

The interface coupling method is a modal approach for both the elastic structure (beam model) and the acoustic fluid (solving the Helmholtz equation using a finite difference model). Coupling is considered through projection of the interaction matrices at the interface.

The numerical results are compared and discussed. The advantages of each method are given, as well as the limits of each type of numerical coupling.

2. Geometry

The geometry is presented in Fig. 1 and represents a two-dimensional rigid duct with an elastic window of length $L = 1$ m. The excitation is located on the open boundaries of the duct. Two types of acoustic excitations are considered: firstly, an incident wave, and secondly a given velocity. Both boundary conditions are symmetric with respect to the elastic window, and vary with time according to frequency $f = 1000$ Hz. The internal fluid as well as the outside one are water; outside radiation takes place in open space.

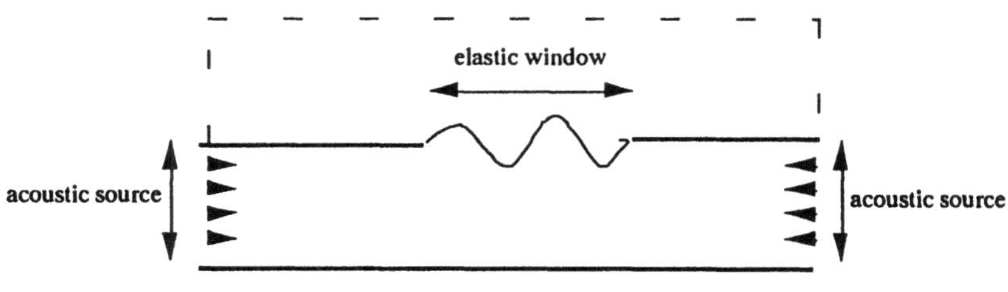

Fig. 1: Geometry

The elastic window is an unbaffled 1 cm high steel beam, assumed to be freely suspended on both edges of the window.

The industrial configuration has been drastically simplified in order to give birth to this test model. It could represent any duct line filled with fluid and being damaged through acoustic-elastic interaction near resonance, or radiating noise.

3. Direct Coupling

The direct coupling of both the elastic equations and the Helmholtz equation yields to non-symmetric formulation. The mixed finite element (u, ϕ, p) however leads to a symmetric formulation of the energy function used in any finite element computing code. Therefore it has been chosen to use this finite element to discretize both the acoustic fluid and the interaction surface with the plate.

p is the acoustic pressure, ϕ is the potential function of the acoustic displacement, and $\vec{u}$ is the displacement of the structure at the fluid interface. u is the component of $\vec{u}$ which is normal to the interface.

The equations of linear acoustics, written using (p, ϕ) as dependant variables are combined into the acoustic energy function, F_a. It includes a volumic term and a surface term. The latter is related to the displacement of the elastic boundary:

$$\begin{aligned} F_a(p,\phi) &= \int_{\Omega_a} \frac{p^2}{\rho C^2} d\Omega + \int_{\Omega_a} \left(\frac{\phi \ddot{p} + p \ddot{\phi}}{C^2} - \rho \, \overrightarrow{\text{grad}} \, \phi \cdot \overrightarrow{\text{grad}} \, \ddot{\phi} \right) d\Omega \\ & \quad + \int_{\Sigma_a} \rho \phi \, \overrightarrow{\text{grad}} \, \ddot{\phi} \cdot \overrightarrow{ds} \, . \end{aligned} \tag{1}$$

The surface interaction term, coupling elastic and acoustic problems can be rewritten as follows:

$$\int_{\Sigma_a} \rho \phi \, \overrightarrow{\text{grad}} \, \ddot{\phi} \cdot \overrightarrow{ds} = \int_{\Sigma_a} \rho \phi \, \overrightarrow{\ddot{u}} \cdot \overrightarrow{ds} \, . \tag{2}$$

The equation of linear mechanics, using $(\vec{u})$ as dependant variable leads to the following elastic energy function, F_S. It includes a volumic term and a surface term:

$$F_S(\vec{u}) = \int_{\Omega_S} k(\vec{u}, \vec{u}) d\Omega + \int_{\Omega_S} m(\vec{u}, \vec{u}) d\Omega + \int_{\Sigma_S} \vec{u} \cdot [\sigma(\vec{u})] \, \overrightarrow{ds} \, . \tag{3}$$

On the interaction surface Ω_S between the acoustic fluid and the elastic plate, the surface stress is due to the acoustic pressure and it can hence be written:

$$\vec{f} = \sigma(\vec{u})\cdot \overrightarrow{ds} = -p\ \overrightarrow{ds} = \rho\ddot{\phi}\ \overrightarrow{ds} \tag{4}$$

and:

$$\int_{\Sigma_S} -p\vec{u}\cdot \overrightarrow{ds} = \int_{\Sigma_S} \rho\ddot{\phi}\vec{u}\cdot \overrightarrow{ds}\ .$$

From equations (2) and (4) one sees that the coupling formulation is symmetric.

The final expression of the direct coupling method is then:

$$\begin{bmatrix} Ku^2 + Mu\ddot{u} & \rho u\ddot{\phi} & 0 \\ \rho\phi\ddot{u} & -\rho\nabla\phi\nabla\ddot{\phi} & \dfrac{\phi\ddot{p}}{C^2} \\ 0 & \dfrac{p\ddot{\phi}}{C^2} & \dfrac{p^2}{\rho C^2} \end{bmatrix} = \begin{bmatrix} 0 \\ \rho\ddot{u}_0 \\ 0 \end{bmatrix}$$

where u_0 is the normal displacement of the acoustic fluid on the open boundaries $(\sum_a - (\sum_S \cap \sum_a))$. It is given by the acoustic source, which is a data of the problem.

4. External Coupling

4.1. The acoustic code

The external coupling method relies on a computing package which solves the linearized Euler equations and the problem of the propagation of an acoustic source in a fluid with any given steady two-dimensional flow.

A semi-Lagrangian approach combined to a direction splitting method is used: following successively both cartesian directions, the hyperbolic Equation (5) is solved on a finite difference grid:

$$\frac{\partial w}{\partial t} + A\cdot\frac{\partial w}{\partial x} + B\cdot w = 0 \tag{5}$$

where matrices w, A, B are respectively equal to:

$$w = \begin{pmatrix} p \\ u \end{pmatrix} \qquad A = \begin{pmatrix} \overline{V} & \overline{\rho}\,\overline{C}^2 \\ \frac{1}{\overline{\rho}} & \overline{V} \end{pmatrix}$$

$$B = \begin{pmatrix} \frac{\gamma}{S}\frac{\partial \overline{V} S}{\partial x} & \frac{\overline{C}^2}{S}\frac{\partial \overline{\rho} S}{\partial x} \\ -\frac{1}{\overline{\rho}^2 \overline{C}^2}\frac{\partial \overline{P}}{\partial x} & \frac{\partial \overline{V}}{\partial x} \end{pmatrix}.$$

In these formulae, p and u denote the acoustic unknown values, and $\overline{\rho}, \overline{V}, \overline{C}, \overline{P}$ characterize the mean flow which is given.

A temporal Galerkin formulation leads to a problem with the test functions ψ_j as unknowns in the time interval $[t^n, t^{n+1}]$:

$$\begin{cases} \frac{\partial_{\psi_j}}{\partial t} +^T A \cdot \frac{\partial_{\psi_j}}{\partial x} -^T B \cdot \psi_j = 0 \\ \psi_j(t = t^{n+1}) = \phi_j \end{cases} \tag{6}$$

where ϕ_j denotes the elementary hat function used as test function (equal to 1 at the grid point j, and equal to 0 anywhere else). j varies from 1 to $2N$ where N is the number of grid points in the x direction:

$$\phi_j \equiv \begin{pmatrix} \phi_j \\ 0 \end{pmatrix} \qquad \text{for } 1 \leq j \leq N$$

and:

$$\phi_j \equiv \begin{pmatrix} 0 \\ \phi_j \end{pmatrix} \qquad \text{for } N+1 \leq j \leq 2N.$$

Equation (6) defines the characteristic curve. Along this curve the Riemann integral is constant:

$$\begin{pmatrix} 1 \\ +/- \overline{\rho}\,\overline{C} \end{pmatrix} \cdot {}^T w = \text{constant}$$

expression which is valid for respectively the backward and forward propagation in the particular case of linear acoustic propagation in a duct with constant section and uniform flow.

4.2. The fluid/structure coupling

The fluid/structure interaction is entirely defined by the boundary condition along the elastic window.

Its numerical treatment follows an explicit time scheme. It actually transfers information from the structure modal space to the acoustic fluid temporal space and vice versa. It involves a Duhamel's integral.

The practical implementation of the external coupling method is described hereafter:

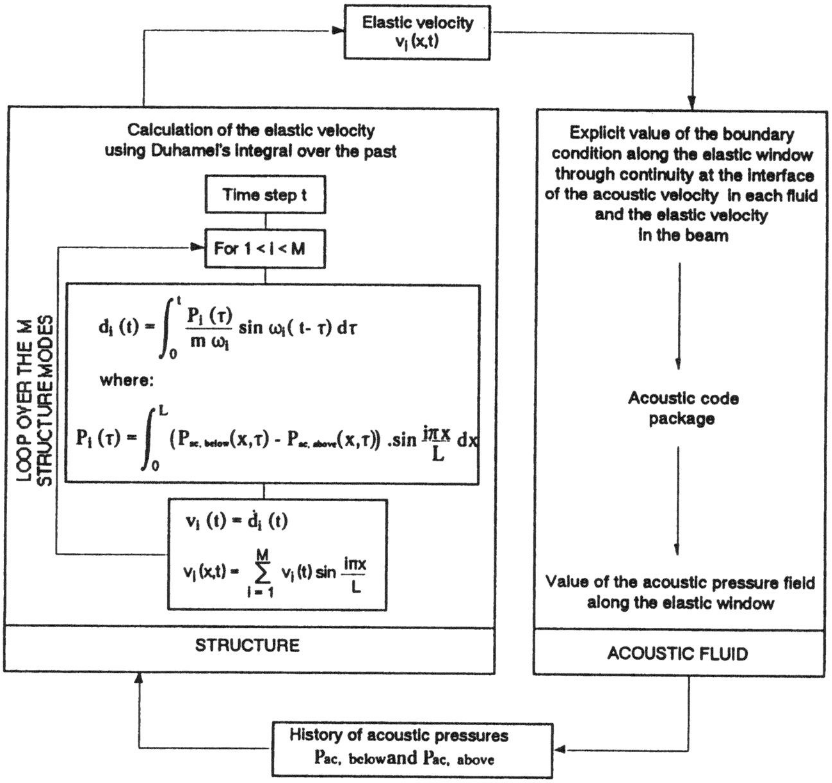

5. Interface Coupling

Both the Helmholtz equation for the acoustic fluids and the bending dynamic equation for the beam are written in frequency space. After discretization on a regular finite difference grid in both the internal and the external fluid and after decomposition of the response of the beam upon its modes, one can write the coupled problem as:

$$\begin{pmatrix} H_1 & \rho\omega^2 C_1 & 0 \\ -{}^tC_1 & L & {}^tC_2 \\ 0 & -\rho\omega^2 C_2 & H_2 \end{pmatrix} \begin{pmatrix} P_1 \\ U \\ P_2 \end{pmatrix} = \begin{pmatrix} 0 \\ 0 \\ S \end{pmatrix} \tag{7}$$

where H_1, L and H_2 respectively denote the acoustic operator in the upper fluid cavity, the vibration operator in the beam responding on bending modes and the acoustic operator in the lower fluid cavity. The vectors P_1, U and P_2 respectively represent the acoustic pressure at the grid points of the upper fluid cavity, the normal displacement coefficients upon the modal base of the beam, and the acoustic pressure at the grid points of the lower fluid cavity. The vector S denotes the acoustic excitation given in the lower cavity as boundary condition.

The coupling matrices are C_1 and C_2. They respectively characterize the interaction between the upper/lower acoustic cavity and the beam:

$$C_{1,ij} = \int_{x_i-\frac{h}{2}}^{x_i+\frac{h}{2}} \sin(j\pi\xi)\delta_1(i)d\xi$$

$$C_{2,ij} = \int_{x_i-\frac{h}{2}}^{x_i+\frac{h}{2}} \sin(j\pi\xi)\delta_2(i)d\xi$$

with $\delta_1(i) = 1$ if the grid point x_i belongs to the upper interface, and $\delta_2(i) = 1$ if the grid point x_i belongs to the lower interface, and $\delta_1(i) = 0$ and $\delta_2(i) = 0$ if not.

Following (7), the algorithm of the interface coupling method is:

- Calculation of the acoustic pressures at the interface, due to an elementary dynamic source corresponding to an elementary bending mode of the beam:

$$P_1|_{\text{interface}} = -\rho\omega^2 H_1^{-1}|_{\text{interface}} C_1$$

$$P_2|_{\text{interface}} = +\rho\omega^2 H_2^{-1}|_{\text{interface}} C_2$$

- Calculation of the response of the beam: the dynamic sources due to the coupled problems are consequently given by $(P_1|_{\text{interface}} \cdot U)$ and $(P_2|_{\text{interface}} \cdot U)$, hence:

$$\{L - C_1|_{\text{interface}}P_1|_{\text{interface}} +^t C_2|_{\text{interface}}P_2|_{\text{interface}}\} \cdot U$$
$$=^t C_2|_{\text{interface}}H_2|_{\text{interface}} \cdot S.$$

- Calculation of the acoustic pressures P_1 and P_2 in both cavities using:

$$H_1P_1 = -\rho\omega^2 C_1 U$$
$$H_2P_2 = \rho\omega^2 C_2 U + S.$$

6. Numerical Results

All the approaches presented in this paper are linear ones. The boundary conditions used for the acoustic excitation are of two types and successively have following values in the computations:

- given velocity:

$$u_0 = 1m/s,$$

or

- incident wave of given amplitude:

$$P_0 = \rho C u_0 = 1.5MPa.$$

The following parameters are used for the comparison of the three numerical coupling methods discussed previously in this paper:

- min and max values of the acoustic level:

$$L_p = 10 \log \frac{p\, p^*}{(2.10^{-5})^2}$$

- min and max values of the acoustic intensity:

$$\vec{I} = \frac{1}{2}\text{Re}(p\vec{v}^*) \quad \text{and} \quad L_I = 10 \log \frac{|\vec{I}|}{10^{-12}}$$

- values of the acoustic power E on the different boundaries:

(1) acoustic source boundary

(2) lower surface of the beam

(3) upper surface of the beam

(4) acoustic absorbing boundary.

These four values should be equal as there is no damping in the system considered. The comparison between the values (2) and (3) is a validation of the coupling method, whereas the comparison of the values (1) and (2), or (3) and (4) is a validation of both the coupling method and the method solving the acoustic propagation problem.

Tables 1 and 2 summarize the numerical results according to the acoustic boundary condition which is considered. The check values appear as they have been previously described, for each of the three coupling methods.

Table 1: Numerical results for an acoustic incident wave.

L_I min	142	142	149
L_I max	173	172	173
L_P min	181	165	182
L_P max	219	216	219
$E(1)$	162	166	165
$E(2)$	162	165	165
$E(3)$	162	165	165
$E(4)$	162	165	164
(dB)	Direct coupling	External coupling	Interface coupling

In the case of an acoustic incident wave, the conservation law is respected with a pretty good accuracy, better than 1%. The acoustic power global check makes a 1 dB lost in the cavity with the acoustic sources, appear for the external coupling method, whereas the direct coupling method is exactly conservative, and the interface coupling method ends up with a 1 dB lost in the cavity with the free radiative acoustic boundaries.

The comparison of the maximum values of both the acoustic level and acoustic intensity makes a good agreement over the three methods clearly appear. The accuracy is better than 3 dB.

Table 2: Numerical results for an acoustic given velocity.

L_I min	164	156	171
L_I max	194	197	204
L_P min	201	190	208
L_P max	250	243	257
$E(1)$	184	194	196
$E(2)$	184	189	195
$E(3)$	184	189	195
$E(4)$	184	189	194
(dB)	Direct coupling	External coupling	Interface coupling

In the case of an acoustic rigid piston however, the global power check is not as good as in the case of an acoustic incident wave. The direct coupling method remains exactly conservative, whereas the interface coupling method looses 2 dB, and the external coupling method 5 dB.

It has clearly appeared during the computations, that the case of a rigid piston as an acoustic source, is a much more severe test for the numerical stability of the methods considered, as it corresponds to a resonance of the lower acoustic cavity, around which the classical acoustic approaches do not manage to find the acoustic solution to the Helmholtz problem.

The comparison of the maximum values as well, is not good, because of non convergence of the interface and external coupling methods. For instance, let us consider the external approach, which is a temporal one. For the frequency $f = 1000$ Hz, the time step is $\Delta t = 0.000001\ s$ and the space step is $\Delta x = 0.05\ m$. In the case of an acoustic incident wave, steady values are obtained after 7 periods of time, whereas steady values are not yet obtained after 25 periods of time in the case of an acoustic given velocity.

All the computations have been run on a CRAY YMP 4-64. The cpu time used by each method varies a lot:

$4s$ cpu time for the direct coupling method

$800s$ cpu time for the external coupling method

$1600s$ cpu time for the interface coupling method.

The Figures 2 and 3 represent a map of the acoustic intensity and acoustic pressure, in the case of an acoustic incident wave, obtained with the interface coupling method.

A computation has been also run in this configuration, adding a two step flow in the duct (around Mach = 0.1 in the lower 66% part of the lower acoustic cavity, and around Mach = 0.01 in the upper 33% part of the lower acoustic cavity), which is a very rough attempt to simulate the influence of a turbulent boundary layer. One can notice on the Figures 4 and 5 that the pattern of both acoustic intensity and pressure is totally different, as well as the localization of the maximum values. The maximum values however remain roughly the same.

7. Conclusion

This paper has discussed the results obtained with three numerical tools which have a totally different approach in order to solve the problem of the interaction of acoustic cavities (with or without fluid in motion in them), and an elastic structure.

The geometry considered is a rough simulation of a pipe with an elastic window in it. It is the first step to the simulation of problems in which Electricité de France is involved, as for instance the noise radiated by vibrating complex structures, or the damage and eventually fracture of some parts of other structures which are excited by an acoustic source.

Two types of acoustic sources have been considered. In the case of incident waves, there is a good agreement between the three methods. It is not as good in the case of given velocities.

The advantages and limitations of each numerical approach are following ones:

- the direct coupling method is applicable to any geometry, but it gives no access to transient events as it is an harmonic approach,
- the external coupling method makes the transient effects appear, and it allows the fluid to be in any steady motion, but it is rather expensive due to space and time steps which values have to be little enough to capture the high frequency phenomena,
- the interface coupling is issued from acoustic transparency theory and makes it easy to run parametric studies for the struc-

ture, but it is not accurate near resonant frequency of the acoustic cavities.

Further work on this subject will be on one hand a comparison between analytical and numerical approaches, and on the other hand, a parametric study. The aim of such study is to be able to characterize the most damageable configuration, and find a set of parameters which minimizes the risk of exciting resonant frequency of both the structure and the acoustic cavities.

References

Courtier-Arnoux, S. et al., 1991. "Confrontation d'outils numeriques our un test de couplage fluide/structure," EDF/DER Internal Report, HP61/ 91.012.

Scott, J. et al., 1991. *Acoustic Radiation of Fluid Loaded Elastic Plates*, TOPEXPRESS Ltd.

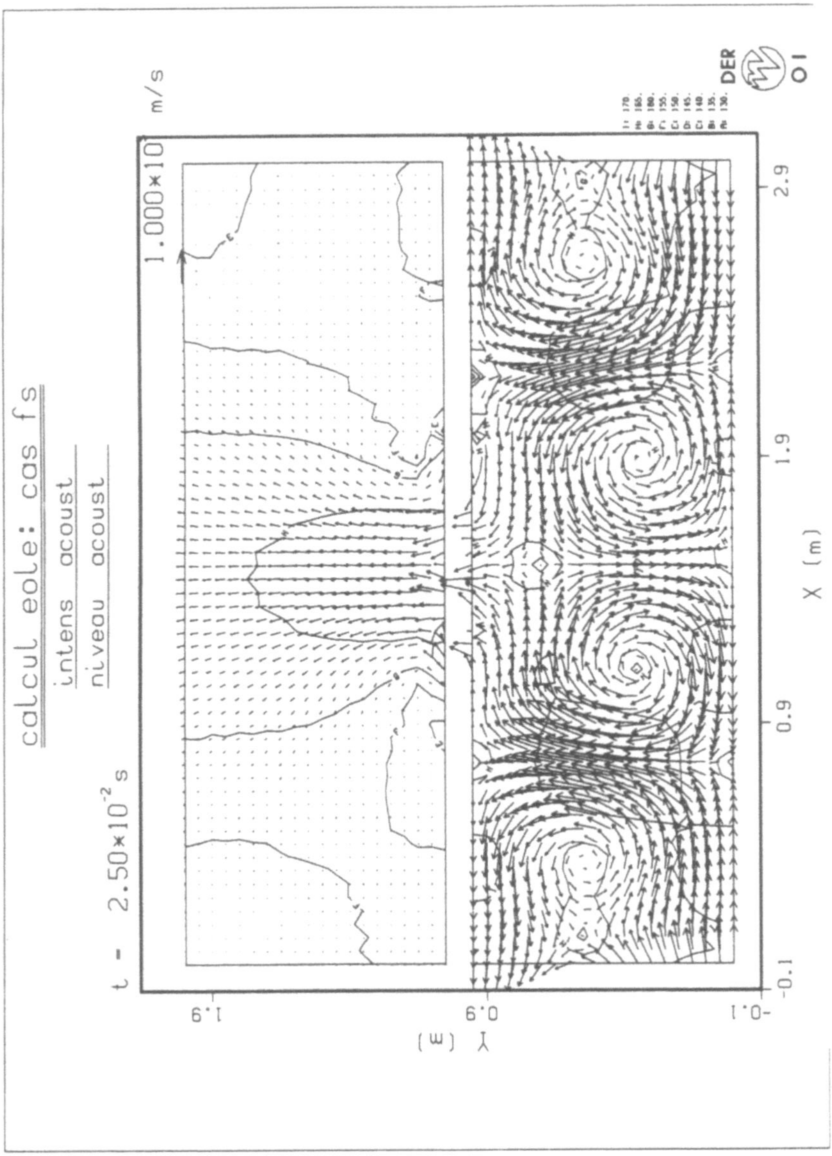

fig.2: Map of the acoustic intensity in the case of acoustic incident waves.

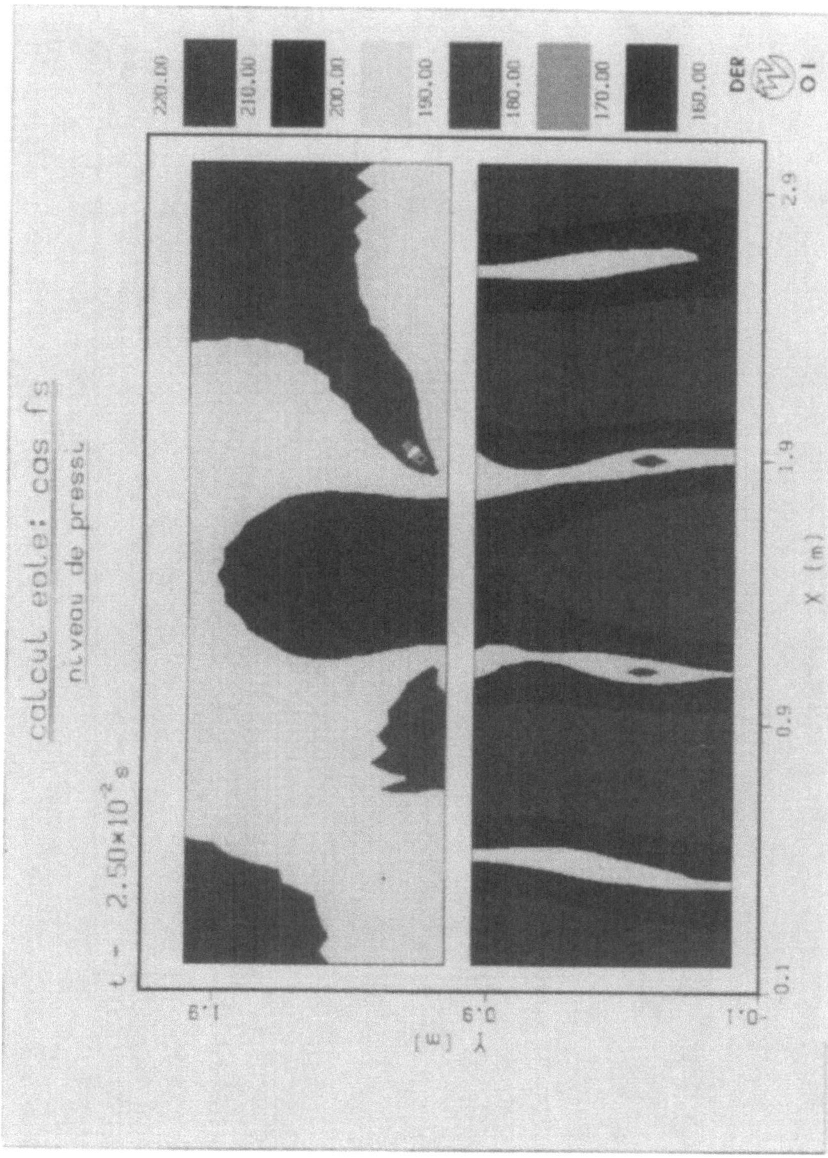

fig.3: Map of the acoustic pressure level in the case of acoustic incident waves.

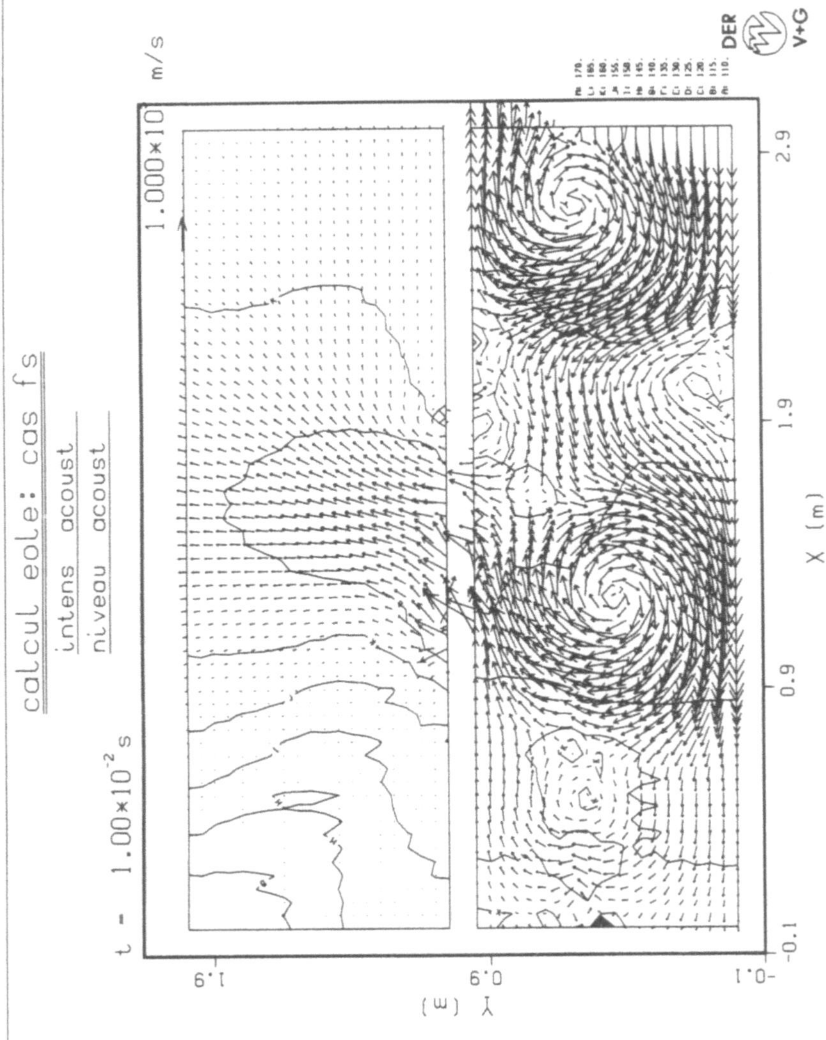

fig.4: Map of the acoustic intensity in the case of a two step flow.

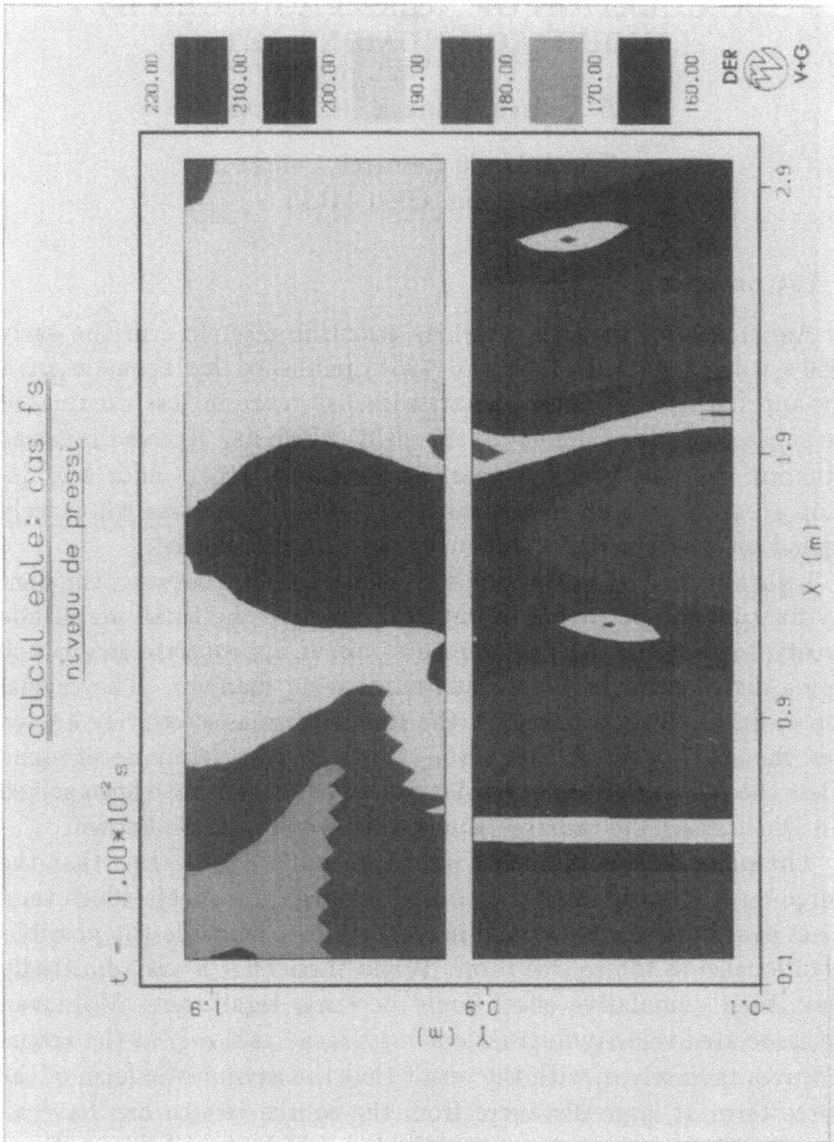

fig.5: Map of the acoustic pressure level in the case of a two step flow.

CALCULATION OF NOISE PRODUCED BY HIGH MACH NUMBER JETS

M. E. Goldstein and R. R. Mankbadi

NASA Lewis Research Center
Cleveland, Ohio 44135

1. Introduction

Aeroacoustics became a serious scientific discipline in the early 1950's when Lighthill (1952 and 1954) published his acoustic analogy approach to jet noise. That work has more or less dominated the subsequent development of this field, which has received renewed attention over the past few years as a result of recent efforts to develop an updated high speed transport – whose successes will largely depend on whether its jet exhaust noise can be reduced.

Lighthill showed that there is an exact analogy between the density fluctuations occurring in any real flow and the small amplitude density fluctuations that would result from an appropriate quadrupole source distribution in a fictitious nonmoving medium. The crucial step in his analysis is to regard the source term as effectively known even though, in reality, that term cannot be completely determined unless the full compressible Navier-Stokes equations have been solved – in which case, the radiated sound field would also be known.

The most serious difficulty with Lighthill's approach is that the source term actually involves acoustic propagation effects which tend to act over large distances and may, therefore, preclude the possibility of localizing the source term. While these effects are admittedly weak, their cumulative effect could be fairly significant. Moreover, the associated velocity fluctuations may decay as slowly as the acoustic waves themselves, with the result that the asymptotic form of the source term at large distances from the source (which can have an important influence on the sound field) could be largely determined by intricate cancellations between its various components. These effects tend to be unimportant at low Mach numbers, but could be quite significant at the high Mach numbers that are of interest for the high speed transport.

2. Extensions of the Acoustic Analogy

These and related considerations led other researchers to introduce various generalizations of the acoustic analogy that could potentially lead to more accurate estimates of the sound field. These approaches all amount to rearranging the Navier-Stokes equation into the form

$$L\Pi = S \tag{1}$$

where L is a differential operator that reduces to the stationary medium wave operator $\Box^2$ at large distances from the source, Π is a scaler variable that reduces to the pressure or density fluctuations at large distances from the source, and S is a, hopefully, localized term that then acts as a source for the acoustic pressure fluctuations. The expectation is that the sound field will be more or less independent of the details of S when S is sufficiently localized. It is plausible that S will be more localized when more of the acoustic propagation effects are included in L.

In the Howe/Powell (Howe, 1975) analogy, L is taken to be the operator describing the irrotational motion of an inviscid fluid, i.e., $L\Pi = 0$, is the equation for the single scaler potential that characterizes the unsteady irrational flow of an inviscid fluid. This equation should be capable of dealing with the most important aspects of the relevant nonlinear wave-phenomena, e.g., wave-steeping effects, and these effects should therefore not be present in S. However, Howe/Powell adopt a somewhat different point of view in arguing for (1). They merely point out that S turns out to be proportional to the vorticity field, which is known to be a highly localized quantity.

Lilley (1974) argues that the most important propagation effects are the acoustic mean flow interactions (i.e., mean flow refraction, mean flow shrouding, etc.) and that $L\Pi = 0$ should therefore reduce to the equation for the propagation of acoustic waves on transversely sheared means flow when the generalized pressure fluctuations Π are sufficiently small. The qualitative features of the experimentally observed sound fields seem to bear this out, and the solution to Lilley's equation, with S replaced by a convecting point source, do seem to predict the experimentally observed 1/3-octave directivity patterns quite well.

3. Direct Computation of Sound Field

An alternative approach to the jet noise problem that is becoming increasingly popular in recent years is to try to solve the compressible Navier-Stokes equations directly and obtain the radiated sound field as a by-product of the solution. Needless to say, the current generation of computers is far from the stage where they can be used to calculate the complete turbulent flow field of a jet at the high Reynolds numbers that are of technological interest. It is therefore necessary to introduce some sort of approximation in order to make progress with this approach.

3.1. Linearization of equations

One could attempt to linearize the equations about an appropriate mean flow, as one does in the rapid distortion theory of turbulence. For jets, it is natural to take the mean flow to be a unidirectional transversely sheared mean flow. This amounts, then, to doing compressible rapid distortion theory on this flow (Moffatt, 1981).

The validity of rapid distortion theory depends upon the following conditions being satisfied (Hunt, 1973): (1) $u'/U << 1$, where u' is the rms turbulence velocity fluctuation and U is the local mean-flow velocity, and (2) the interaction or change being calculated is completed in a time, say τ_I, that is short compared with τ_{decay}, where $\tau_{\text{decay}} = 0(\ell/u')$ is the decay time or lifetime of a turbulent eddy, ℓ being the characteristic size of that eddy. Rapid-distortion calculations are usually based on the inviscid equations – an approximation that is justified when both the mean flow and turbulence Reynolds numbers are large.

The ratio of rms turbulence velocity to local mean flow velocity is reasonably small in the region of maximum turbulence, so that the first requirement for the validity of rapid distortion theory is satisfied (Tennekes and Lumley, 1972). However, the interaction time τ_I, which, in the present context, should be taken as the time for the sound generation to occur, is equal to the decay time τ_{decay} of the turbulence, and the second requirement of the theory is therefore not satisfied. But in the absence of another completely satisfactory alternative, one might still attempt to use u'/U as a small parameter and carry the corresponding asymptotic expansion to its logical conclusion.

It is encouraging to note that Townsend (1976) achieved considerable success using this type of rapid distortion theory to predict the essential anisotropic features of the relatively fine-grained turbulence in free shear flows. The characteristic time τ_I in the Townsend analysis is the reciprocal mean shear $(\partial U/\partial x_2)^{-1}$, where x_2 is a Cartesian coordinate in a direction perpendicular to U. Prandtl's hypothesis, namely $u' = \ell\partial U/\partial x_2$, suggests that these two times, namely τ_I and τ_{decay}, are again equal. If rapid distortion theory can succeed here, it certainly should be capable of predicting the weak sound fields generated by the turbulence.

The lowest order equations are the compressible Rayleigh's (or linearized Euler) equations. It is well known (Betchov and Criminale, 1967) that the velocity components can be eliminated between these equations to obtain a single equation for the normalized first-order isentropic density fluctuation Π_1. This equation can be written symbolically as

$$L\Pi_1 = 0. \tag{2}$$

As is usual in such perturbation equations, the next order density fluctuations Π_2, Π_3, et cetera satisfy homogeneous versions of this equation

$$L\Pi_2 = S_1(\Pi_1, q_1) \tag{3}$$

$$L\Pi_3 = S_2(\Pi_1, \Pi_2, q_1, q_2) \tag{4}$$

where the source term S_1 is quadratic in the first-order quantities and q_i for $i = 1, 2$ denote generic first- and second-order variables.

Equation 2 possesses two classes of homogeneous solutions. One of them consists of the linear instability waves that will exist when the mean velocity profile is inflexional. The other corresponds to the continuous spectrum and may somehow be linked with the fine-grained turbulence. At subsonic speeds, these solutions decay exponentially fast outside the jet and, therefore, possess no radiation field. However, the solutions are only local, as evidenced by the fact that the expansion

$$\Pi = \Pi_1 + \Pi_2 + \Pi_3 + \ldots \tag{5}$$

is nonuniformly valid in the sense that Π_2 and/or Π_3 eventually become larger than Π_1 when either the streamwise or transverse coordinate becomes sufficiently large.

In certain cases, it is possible to use standard singular perturbation techniques (usually together with some additional ad hoc approximations) to render the solution uniformly valid in both the streamwise and transverse directions (Nayfeh, 1973), and the resulting solutions will then have an acoustic radiation field. This approach goes beyond the usual rapid distortion theory and allows one to follow the flow over time scales that are long enough to properly evaluate the radiated sound field (i.e., time scales that are much longer than the sound generation process itself). It has been pioneered by Tam and Morris (1980).

However, it is also worth noting that the global solution of

$$L(\Pi_1 + \Pi_2) = S_1(\Pi_1, q_1), \tag{6}$$

obtained by adding (2) and (3), will possess a radiation field when Π_1, q_1 are rendered uniformly valid in the streamwise direction, but not in the transverse direction. This approach has been pioneered by Crighton and Huerre (1983). Further study of the relation between these two approaches could prove to be highly beneficial.

3.2. Large eddy simulation

Another approach to simplifying the full compressible Navier-Stokes equation is to realize that a major obstacle to obtaining numerical solutions at high Reynolds numbers is due to the large range of scales that need to be resolved, while conventional wisdom suggests that only the larger, more coherent motions are major producers of sound. This suggests that one ought to calculate only the larger scale motion and to simply model the effect of the smaller scales on that motion. This amounts to using a compressible version of the so-called "Large Eddy Simulation" or "Subgrid Scale Modeling" theory. This subject has been highly developed over the past few years and may now be ready for use as a jet noise prediction tool. It is, in a certain sense, a natural extension of the acoustic analogy approach, which separates Navier-Stokes equations into acoustic propagation terms and source terms and models the resulting Reynolds stress-type source terms. The large eddy simulation begins with the filtered Navier-Stokes equation and models the filtered Reynolds stress terms. It therefore represents a natural progression in complexity that allows fuller exploitation of currently available computational capabilities.

There are, however, a number of issues that have to be dealt with before this can be carried out in practice. The most important of these is related to reducing the storage requirements by limiting the numerical computation to the near-field region where nonlinear effects are important, while using analytical procedures to determine the flow in the far field where the motion is presumably governed by the linear acoustic equations to a close approximation. It would not just be a wasteful use of valuable computational resources to solve the full Navier/Stokes equations in this latter region, but would actually increase the storage requirements beyond the capability of any currently available machine.

Two approaches are currently being developed to deal with this problem – both of which are direct extensions of the approaches developed for the linear theory. The first (which is an extension of the Crighton/Huerre [Crighton and Huerre, 1983] approach) involves the use of Lighthill's (or some other) acoustic analogy equation with its source term evaluated from near-field compressible Navier-Stokes solutions. The major difficulty with this approach stems from the fact that some acoustic propagation and other compressibility effects probably remain embedded in the source term and may be very difficult to estimate from a numerical simulation to a sufficient degree of accuracy.

The other approach (which is an extension of the Tam/Morris [Tam and Morris, 1980] approach) involves patching the appropriate far-field solution to the linear wave equation to the near-field Navier-Stokes solution on some large surface that completely encloses the flow. Here the major problems arise because the available computational resources are not great enough to obtain accurate solutions of the Navier-Stokes equations at the large distances where the linear acoustics solutions become valid.

This approach also bears some resemblance to an analytical procedure introduced by Dowling et al. (Dowling, Ffowcs, and Goldstein, 1978) in an effort to incorporate mean-flow shrouding effects into Lighthill's theory. This analysis, which also involved patching a near-field solution to the solution of an acoustic wave equation, showed that the patching tends to produce extraneous acoustic sources which can be minimized if the patching is carried out on a surface that moves with the flow, rather than on a fixed surface – which might seem to be the more obvious choice. Further study of the linear theory might shed some light on these issues.

Another difficulty is associated with finding appropriate outflow boundary conditions for the near-field solution. The true boundary conditions should be imposed at downstream infinity where the jet's momentum has been dissipated by viscous effects, but available computer resources require that the boundary condition be imposed fairly close to the nozzle exit where the momentum is still fairly large. Any artificial reflections resulting from even small errors in modeling this boundary condition could cause large errors in the predicted acoustic radiation.

The filtered Navier-Stokes equations can be written (in the cylindrical coordinate system that would be appropriate to a round jet) as

$$\frac{\partial Q}{\partial t} + \frac{\partial F}{\partial x} + \frac{1}{r}\frac{\partial(rG)}{\partial r} + \frac{1}{r}\frac{\partial H}{\partial \theta} = R, \tag{7}$$

$$Q = \begin{bmatrix} \rho \\ \rho u \\ \rho v \\ \rho w \\ \rho e \end{bmatrix}, \tag{8}$$

$$F = \begin{bmatrix} \rho u \\ \rho u^2 - \tau_{xx} \\ \rho uv - \tau_{xr} \\ \rho uw - \tau_{x\theta} \\ \rho ue - u\tau_{xx} - v\tau_{xr} - w\tau_{x\theta} - kT_x \end{bmatrix}, \tag{9}$$

$$G = \begin{bmatrix} \rho v \\ \rho uv - \tau_{xr} \\ \rho v^2 - \tau_{rt} \\ \rho vw - \tau_{r\theta} \\ \rho ve - u\tau_{r\theta} - v\tau_{rt} - w\tau_{r\theta} - kT_r \end{bmatrix}, \tag{10}$$

$$H = \begin{bmatrix} \rho w \\ \rho uw - \tau_{x\theta} \\ \rho vw - \tau_{r\theta} \\ \rho w^2 - \tau_{\theta\theta} \\ \rho we - u\tau_{x\theta} - v\tau_{r\theta} - w\tau_{\theta\theta} - kT_\theta \end{bmatrix}, \tag{11}$$

$$R = \frac{1}{r}\begin{bmatrix} 0 \\ 0 \\ -\tau_{\theta\theta} \\ 0 \\ 0 \end{bmatrix}. \tag{12}$$

Here Q is the unknown vector; F, G, and H are the fluxes in the x, r, and θ direction, respectively; R is a source term that arises when cylindrical polar coordinates are used; and τ_{ij} are the shear stresses, including the filtered Reynolds stresses.

Needless to say, highly accurate numerical computations are required to obtain even relatively crude approximations to the radiated sound field, since only a small fraction of the near-field flow energy actually gets radiated as sound – presumably as a result of very subtle phase cancellations within the flow. It might therefore seem that spectral methods, which have infinite order accuracy, would be ideally suited for the computations – especially for round jets where Fourier spectral methods can be used in the circumferential direction. This will probably turn out to be the case for subsonic flows, but, as already indicated, the major technological interest is currently in supersonic flows. Spectral methods cannot resolve the shock waves in these flows, and higher order finite-difference methods are then required. It is felt that the chosen method should be at least fourth-order accurate in space and second-order accurate in time.

An attractive sixth-order accurate compact finite-difference scheme for space discretization, which has been extensively developed by Lele (1990) at Stanford/Ames and is currently being developed at Lewis by S. T. Yu, is to determine the first derivatives at the ith grid location, say u'_i, by inverting the matrix equation (Collatz, 1966)

$$u'_{i-1} + 3u'_i + u'_{i+1} = \frac{1}{12\Delta x}(u_{i+2} + 28u_{i+1} - 28u_{i-1} + u_{i-2}) + 0(\Delta x^6). \tag{13}$$

Only three points are used to achieve sixth-order accuracy by solving for the derivatives simultaneously at all locations. A compact-storage, four-step Runge-Kutta scheme is used for time-discretization.

Another approach, which is being developed at Lewis by M. E. Hayder, is the Turkel/Gottlieb (Gottlieb and Turkel, 1976; Bayliss, Paresh, Mastrello, and Turkel, 1985) predictor/corrector method.

Here the predictor step is

$$\overline{Q_i} = Q_i^n + \frac{\Delta t}{6\Delta x}[7(F_{i+1}^n - F_i^n) - (F_{i+2}^n - F_{i+1}^n)] + \Delta t R_i \qquad (14)$$

while the corrector step is

$$Q_i^{n+1} = \frac{1}{2}\left\{\overline{Q}_i + Q_i^n + \frac{\Delta t}{6\Delta x}[7(\overline{F}_i - \overline{F}_{i-1}) - (\overline{F}_{i-1} - \overline{F}_{i-2})] + \Delta t R_i\right\}. \qquad (15)$$

The time step is second-order accurate.

References

Bayliss, A., Paresh, P., Mastrello, L., and Turkel, E., 1985. "A fourth-order scheme for unsteady compressible Navier-Stokes equations," ICASE Report No. 85-44.

Betchov, R. and Criminale, W. O., Jr., 1967. *Stability of Parallel Flows*, Academic Press, Inc.

Collatz, L., 1966. *The Numerical Treatment of Differential Equations*, Springer-Verlag, NY, p. 538.

Dowling, A. P., Ffowcs Williams, J. E., and Goldstein, M. E., 1978. "Sound production in a moving stream," *Philos. Trans. Roy. Soc. London Ser. B* **288**(1353), pp. 321-349.

Gottlieb, D. and Turkel, E., 1976. "Dissipative two-four methods for time dependent problems," *Math. Comp.* **30**, pp. 703-723.

Howe, M. S., 1975. "Contributions to the theory of aerodynamic sound with application to excess jet noise and the theory of the flute," *J. Fluid Mech.* **71**(4), pp. 625-673.

Huerre, P. and Crighton, D. G., 1983. "Sound generation by instability waves in a low Mach number flow," AIAA-83-0661.

Hunt, J. C. R., 1973. "A theory of turbulent flow round 2-dimensional bluff bodies," *J. Fluid Mech.* **61**(4), pp. 625-706.

Lele, S. K., 1990. "Compact finite difference schemes with spectral-like resolution," CTR Manuscript 107, submitted to *J. Comput. Phys.*

Lighthill, M. J., 1952. "On sound generated aerodynamically: I – General theory," *Proc. Roy. Soc. London Ser. A* **211**(1107), pp. 564-587.

Lighthill, M. J., 1954. "On sound generated aerodynamically: II – Turbulence as a source of sound," *Proc. Roy. Soc. London Ser. A* **222**(1148), pp. 1-32.

Lilley, G. M., 1974. "On the noise from jets," *Noise Mechanisms*, AGARD-CP-131, pp. 13.1-13.12.

Moffatt, H. K., 1981. "Some developments in the theory of turbulence," *J. Fluid Mech.* **106**, pp. 27-47.

Nayfeh, A. H., 1973. *Perturbation Methods*, John Wiley & Sons, Inc.

Tam, C. K. W. and Morris, P. J., 1980. "The radiation of sound by the instability waves of a compressible plane turbulent shear layer," *J. Fluid Mech.* **98**(2), pp. 349-381.

Tennekes, H. and Lumley, J. L., 1972. *A First Course in Turbulence*, M.I.T Press.

Townsend, A. A., 1976. *The Structure of Turbulent Shear Flow*, Cambridge University Press.

REGARDING NUMERICAL CONSIDERATIONS FOR COMPUTATIONAL AEROACOUSTICS

Jay C. Hardin

NASA Langley Research Center
Mail Stop 462
Hampton, Virginia 23681-0001

ABSTRACT

This paper presents a potpourri of factors which are of concern in the numerical analysis of sound generation and propagation by fluid flows. It is shown that successful computational aeroacoustics (CAA) solutions require that careful attention be given to numerical algorithms, stability considerations, and gridding. In particular, difficulties inherent in high Mach and Reynolds number viscous calculations are discussed.

1. Introduction

With the development of larger and faster computers, the field of Computational Aeroacoustics, where the sound produced by a flow as well as the underlying flowfield itself is computed from first principles, has become a practical reality. However, such calculations are not trivial extensions of compressible computational fluid dynamics codes, which are usually developed for steady flows and are thus designed to damp out oscillations, but must be carefully tailored to extract the acoustic quantities of interest. Successful CAA calculations must overcome several challenges:

1. The small size of the quantities of interest. A fairly intense acoustic wave of 100 decibels contains pressure fluctuations which are three orders of magnitude smaller than the pressure fluctuations in a typical turbulent flow. Thus, computational inaccuracies can swamp the quantity of interest in numerical error and, in some cases, introduce nonphysical intense acoustic sources.

2. The high frequencies of the quantities of interest. The human ear responds to frequencies in the range 50 Hz - 20 kHz with peak sensitivity near 1-2 kHz. These frequencies are much

higher than those considered in a typical unsteady aerodynamics calculation and place significant demands on numerical algorithms and grids.

3. The need for long time solutions. The result typically desired from CAA calculations is the spectral density of the farfield sound. Due to data analysis requirements, adequate resolution in the spectral analysis can only be achieved through long data records, with concurrent demands on numerical accuracy and computer resources.

4. The need to minimize damping and dispersion of oscillations. Since acoustic waves occur at high frequency and the acoustic field is sensitively dependent upon phase, damping and dispersion of high frequency oscillations, which are typically introduced by numerical algorithms, must be reduced.

5. The need for nonreflecting boundary conditions. Unbounded aeroacoustic phenomena must be calculated on a bounded computational domain. Since transitional flows can be very sensitive to sound and linear wave fields superimpose, it is necessary that nonphysical boundary reflections be avoided.

6. Interest in high Mach and Reynolds number flows. Most of the CAA applications in aeronautics involve both high Mach and Reynolds numbers. High Mach number causes complicated convective effects in acoustic fields and may induce new nonlinear source mechanisms, while high Reynolds number introduces multiple scale difficulties due to the disparity between the acoustic wavelength and the size of the energy dissipating eddies.

This paper discusses the implications of these challenges in attempts to compute the sound produced by a fluid flow using adaptions of classical computational fluid dynamics techniques.

2. Numerical Considerations

The complete set of equations (Batchelor, 1967) governing air flow at normal temperatures and pressures consists of six equations for the six unknowns, the density, ρ, pressure, p, temperature, T, and three velocity components, u_i. These governing equations may, in theory, be solved numerically for the sound produced by any fluctuating air flow. Such "direct simulations" have been attempted by several authors, including Brentner (1990), Watson (1991), and Mitchell, Lele, and Moin (1992), with varying degrees of success. These mixed reviews are not surprising when one considers the difficulties inherent in the calculation, especially for viscous flows. In such numerical simulations, the continuous governing partial differential equations are replaced by some sort of discrete approximation and the question arises as to how faithfully the solutions of the discrete analog represent those of the continuous equations. Consider the first order forward difference.

$$\frac{d\hat{f}}{dt} = \frac{f(t+\Delta t) - f(t)}{\Delta t} + 0(\Delta t)$$

which is often applied to estimate a first derivative. If $f(t) = e^{i\omega t}$, then

$$\frac{d\tilde{f}}{dt} = \left\{ \frac{\sin \frac{\omega \Delta t}{2}}{\frac{\omega \Delta t}{2}} \right] e^{\frac{i\omega \Delta t}{2}} \frac{df}{dt}$$

where ω is the temporal frequency and it can be seen that the discrete approximation is the product of a frequency dependent dissipative term in the square brackets and a frequency dependent phase shift (dispersion) times the actual derivative of the function. Since acoustic waves are very sensitive to phase (a 180° phase shift can make the difference between complete cancellation and doubling of two waves of the same frequency), this dispersive phase shift is a critical disadvantage.

The phase shift can be eliminated by going to a second order central difference approximation

$$\frac{d\hat{f}}{dt} = \frac{f(t+\Delta t) - f(t-\Delta t)}{2\Delta t} + 0(\Delta t^2)$$

in which case

$$\frac{d\hat{f}}{dt} = \left[\frac{\sin \omega \Delta t}{\omega \Delta t} \right] \frac{df}{dt}.$$

However, to use such an approximation requires storage of the field information at the previous time step $t - \Delta t$. Further, as can be seen in Figure 1, the dissipation of the scheme has actually increased over that of the first order scheme. This inherent dissipation places severe restrictions on the time step in order to preserve frequencies of acoustic interest in the solution, particularly when one realizes that this error occurs every time step and one desires a long time solution.

Similar considerations apply to numerical approximations of space derivatives. Here, central differences are usually employed since no additional storage is required. Consider the second order central difference approximation to a second space derivative,

$$\frac{d^2\hat{f}}{dx^2} = \frac{f(x+\Delta x)+f(x-\Delta x)-2f(x)}{(\Delta x)^2} + 0(\Delta x^2). \qquad (1)$$

If $f(x) = e^{ikx}$,

$$\frac{d^2\hat{f}}{dx^2} = \left[\frac{\sin\frac{k\Delta x}{2}}{\frac{k\Delta x}{2}}\right]^2 \frac{d^2 f}{dx^2}$$

where k is the wavenumber. Again it can be seen that the actual second derivative is multiplied by a dissipation function which depends upon $k\Delta x$ and places a restriction on the grid spacing Δx in order to resolve various wave numbers k. Since the wavelength $\lambda = 2\pi/k$ of acoustic waves is generally long ($\lambda \simeq 1$ foot at $f = 1$kHz) compared to the scale of energy dissipating eddies in the flow, acoustic waves tend to occur a low wavenumber and thus this restriction is more critical for the viscous flow calculation than for the acoustic waves.

In order to resolve the higher wavenumbers, higher order difference approximations such as

$$\frac{d^2\hat{f}}{dx^2} = \frac{16f(x+\Delta x)-f(x+2\Delta x)-30f(x)-f(x-2\Delta x)+16f(x-\Delta x)}{12\Delta x^2}$$

$$+0(\Delta x^4)$$

are often utilized in computational fluid dynamics (CFD). Here, if $f(x) = e^{ikx}$,

$$\frac{d^2\hat{f}}{dx^2} = \left[\frac{30+2\cos 2k\Delta x - 32\cos k\Delta x}{12(k\Delta x)^2}\right]\frac{d^2 f}{dx^2}$$

and it can be seen in Figure 2 that the wavenumber characteristics of the fourth order scheme are much better than those of the second order scheme. However, implementation of boundary conditions is more difficult.

Evaluation of the effect of numerical approximation on individual derivatives is not sufficient to answer the question of how the numerical solutions compare to those of the continuous equations. For that, the equations themselves must be employed. Consider the one dimensional wave equation $p_{tt} = c_0^2 p_{xx}$ where c_0 is the ambient sound speed. This equation is much simpler than the equations of interest, but is the equation to which they reduce for small perturbations in an inviscid medium. Further, it has the characteristic wave behavior of concern in that it admits the solution $p(x,t) = f(x - c_0 t)$ where $f(\cdot)$ is an arbitrary function. This solution can be interpreted as a wave travelling in the positive x-direction at speed c_0.

If both the time and space derivatives in this wave equation are differenced using the second order numerical approximation given by Eqn. (1), the continuous solution $f(x - c_0 t)$ will also be a solution to the discrete analog equation if and only if

$$\begin{aligned} &f(x + c_0\Delta t - c_0 t) + f(x - c_0\Delta t - c_0 t) - 2f(x - c_0 t) \\ &= \beta^2[f(x + \Delta x - c_0 t) + f(x - \Delta x - c_0 t) - 2f(x - c_0 t)] \end{aligned} \tag{2}$$

where

$$\begin{aligned} \beta &= \frac{c_0 \Delta t}{\Delta x} = \frac{\text{distance wave travels in one time step}}{\text{space step}} \\ &= \frac{c_0}{\Delta x/\Delta t} = \frac{\text{wave speed}}{\text{numerical information speed}} \end{aligned}$$

is called the CFL number. The only nontrivial solution to Eqn. (2) is $\beta = 1$. In CFD, it is shown that $\beta = 1$ is the stability limit for hyperbolic solvers and thus calculations usually take $\beta < 1$. However, accurate CAA solutions need to be run at the stability limit, $\beta = 1$, which says physically that the next grid point should be at exactly the distance the wave will travel in one time step. If not, error will appear and the discrete solution will not match that of the continuous equation.

The CFL criterion becomes more complicated in two dimensions. Consider the two-dimensional wave equation $p_{tt} = c_0^2(p_{xx} + p_{yy})$. This equation admits the solution $p(x,y,t) = \exp i[\omega t - k_1 x - k_2 y]$ where $k_1^2 + k_2^2 = k^2$ and $k = \omega/c_0$ is the acoustic wavenumber, which may be interpreted as a plane wave propagating at the angle $\theta = \tan^{-1}(k_2/k_1)$ with respect to the positive x-axis. If this two dimensional wave equation is finite differenced, again using Equation (1) for both time and space derivatives, it can be shown that the continuous solution will also be the solution to the discrete version if and only if

$$\beta^2\left(1-\cos\frac{k_1c_0\Delta t}{\beta}\right)+\gamma^2\left(1-\cos\frac{k_2c_0\Delta t}{\gamma}\right)=1-\cos kc_0\Delta t \quad (3)$$

where now two CFL numbers, $\beta = c_0\Delta t/\Delta x$ and $\gamma = c_0\Delta t/\Delta y$ are introduced. The only nontrivial solution to Eqn. (3) is $\beta = k_1/k$ and $\gamma = k_2/k$. Thus, the optimum CFL numbers now depend upon the angle of propagation as shown in Figure 3.

The situation described above is tolerable in problems where there is only one, known propagation direction. However, a much more common situation is shown in Figure 4 which is an instantaneous plot of the pressure field radiated by a circular piston (radius $= a$) in a infinite plate. Clearly here sound is propagating in all directions. Yet one can only optimize the numerical scheme in one.

The phenomenon described above may be called anisotropy. Whereas the continuous wave equation allows waves to propagate in all directions at the same speed, the discrete approximation will cause waves to propagate in different directions at different speeds, dependent upon the numerical algorithm and grid. In order to minimize this effect, Tam et al. (1992) have proposed a class of dispersion-relation-preserving (DRP) finite difference techniques. The spirit of these techniques may be understood by noting that the square error in the numerical scheme for solving the two dimensional wave equation described by Eqn. (3) is

$$\begin{aligned} e^2(\omega,\theta) \;=\; & \left[\beta^2\left(1-\cos\left[\frac{\omega\Delta t\cos\theta}{\beta}\right]\right)+\gamma^2\left(1-\cos\left[\frac{\omega\Delta t\sin\theta}{\gamma}\right]\right]\right. \\ & \left.-(1-\cos\omega\Delta t)\right]^2 \end{aligned} \quad (4)$$

which depends upon the frequency ω and the direction of propagation θ. If one chooses a frequency range and angle of propagation range of interest, the CFL numbers β and γ may be chosen to minimize Eqn. (4) over the ranges of interest under the constraint that the amplification $R(\omega,\theta)$ satisfy

$$\begin{aligned} R^2(\omega,\theta) = & \left[2\beta^2\cos\left(\frac{\omega\Delta t\cos\theta}{\beta}\right)+2\gamma^2\cos\left(\frac{\omega\Delta t\sin\theta}{\gamma}\right)\right. \\ & \left.+2(1-\beta^2-\gamma^2)-\cos\omega\Delta t\right]^2+\sin^2\omega\Delta t \le 1 \end{aligned}$$

Such techniques appear to offer much promise for optimal CAA calculations.

It should be mentioned that, while all work described in this paper has been for rectangular grids, non rectangular grids are often required for body fitting and much effort has gone into the development of optimal grids for various purposes in CFD. Such grids can also be useful in CAA, especially if the primary propagation direction is along a coordinate line. However, it should be cautioned that a CFL analysis must be carried out for each grid and dissipation and dispersion are much more difficult to minimize.

3. Effects of Mach and Reynolds Numbers

As mentioned in the introduction, most of the problems of interest to CAA occur at high Mach and Reynolds numbers. Thus, some discussion of the effects of these two parameters is warranted.

With regard to Mach number, it should first be understood that there are actually two Mach numbers of interest to CAA, the standard $M = U/c_0$, where U is a characteristic convective velocity, and the acoustic Mach number $M' = u'/c_0$, where u' is a characteristic acoustic fluctuation velocity. The first of these produces convective effects, since acoustic waves propagate at the speed $U \pm c_0$ in the presence of a convective flow, and can introduce new acoustic sources if $M > 1$ where shocks can appear in the flow. The second of these is a measure of the linearity of the acoustic field, with nonlinearity to be suspected if $M' > 0.3$. This condition often occurs in the nearfield of turbulent flows.

The situation with regard to Reynolds number is more curious. Although most sound generation by flows is ultimately due to viscous action, (in the sense that if there is no viscosity, there is no vorticity, which implies that the most common source mechanism, acceleration of vorticity, is absent), it is difficult to ascertain much effect of Reynolds number on aeroacoustics except in transitional flows. Spectra display some sensitivity as illustrated by Figure 5 which portrays spectra of farfield sound radiated by three supersonic ($M \simeq 2.1$) jets whose Reynolds numbers vary over a range of 1000. Note that at low Reynolds number, the spectrum is narrowband, while as the Reynolds number increases, the spectra broaden. However, all of these spectra peak at the same Strouhal number, $St = fL/U$, where L is a characteristic length scale, in this case the variable diameter of the jets. This Strouhal number invariance is typical of the behavior of aeroacoustic sources. Further, the total acoustic power (integral under the spectrum) and directivity patterns of these jets are essentially the same at equivalent scaled distances, varying only with the Mach number of the flow.

This relative invariance of aeroacoustic sources with Reynolds number may help to alleviate some of the multiple scale problems faced by CAA in viscous flows. It suggests that CAA calculations may be accomplished at lower than the full scale value and frequencies and levels scaled to full scale by Strouhal and Mach number factors. Even spectra are invariant over some range (compare the plots on Figure 5 at $Re = 7 \times 10^4$ and $Re = 5.2 \times 10^6$).

Figure 6 presents a schematic of the various approaches that have been or are being employed in CAA to go from the governing compressible Navier-Stokes equations of viscous fluid flow to calculate the spectra of the farfield sound radiation. Since the size of the energy dissipating eddies scales like (Reynolds number)$^{-1}$, one of the big problems with the Direct Simulation or even Large Eddy Simulation (LES), where only the smaller wavenumbers are computed and the energy drain to higher wavenumbers is modelled, approaches is that such a fine grid is required to resolve the flowfield that one does not have the computer resources to carry the computation to the acoustic farfield. This is true even in two dimensions while many aeroacoustic phenomena of interest are fully three dimensional. For this reason, utilization of the Kirchhoff theory has been proposed to confine the extent of the computational domain. However, if the bounds of applicability of the idea of Reynolds number invariance of aeroacoustic radiation can be fully delineated and computations performed at lower than full scale Reynolds numbers, this problem will be greatly alleviated.

4. Conclusion

This paper has shown that numerical approximations to the computation of continuous wave fields must be carefully tailored in order to extract the small quantities of interest in the presence of limitations on frequency, wavenumber, and stability of the calculation. In addition, anisotropy due to the computational grid must be minimized. Criteria to optimize the numerical approximation are developed.

The idea of Reynolds number invariance of aeroacoustic phenomena is also advanced as a way of overcoming the multiple scale problem of viscous CAA. If calculations may be carried out at lower than full scale Reynolds number and then scaled by Strouhal and Mach number factors, direct simulation of aeroacoustic sound generation, even in three dimensions, may prove viable.

References

Batchelor, G. K., 1967. *An Introduction to Fluid Dynamics*, Cambridge University Press.

Brentner, K. S., 1990. *The Sound of Moving Bodies*, Cambridge University Thesis.

Mitchell, B. E., Lele, S. K., and Moin, P., 1992. "Direct computation of the sound from a compressible co-rotating vortex pair," AIAA Paper No. 92-0374.

Tam, C. K., Webb, J. C., and Zhong, D., 1992. "A dispersion relation preserving finite difference scheme for computational acoustics," presented at the DGLR/AIAA 14th Aeroacoustics Conference, Aachen, Germany, May 11-14, 1992, AIAA Paper No. 92-02-033.

Watson, W. R., 1991. *A Time Domain Numerical Theory for Studying Steady-State Acoustic Disturbance in Flow*, George Washington University Thesis.

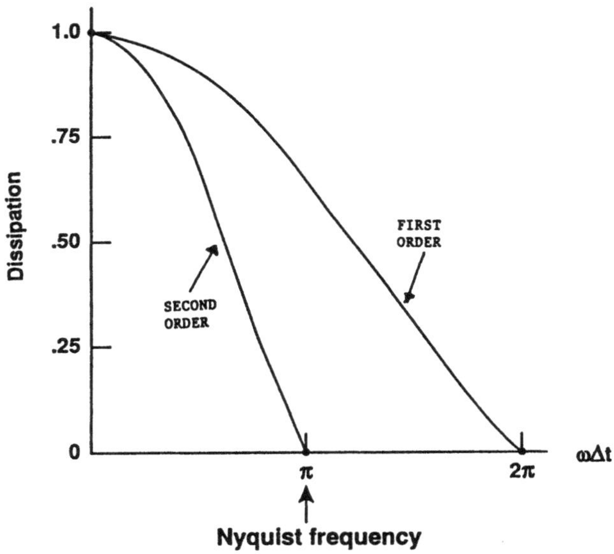

Figure 1: Frequency/Time Step Restriction

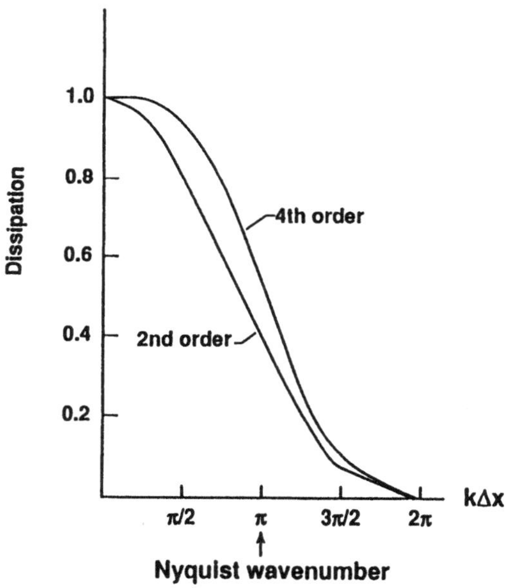

Figure 2: Comparison of Orders

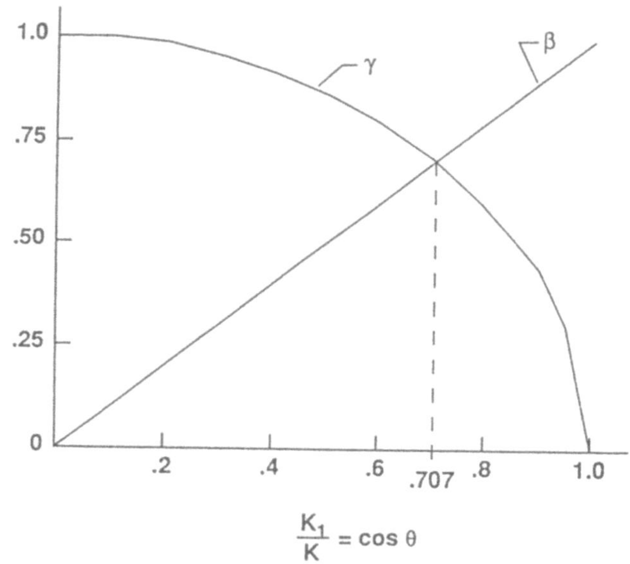

Figure 3: Optimum CFL Numbers

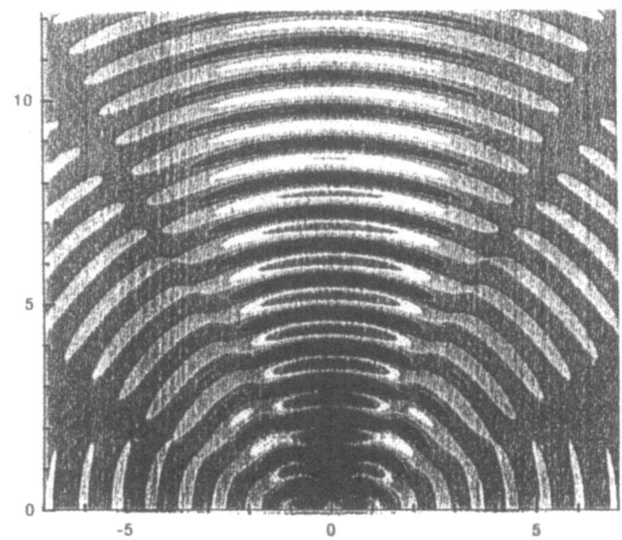

Figure 4: Radiation Field Produced by Piston in Plate (ka=7.5)

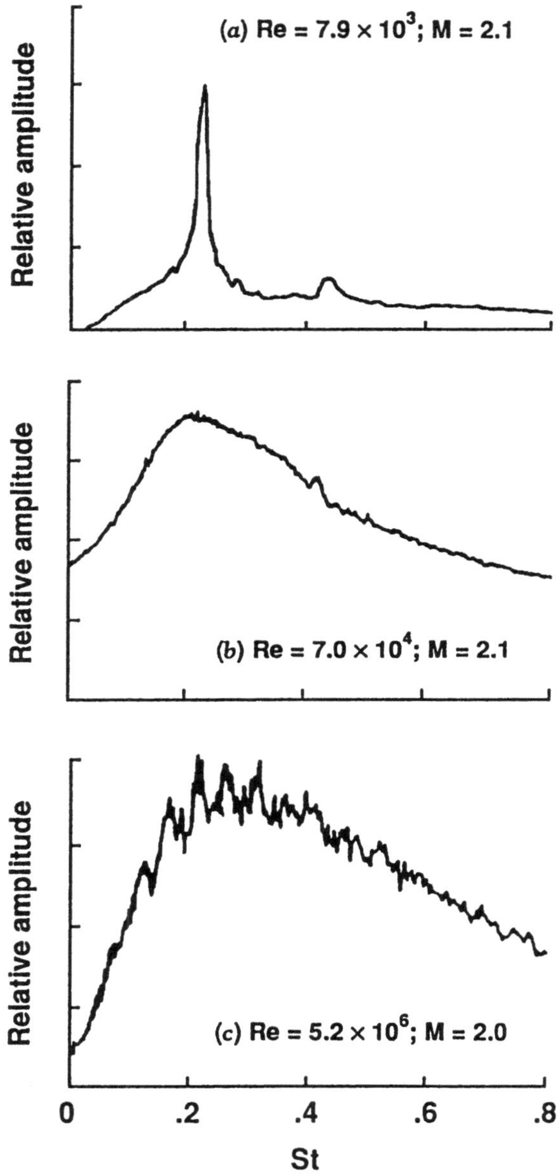

Figure 5: Supersonic Jet Noise Spectra

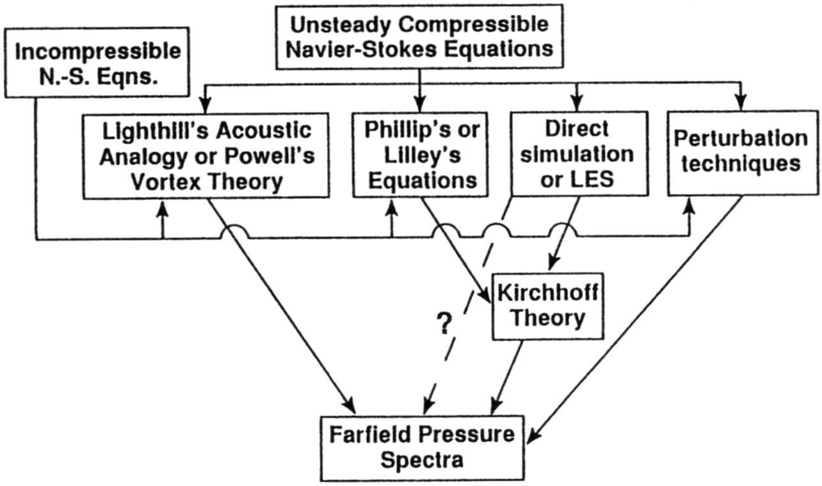

Figure 6: Approaches to CAA

OBSERVED AND COMPUTED WAVES OF AERODYNAMIC SOUND

T. Kambe

Department of Physics, University of Tokyo
Hongo, Bunkyo-ku, Tokyo 113, Japan

ABSTRACT

Evidences are accumulating to show that the theory of vortex sound in the aeroacoustics can describe characteristic features of the waves radiated by a localized vortex motion at low Mach numbers. Examples are given for collisions of two vortex rings and interaction of a vortex ring with a cylinder. An unresolved problem in our study so far is whether there exists a monpolar (*i.e.* isotropic) component in the radiated waves which can be attributed to the viscous dissipation of kinetic energy. A discussion is given about implication of the octapole component observed in the experiment of oblique collision of two vortex rings.

1. Introduction

In the theory of vortex sound (Kambe, 1986; Müller & Obermeier, 1988) at low Mach number M much less than unity, the vortex motion scaled on a length l of vortex size is regarded as a flow in *inner* region. This flow field is surrounded by *outer wave* field scaled on $\lambda \approx c\tau = O(l/M)$, where $\tau = l/u$, u being a typical velocity of the flow field. The sound speed and density in the undisturbed medium are denoted by c and ρ_0, respectively.

At large distance r away from the vortex motion, the inner pressure $p^{(i)}$ at a point $\boldsymbol{x}$ in the inner flow region can be expanded asymptotically (as $r/l \to \infty$, where $r = |\, \boldsymbol{x} \,|$) in the form (Kambe *et al.*, 1990),

$$p^{(i)}(\boldsymbol{x}, t) = -\rho_0 \dot{Q}_i \, \partial_i \frac{1}{r} - \rho_0 \dot{Q}_{ij} \, \partial_i \partial_j \frac{1}{r} - \rho_0 \dot{Q}_{ijk} \, \partial_i \partial_j \partial_k \frac{1}{r} + O(r^{-5}) \tag{1}$$

where the summation convention is assumed and an over-dot denotes a time differentiation, and

$$Q_i(t) = \frac{1}{8\pi} \int (\boldsymbol{y} \times \boldsymbol{\omega})_i d^3\mathbf{y} \quad , \tag{2}$$

$$Q_{ij}(t) = -\frac{1}{12\pi}\int (y \times \omega)_i y_j d^3\mathbf{y} \ , \tag{3}$$

$$Q_{ijk}(t) = \frac{1}{32\pi}\int (y \times \omega)_i y_j y_k d^3\mathbf{y} \ . \tag{4}$$

It is almost trivial to see the properties,

$$Q_{ii}(t) = 0 \ , \quad Q_{iik}(t) = 0 \ , \quad Q_{iji}(t) = 0 \tag{5}$$

by the orthogonality of the two vectors y and $y \times \omega$. The resultant impulse of the vortex system defined by $4\pi Q_i$ characterizes a dynamical aspect of the vortex motion.

The wave pressure $p^{(w)}$ matching to the inner solution asymptotically as $| x | / \lambda \to 0$ is written as

$$\begin{aligned} p^{(w)}(x,t) \ &= \ -\rho_0\, \partial_i \left[\frac{\dot{Q}_i(t_r)}{r}\right] - \rho_0\, \partial_i\partial_j \left[\frac{\dot{Q}_{ij}(t_r) + q_0(t_r)\delta_{ij}}{r}\right] \\ &\quad -\rho_0\, \partial_i\partial_j\partial_k \left[\frac{\dot{Q}_{ijk}(t_r) + q_k(t_r)\,\delta_{ij}}{r}\right] + \cdots \end{aligned} \tag{6}$$

where $t_r = t - r/c$ is the retarded time. Applying the space derivatives to the factor $1/r$ and retaining only such terms, we obtain the expression (1) since $q_0\delta_{ij}\partial_i\partial_j(1/r) = q_0\nabla^2(1/r) = 0$ and $q_k\delta_{ij}\partial_i\partial_j\partial_k (1/r) = q_k\partial_k\nabla^2(1/r) = 0$ for $r \neq 0$, the other terms being higher order in an intermediate region as $| x | / \lambda \to 0$ but $O(| x | / l)$ being sufficiently large (existence of such a region is assumed). The origin of the arbitrary term $q_0(t)\delta_{ij}$ in the second parenthesis is considered in Kambe (1984) and Kambe & Minota (1983). It is found from the dynamical equation of motion (Navier-Stokes equation) that

$$q_0(t) = -\frac{5-3\gamma}{12\pi}\, K(t) \ , \quad K(t) = \frac{1}{2}\int v^2(\mathbf{y},t)\, d^3\mathbf{y} \tag{7}$$

where K is the total kinetic energy and γ the ratio of the specific heats ($\gamma = 7/5$ for the air). Another arbitrary term $q_k\delta_{ij}$ in the third parenthesis will be considered below. The first term of (6) represents a dipole emission due to the change of total impulse $4\pi Q_i$.

We first consider the vortex sound in an unbounded fluid with no solid body. In this case the impulse $4\pi Q_i$ does not change and the first term disappears. The pressure in the acoustic far-field takes simpler form since space derivatives applied to r^{-1} become higher order of smallness than those applied to the functions of t_r. Thus the pressure observed at a point $x = (x_1, x_2, x_3)$ in the acoustic far-field is given as

$$\begin{aligned} p^{(f)}(x,t) &= -\frac{\rho_0}{c^2} q_0^{(2)}(t_r)\frac{1}{r} - \frac{\rho_0}{c^2} Q_{ij}^{(3)}(t_r)\frac{x_i x_j}{r^3} \\ &\quad + \frac{\rho_0}{c^3} Q_{ijk}^{(4)}(t_r)\frac{x_i x_j x_k}{r^4} + \frac{\rho_0}{c^3} q_k^{(3)}(t_r)\frac{x_k}{r^2} + \cdots \end{aligned} \quad (8)$$

where superscript (n) denotes the n-th time derivative. The second quadrupole term (Möhring's quadrupole: Möhring, 1978) derives from the non-isotropic part of the Reynolds stress $\rho_0 v_i v_j$ (Kambe, 1984). The conservation of the resultant moment of impulse (angular impulse) leads to the symmetry property $\dot{Q}_{ij} = \dot{Q}_{ji}$. The first isotropic (monopole) term arises when the total kinetic energy K of the system changes, but vanishes identically in an inviscid fluid. Here we have written up to the third order terms. An experimental observation (described in § 3) shows existence of this order. The fourth term is again of dipole character, which is fixed by the requirement of (say) no dipole emission in the absence of external force.

2. Collision of Two Vortex Rings: Head-on Collision

Axisymmetric collision of two vortex rings (Kambe & Minota, 1983) is a particularly simple example of the vortex sound. In an inviscid fluid, the first term of the formula (8) vanishes and the second quadrupole term reduces to

$$p = -\frac{\rho_0}{12c^2} Q'''(t_r)\,\frac{1}{r}(1 - 3\cos^2\theta) \quad (9)$$

where the vorticity is assumed to have only azimuthal ϕ-component $\omega(\sigma, z)$ in the cylindrical coordinate system (z, σ, ϕ), leading to

$$Q(t) = \int \omega(\sigma, z)\,\sigma^2 z\, d\sigma\, dz\ , \quad (10)$$

with $z = x_3$ and $\sigma = \sqrt{x_1^2 + x_2^2}$ and $\theta = \arccos(z/r)$. In this case the pressure profile is described by a single scalar function $Q(t)$.

2.1. Computations

Thin-core rings: Suppose that we have two vortex rings having a common symmetry axis z with one vortex being a mirror image of the other with respect to the plane $z = 0$, and that they approach each other according to the equation of motion. The interacting motion of two vortex rings having very thin cores is described by a system of ordinary differential equations (Kambe & Minota, 1981) originally derived by Dyson (1893). When the vortex is characterized by its strength $-\Gamma$, ring radius $R(t)$ (core radius δ) and the distance $Z(t)$ from the symmetry plane $z = 0$, the governing equations are

$$R\frac{\mathrm{d}Z}{\mathrm{d}t} = -\frac{\Gamma}{4\pi}(\log\frac{8R}{\delta} - \frac{1}{4}) + \frac{\partial U}{\partial R}, \qquad R\frac{\mathrm{d}R}{\mathrm{d}t} = -\frac{\partial U}{\partial Z} \tag{11}$$

where

$$U = \frac{\Gamma}{4\pi}R\left[\left(\frac{2}{k} - k\right)F(k) - \frac{2}{k}E(k)\right], \qquad k = \frac{R}{\sqrt{Z^2 + R^2}} \tag{12}$$

where $F(k)$ and $E(k)$ are the complete elliptic integrals of the first and second kind. The profile function $Q(t)$ is given by $-2\Gamma R^2(t)Z(t)$. The wave pressure form obtained numerically by solving the above system of differential equations is shown in Figure 1.

Finite-core rings: Effect of finite core-size and core deformation on the wave profile is studied in detail by the contour dynamics (Shariff, Leonard & Ferziger, 1989). This analysis suggests that the

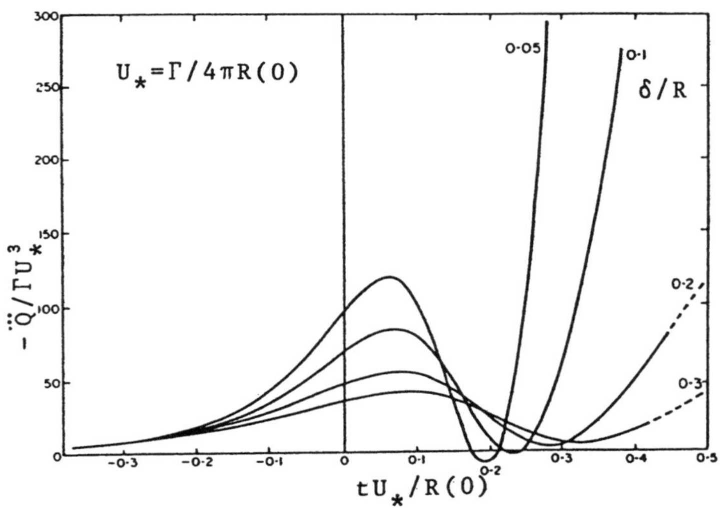

Figure 1. Computed wave pressure emitted by head-on collision of two thin-core vortex rings.

negative peak observed in the wave profile $p_q(t)$ of Figure 2 is due to core deformation, although the core size used in the numerical simulation seems to be larger than in the experiment. Calculations for a thinner core include high-frequency oscillations in the profile. It is interesting to remind the property that observed acoustic signals in the experiments also have such high-frequency oscillations before averaging.

2.2. Observations

Experimental observations of the corresponding acoustic waves due to the vortex collision were carried out by Kambe & Minota (1983) and Minota & Kambe (1986). In the latter, the waves were observed in all directions at $\theta = 10°$ to $350°$ with $10°$ interval in a meridional plane including the symmetry axis x_3. The observed pressures at 35 stations are represented in the truncated Fourier series: $p(\theta, t) = a_0(t) + a_1(t)\cos\theta + b_1(t)\sin\theta + a_2(t)\cos 2\theta + b_2(t)\sin 2\theta$. The two terms $a_0(t)$ and $a_2(t)\cos 2\theta$ are rewritten as

$$p_{\mathrm{mq}}(\theta, t) = p_{\mathrm{m}}(t) + p_{\mathrm{q}}(t)(1 - 3\cos^2\theta) , \tag{13}$$

where p_{m} and p_{q} represent the monopolar and quadrupolar components respectively. Significant amplitudes of $p_{\mathrm{m}}(t)$ and $p_{\mathrm{q}}(t)$ are detected (Figure 2), although non negligible amplitude is also recorded for $a_1(t)$ (*e.g.* Figures 4 and 6 of Minota & Kambe (1986)).

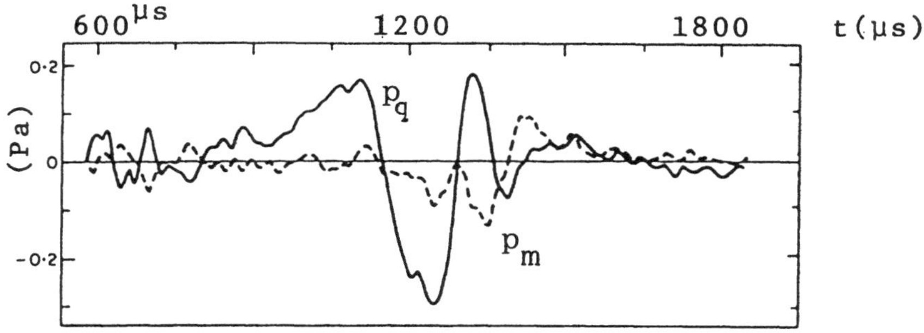

Figure 2. Observed profiles of $p_{\mathrm{m}}(t)$ and $p_{\mathrm{q}}(t)$.

3. Collision of Two Vortex Rings: Oblique Collision

Oblique collision of two vortex rings at right angles is studied numerically and acoustically by Kambe *et al.* (1990). Evidently this oblique collision has no axisymmetry like that in the previous section. This requires an increased amount of data for the analysis. At the collision of vortices, reconnection of vortex lines is expected to occur due to the viscosity effect. This event is immediately followed by violent motion of vorticity and excitation of acoustic waves. Analysis of the wave data provides indirect information of the vortexline reconnection and associated complex vortex motion.

3.1. Expansion in terms of the spherical harmonics

Here we consider the formula (8) up to the third order terms. The formula is rewritten by using the spherical polar coordinates (r, θ, ϕ):

$$x_1 = r\sin\theta\cos\phi, \quad x_2 = r\sin\theta\sin\phi, \quad x_3 = r\cos\theta. \tag{14}$$

Then n-th order form, like the expression $F_2 = C_{ij}x_ix_j/r^2$ for $n = 2$ or $F_3 = C_{ijk}x_ix_jx_k/r^3$ for $n = 3$, can be represented in terms of the n-th order (or lower order) spherical harmonics,

$$P_n^0(\zeta),\ \ P_n^1(\zeta)\,(\cos\phi,\ \sin\phi),\ \ \cdots,\ \ P_n^n(\zeta)\,(\cos n\phi,\ \sin n\phi), \tag{15}$$

where $\zeta = \cos\theta$, $P_n^0(\zeta)$ and $P_n^k(\zeta)$, $(k = 1, \cdots, n)$ are the Legendre polynomials. In fact, we have

$$\begin{aligned} F_2 \;=\; & \frac{1}{3}(-C_{11} - C_{22} + 2C_{33})P_2^0 \\ & + \frac{1}{3}(C_{13} + C_{31})P_2^1\cos\phi + \frac{1}{3}(C_{23} + C_{32})P_2^1\sin\phi \\ & + \frac{1}{6}(C_{11} - C_{22})P_2^2\cos 2\phi + \frac{1}{6}(C_{12} + C_{21})P_2^2\sin 2\phi. \end{aligned} \tag{16}$$

Similarly the third-order form F_3 is represented in terms of (P_3^0, $P_3^1\cos\phi$, $P_3^1\sin\phi$, $P_3^2\cos 2\phi$, $P_3^2\sin 2\phi$, $P_3^3\cos 3\phi$, $P_3^3\sin 3\phi$). In particular, we give explicit expression to the terms to be used later:

$$F_3 = \frac{1}{5}(2Q_{333}^{(4)} - \tilde{Q}_{113} - \tilde{Q}_{223})P_3^0 + \frac{1}{30}(\tilde{Q}_{113} - \tilde{Q}_{223})P_3^2\cos 2\phi \ + \ \cdots \tag{17}$$

where $Q_{ijk}^{(4)}$ are used in place of C_{ijk}, and the tilde symbol denotes

$$\tilde{Q}_{113} = Q_{113}^{(4)} + Q_{131}^{(4)} + Q_{311}^{(4)} \quad \textit{etc.}$$

The coefficient of P_3^0 of (17) reduces to

$$Q^{(4)}_{333} - \frac{1}{5} Q^{(4)}_{3kk} \tag{18}$$

by the relations (5). Here the Legendre functions are

$$\begin{aligned}
P_2^0 &= \frac{3}{2}\cos^2\theta - \frac{1}{2}, \quad P_2^1 = 3\cos\theta\sin\theta, \quad P_2^2 = 3\sin^2\theta, \\
P_3^0 &= \frac{1}{2}(5\cos^3\theta - 3\cos\theta), \quad P_3^1 = \frac{3}{8}(\sin\theta + 5\sin 3\theta), \\
P_3^2 &= 15(\cos\theta - \cos^3\theta), \quad P_3^3 = 15\sin^3\theta\ .
\end{aligned} \tag{19}$$

3.2. Geometrical arrangement of oblique collision

Consider a problem that the initial state is given in such a way that two vortex rings are set to move along the paths intersecting at the origin at right angles and collide with each other (Kambe *et al.*, 1990). The bisecting straight line between the two paths of the vortex center is taken as the polar axis $\theta = 0$ (along the x_3 axis) of the spherical coordinate system. The plane perpendicular to the x_3 axis is the (x_1, x_2) plane on which $\theta = \pi/2$. There are two symmetry planes including the x_3 axis: one includes the trajectories of the vortex centers which is defined as (x_2, x_3) plane and the plane (x_1, x_3) perpendicular to it is also a symmetry plane which bisects the two trajectories. The plane $\phi = 0$ is taken along the positive x_1 axis. Thus the two vortex rings (centers) move toward the origin along the direction of the angles $(\theta, \phi) = (\pi/4, \pi/2)$ and $(\pi/4, 3\pi/2)$ before colliding interaction.

3.3. Observation

The acoustic waves emitted by the 90° collision were detected at 102 different angular positions on the three great circles of radius $r = 620mm$ on the three orthogonal planes: (1) $\phi = \pi/2, 3\pi/2$; (2) $\phi = 0, \pi$; (3) $\theta = \pi/2$. The trajectories of the vortex cores in the (x_2, x_3) plane observed by the photosensor are shown in Figure 3, where the ring diameter $2R(0)$ of the single (unperturbed) vortex is 9.5mm.

From the geometry of the experimental arrangement, the acoustic pressure $p(\theta, \phi, t)$ is characterized by the symmetry:

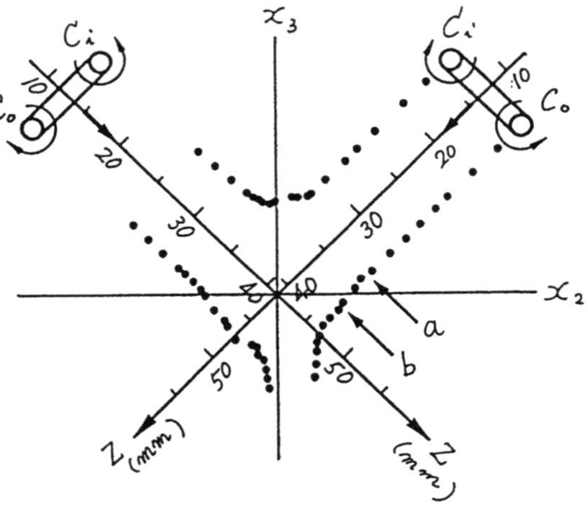

Figure 3. Observed trajectories of two colliding vortices.

$$p(\theta, \phi + \pi, t) = p(\theta, \phi, t) \ , \qquad p(\theta, -\phi, t) = p(\theta, \phi, t) \ . \tag{20}$$

This is found to be consistent with the observed data. In fact, the acoustic pressure detected in the plane $\theta = \pi/2$ can be expanded into Fourier series with respect to the angle ϕ, and it is found that the Fourier coefficients of $\sin m\phi$ ($m = 1 \sim 4$) and $\cos m'\phi$ ($m' = 1, 3, 4, 5$) are negligible. In the decomposition with respect to the angle θ, non-negligible component of $\sin 2\theta$ which is considered to be due to asymmetrical vortex motion is found. But its amplitude is relatively small and neglected. Supposing that the acoustic pressure is represented in terms of the spherical harmonics (15) up to the third order having the symmetries (20), the pressure is represented as

$$\begin{aligned} p(\theta, \phi, t) \ &= \ A_0(t) + A_1(t)P_2^0(\cos\theta) + A_2(t)P_2^2(\cos\theta)\cos 2\phi \\ &\quad + B_1(t)P_3^0(\cos\theta) + B_2(t)P_3^2(\cos\theta)\cos 2\phi \ , \end{aligned} \tag{21}$$

where higher order terms are omitted since the amplitudes are not significant. Three sets of data for $[A_0, A_1, A_2, B_1, B_2]$ can be determined from the observed data. These three are found to be almost coincident. The average of these profiles are shown in Figure 4. The ordinates $[\tilde{A}_0, \tilde{A}_1, \tilde{A}_2, \tilde{B}_1, \tilde{B}_2]$ marked on the right-hand side denotes the scale of the modes A_0, A_1 and A_2 normalized by $(\rho_0/c^2 r)RU^4$ and the modes B_1 and B_2 normalized by $(\rho_0/c^3 r)RU^5$.

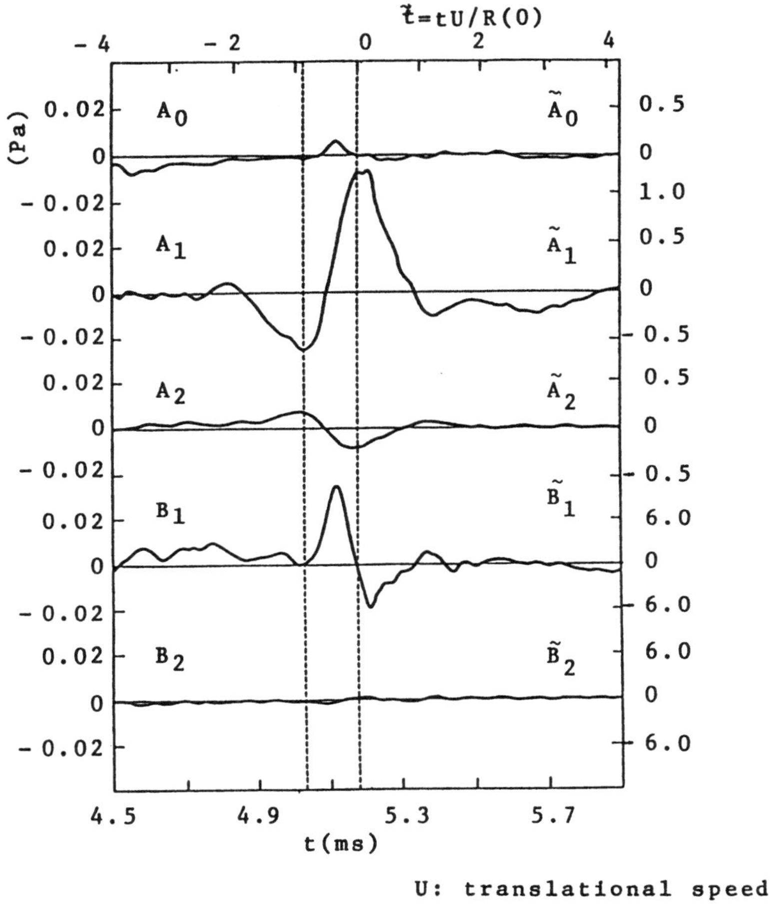

Figure 4. Average mode amplitudes of observed waves.

3.4. Computation

Numerical simulation of the vortex collision at right angles, which corresponds to the experiment, was carried out, using the method of Kida *et al.* (1989, 1990). The incompressible Navier-Stokes equation is solved numerically, using the continuity equation, by the spectral method on 64^3 grid points. Periodic boundary conditions are imposed with the 2π period in the three coordinate directions. The collision of two vortex rings at the 90° initial start was simulated in detail by Takaoka in order to estimate the acoustic emission.

The wave profiles are calculated from the variables of the vortex motion. The isotropic component is proportional to the time derivative of the energy dissipation rate, and the quadrupole components and higher modes are related to the change of moments of vorticity distribution. Using the data from the simulation, we can calculate the tensors $Q_{ij}(t)$ and $Q_{ijk}(t)$ of (3) and (4). Thus we readily obtain the main mode coefficients $[\tilde{A}_0, \tilde{A}_1, \tilde{A}_2, \tilde{B}_1, \tilde{B}_2]$. The amplitudes are shown in Figure 5.

The curves of Figures 4 and 5 are qualitatively similar, but quantitatively different. Nomalized amplitudes of the modes $\tilde{A}_1$ and $\tilde{A}_2$ obtained from the experiment are about twenty times larger than those of the computation, while the observed $\tilde{B}_1$ is larger by two order of magnitude (10^2) than that of computation.

3.5. Discussion

Observed amplitudes A_1 and A_2 of the two quadrupoles P_2^0 and $P_2^2 \cos 2\phi$ are significantly large. The amplitudes $A_1(t)$ and $A_2(t)$ are considered as consisting of three longitudinal quadrupoles Q_{11}, Q_{22},

Q_{33} (to the directions x_1, x_2, x_3), as understood from the form of the coefficients in (16). It appears that there exists non-negligible amplitude in the monopole.

Out of the two third-order components, the amplitude B_2 of $P_3^2 \cos 2\phi$ is negligibly small, but the appearance of the mode P_3^0 is to be remarked. To see the significance of this mode P_3^0, we examine its coefficient B_1 which is given by (18) as $Q_{333}^{(4)}(t) - \frac{1}{5} Q_{3kk}^{(4)}(t)$, where

$$Q_{333} = \frac{1}{32\pi} \int (y \times \omega)_3 y_3^2 d^3\mathbf{y}, \quad Q_{3kk} = \frac{1}{32\pi} \int (y \times \omega)_3 y^2 d^3\mathbf{y} \ ,$$

$$p_3 \equiv (y \times \omega)_3 = y_1\omega_2 - y_2\omega_1 = y_* \times \omega_* \ ,$$

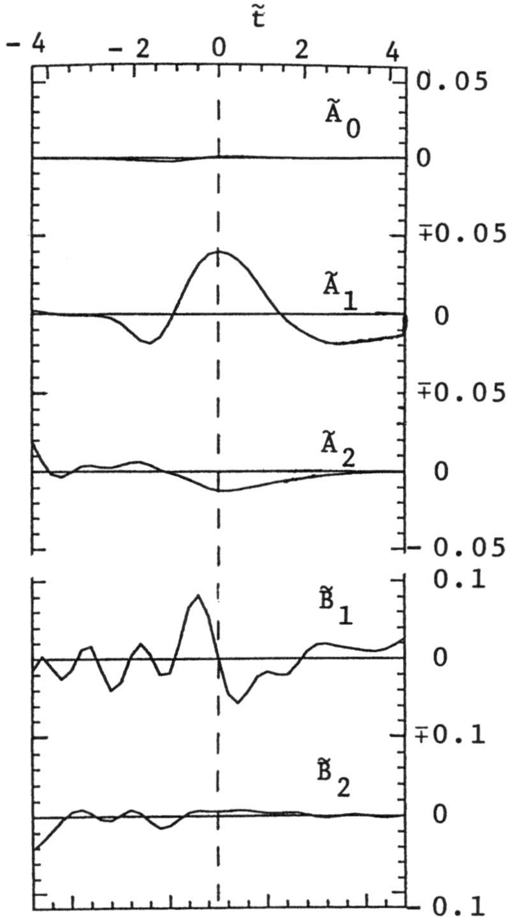

Figure 5. Computed main mode amplitudes.

the vectors $y_* = (y_1, y_2)$ and $\omega_* = (\omega_1, \omega_2)$ being projection of y and ω on the plane (x_1, x_2). Restricting our attention to the location of vortexline reconnection and reminding the geometry of the vortexlines in the sections 3.2 and 3.3, it is expected that the variable p_3 (near $y_* = 0$) changes its sign before and after the reconnection, and that its rapid change gives substantial contribution to the terms $Q^{(4)}_{333}(t)$ and $Q^{(4)}_{3kk}(t)$. Thus it is suggested that the significant amplitude $B_1(t)$ observed in the experiment (see Figure 4) is representing an acoustic signal of vortexline reconnection.

4. Vortex and Cylinder Interaction

Presence of an solid body in a flow can cause a considerable increase in the intensity of generated sound. The increase depends crucially on the boundary conditions associated with the body. If the body is finite with its dimension being small compared with a typical wavelength, then the generated wave field is characterized by a dipole emission whose intensity is proportional to the sixth power of the typical velocity u. On the other hand, the intensity of sound generated by eddies in the vicinity of a sharp edge of a flat semi-infinite plate is proportional to the fifth power of u and the wave field is characterized by a cardioid directivity pattern. [Kambe (1986) for general formulation; Kambe *et al.* (1985) for a semi-infinite plate; Minota & Kambe (1987) for a circular cylinder; Minota *et al.* (1988a) for a sphere ; Minota *et al.* (1988b) for a wedge.]

4.1. General formula

If the eddy and the body without sharp edge are both compact relative to the acoustic wavelength, the acoustic pressure obtained by Curle (1955) is expressed as

$$p(x,t) = -\frac{1}{4\pi c}\frac{x_i}{r^2}\frac{\partial}{\partial t}F_i(t-\frac{r}{c}) \tag{22}$$

in the far-field where $F_i(t)$ is the resultant force exerted on the body by the surrounding fluid.

A different expression of the acoustic pressure is obtained by Obermeier (1980) and Kambe (1986) for the same problem. The pressure generated by the interaction of a solid body and a closed-loop vortex (σ) as

$$p(\boldsymbol{x},t) = \frac{\rho_0}{4\pi c}\frac{x_i}{r^2}\frac{\partial^2}{\partial t^2}\Pi_i(t-\frac{r}{c}) \tag{23}$$

where

$$\Pi(t) = \Gamma \oint_\sigma \Psi_i \cdot d\mathbf{s} = \Gamma \int_S (\nabla \times \Psi_i)\cdot \mathbf{n} dS = \Gamma J_i(t)\ , \tag{24}$$

the function Ψ_i ($i = 1,2,3$) is a vector potential (with div $\Psi_i = 0$) for the velocity of an imaginary potential flow (around the body) of a unit velocity at infinity in the i-th direction. The constant Γ is the strength of the vortex, the infinitesimal variables ds and dS are, respectively, a line element of the closed loop σ and an element of the open surface S bounded by σ, and $\mathbf{n}$ is a unit vector normal to the surface S. The function $J_i(t)$ denotes the volume flux of the potential flow Ψ_i passing through the loop σ.

4.2. A vortex passing by a circular cylinder

Consider a circular vortex ring of radius R passing the side of an infinite rigid circular cylinder of radius a. The x_3 axis is taken along the cylinder axis. The vortex center is assumed to move within the plane (x_1, x_2) perpendicular to x_3. Then the above expression reduces to

$$p(\boldsymbol{x},t+\frac{r}{c}) = \frac{\rho_0}{4\pi c}\Gamma\frac{\sin\phi}{r}[\ddot{J}_1(t)\cos\theta + \ddot{J}_2(t)\sin\theta] \tag{25}$$

$$= \frac{\rho_0}{4\pi c}\Gamma\frac{\sin\phi}{r}\ddot{J}(t)\cos(\theta - \Theta(t)) \tag{26}$$

where $\ddot{J}^2 = \ddot{J}_1^2 + \ddot{J}_2^2$ and $\tan\Theta = \ddot{J}_2^2/\ddot{J}_1^2$, since J_3 is taken to be zero and Ψ_i has only the x_3 component: (0, 0, Ψ_i). The angle θ denotes the azimuthal angle of the projection of the position vector $\boldsymbol{x}$ of observation on the (x_1, x_2) plane, measured from the x_1 axis. The angle ϕ is the polar angle of the direction $\boldsymbol{x}$ from the x_3 axis.[1] If the vortex is sufficiently distant from the cylinder, the vortex path is regarded as rectilinear. Evidently the directivity of the acoustic emission (26) is of dipole character with its axis in the direction $\Theta(t)$ determined by $\ddot{J}_1$ and $\ddot{J}_2$. The coefficient $\ddot{J}(t)$ is the second time

[1]In this section, the angles θ and ϕ are interchanged with those of previous sections.

derivative of the volume flux passing through the vortex ring of the flow around the cylinder toward the observation direction θ (Minota & Kambe, 1987).

The curves in Figure 6 show normalized amplitude functions $h_1 = \ddot{J}_1$ and $h_2 = -\ddot{J}_2$ for the vortex ring moving along a straight line parallel to the x_2 axis with a constant speed.

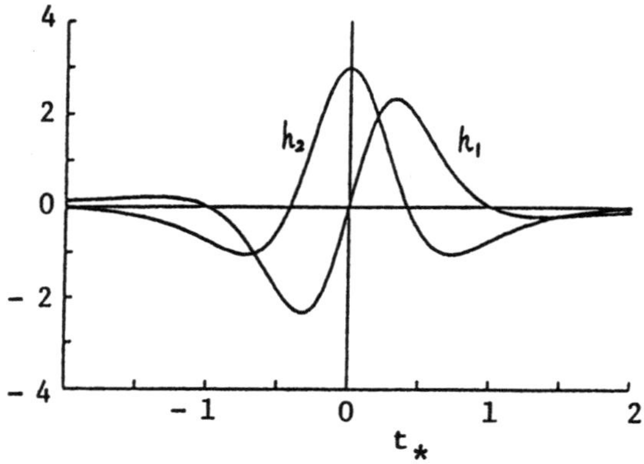

Figure 6. Computed amplitudes $h_1 = \ddot{J}_1$ and $h_2 = -\ddot{J}_2$ for rectilinear vortex motion of constant speed.

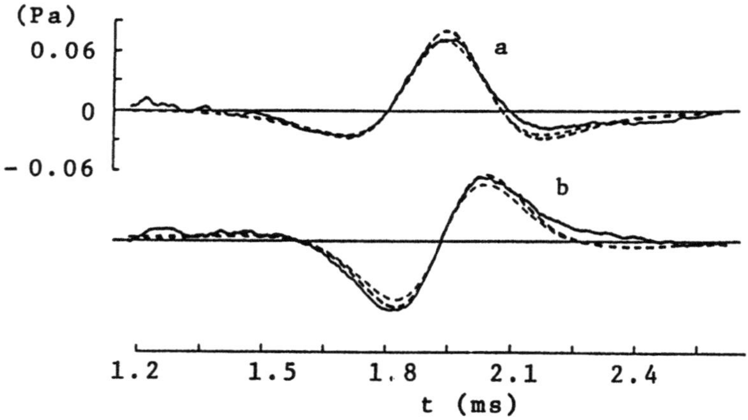

Figure 7. Observed amplitudes of $a(t)$ and $b(t)$ compared with computed ones for slowing-down vortices.

4.3. Observed waves of vortex and cylinder interaction

Corresponding acoustic waves were detected at various angles θ in the plane $\phi = 90°$ and various angles ϕ in the plane $\theta = 0°$ and 180°. Average wave observed at every 10° position (in the plane $\phi = 90°$) is expanded into Fourier series with respect to θ. It is found that the main component is given in the form,

$$p_{\mathrm{obs}}(\theta, t) = a(t) \sin\theta - b(t) \cos\theta \ ,$$

in accordance with the expression (25). The solid curves a, b in Figure 7 are the observed profiles. The broken curves are computed ones, taking account of the observed variation of the velocity. Agreement in absolute values between the observed and predicted profiles are fairly good.

Acknowledgement

The author wishes to thank Dr. Toyoko Minota for collaborative work on the present subject and Dr. Masanori Takaoka on the computational aspect of the oblique vortex collision.

References

Curle, N., 1955. "The influence of solid boundaries upon aerodynamic sound," *Proc. Roy. Soc. London A* **231**, pp. 505-514.

Dyson, F. W., 1983. "The potential of an anchor ring, Part II," *Philos. Trans. Roy. Soc. London A* **184**, pp. 1041-1106.

Kambe, T., 1984. "Influence of viscosity on aerodynamic sound emission in free space," *J. Sound Vibration* **95**, pp. 351-360.

Kambe, T., 1986. "Acoustic emissions by vortex motions," *J. Fluid Mech.* **173**, pp. 643-666.

Kambe, T. and Minota, T., 1981. "Sound radiation from vortex systems," *J. Sound Vibration* **74**, pp. 61-72.

Kambe, T. and Minota, T., 1983. "Acoustic wave radiated by head-on collision of two vortex rings," *Proc. Roy. Soc. London A* **386**, pp. 277-308.

Kambe, T., Minota, T., and Ikushima, Y., 1985. "Acoustic wave emitted by a vortex ring passing near the edge of a half-plane," *J. Fluid Mech.* **155**, pp. 77-103.

Kambe, T., Minota, T., Murakami, T., and Takaoka, M., 1990. "Oblique collision of two vortex rings and its acoustic emission," in *Topological Fluid Mechanics* (eds: H. K. Moffatt and A. Tsinober, Cambridge University Press), pp. 515-524.

Kida, S., Takaoka, M., and Hussain, F., 1989. "Reconnection of two vortex rings," *Phys. Fluids* **A1**, pp. 630-632; and 1991, "Collision of two vortex rings," *J. Fluid Mech.* **230**, pp. 583-646.

Minota, T. and Kambe, T., 1986. "Observation of acoustic emission from head-on collision of two vortex rings," *J. Sound Vibration* **111**, pp. 51-59.

Minota, T. and Kambe, T., 1987. "Acoustic waves emitted by a vortex ring passing near a circular cylinder," *J. Sound Vibration* **119**, pp. 509-528.

Minota, T., Kambe, T., and Murakami, T., 1988a. "Acoustic emission from interaction of a vortex ring with a sphere," *Fluid Dyn. Res.* **3**, pp. 357-362.

Minota, T., Murakami, T., and Kambe, T., 1988b. "Acoustic waves emitted by a vortex ring passing near a wedge-like plate," *Fluid Dyn. Res.* **4**, pp. 57-71.

Möhring, W., 1978. "On vortex sound at low Mach number," *J. Fluid Mech.* **85**, pp. 685-691.

Müller, E.-A. and Obermeier, F., 1988. "Vortex sound," *Fluid Dyn. Res.* **3**, pp. 43-51.

Obermeier, F., 1980. "The influence of solid bodies in low Mach number vortex flow," *J. Sound Vibration* **72**, pp. 39-47.

Shariff, K., Leonard, A., and Ferziger, J. H., 1989. "Dynamics of a class of vortex rings," NASA Tech. Mem. 102257.

VALIDATION OF COMPUTATIONAL AEROACOUSTICS ALGORITHMS

Philip J. Morris

Department of Aerospace Engineering
The Pennsylvania State University, 153D Hammond Bldg.
University Park, Pennsylvania 16802

ABSTRACT

In most cases of aerodynamically generated noise there is no back reaction of the radiated noise on the flow. However, in some situations this is not the case. These cases can serve as valuable tests of computational aeroacoustic algorithms and boundary conditions. The simple class of problems considered in the present paper involves the radiation of sound by vortex flows. The interesting feature of these flows is that the loss of angular momentum and energy by the flow due to acoustic radiation alters the flow evolution. The general solution involves the matching of a potential flow surrounding a rotational region with an outer acoustic solution. Examples are given of the effects of the acoustic back reaction for point vortex and distributed vorticity solutions. In the latter case the inner solution is described by a general set of hypocycloidal vortex regions. The solutions are scaled in terms of parameters that are time invariant. The applicability of these solutions as tests of computational aeroacoustic algorithms is discussed. Finally, some discussion is given of the computational resources required for the solution of aeroacoustic problems.

1. Introduction

In the forty years since the original introduction of Lighthill's theory of aerodynamic noise incredible increases in the power of digital computers have occurred. This now makes possible the direct solution of aeroacoustic problems. Previously, the solution to such problems depended on numerous approximations: usually in the description of the aerodynamic noise sources. It should be recalled that, though Lighthill's equation is an exact manipulation of the equations of motion, the prediction of the radiated noise depends on a detailed knowledge of the turbulence itself. Relationships between the statistical properties of the turbulence and the far-field noise can be derived and propagation effects may be included. Examples of such an analysis are given in the present volume. For an example of a jet noise model see (Lilley, 1992). However, for the prediction of the radiated noise a number of empirical factors must be specified. These include, for example, the length and time scales of the turbulence, the convection velocity of the noise-producing structures, and the level of the turbulent fluctuations. Such values may be obtained from experiments and the results of the predictions are satisfactory. However, if the flow is modified, perhaps the jet exit geometry is altered, the various empirical factors must be reevaluated.

Thus, it is desirable, and probably essential, that a three-dimensional time-dependent solution for the unsteady flow be obtained. Such a simulation is feasible with present computational power. Some estimates of the computational resources required are discussed below. However, it is not clear that there are any advantages in the computation of the flow field alone, to provide the *sources*, if the entire acoustic field is required. This is especially true if both the near and far field are of interest. Such would be the case if the unsteady acoustic loads on nearby structures were important. Thus, the challenge of computational aeroacoustics is the direct computation of both the turbulent flow and the sound it produces. Though the simulation of high Reynolds number turbulent flows will certainly remain beyond our computational resources for the foreseeable future it may not be necessary to simulate all the scales of motion, or to perform high Reynolds number calculations, to make meaningful noise predictions.

If a direct simulation of the turbulent flow and its radiated noise are to be achieved many numerical problems have to be overcome. These are discussed in other papers in this volume. Among the most important are the minimization of numerical dispersion and dissipation, and the implementation of a non–reflecting boundary treatment. These new numerical methods must be validated. This could be accomplished in a number of ways. Comparisons could be made with carefully controlled experiments. Examples of experiments of this kind are described by (Kambe, 1992). However, though the experiments capture all the real physics, the specification of exact initial and boundary conditions is very difficult and the signal to noise ratio may be very low particularly for low Mach number flows. Alternatively, the computations could be compared with analytical solutions. These could either be linear or non–linear. Such problems provide the exact initial and boundary conditions. However there are a limited number of exact solutions in the non–linear case. Finally, checks for internal consistency could be made. For example, (Brentner, 1992) showed how entropy generation in his aeroacoustic calculations could be traced to numerical damping.

In this paper analytical solutions for the sound radiation by regions of distributed vorticity are presented. It is shown how the acoustic radiation can alter the evolution of the flow and how this alteration depends on the transfer of angular momentum and energy to the radiated field. The use of solutions of this type as validation problems for computational aeroacoustic algorithms is discussed. In addition, the computational resources needed to predict both the turbulent flow and the noise it radiates are described. This includes methodologies based on the *acoustic analogy* and direct calculations of the noise field.

2. Analysis and Computations

As a simple example of a vortex flow whose sound radiation alters its behavior consider a pair of rectilinear vortices. The complex potential may be written (Milne-Thomson, 1967)

$$\Phi = -\mathrm{i}\kappa_1 \ln(re^{\mathrm{i}\theta} - a_1 e^{\mathrm{i}\omega t}) - \mathrm{i}\kappa_2 \ln(re^{\mathrm{i}\theta} - a_2 e^{\mathrm{i}\omega t}) \qquad (1)$$

where, $a_1 = a\kappa_2/(\kappa_1 + \kappa_2)$, $a_2 = -a\kappa_1/(\kappa_1 + \kappa_2)$, and $\omega = (\kappa_1 + \kappa_2)/a^2$. κ_1 and κ_2 are the strengths of the vortices, a is their separation, and $z = e^{\mathrm{i}\theta}$. Now, in the limit as $a/\lambda = a\omega/c \ll 1$ the unsteady part of the complex potential in the far field may be written

$$\Phi_s = -\frac{\mathrm{i}}{2}\frac{\kappa_1\kappa_2}{(\kappa_1 + \kappa_2)}\left(\frac{a}{r}\right)^2 e^{2\mathrm{i}(\omega t - \theta)}, \quad \lambda \gg r \gg a. \qquad (2)$$

The acoustic potential satisfies the wave equation and takes the form

$$\phi(r, \theta, t) = \sum_{n=0}^{\infty} A_n H_n^{(2)}\left(\frac{n\omega}{c} r\right) e^{\mathrm{i}n(\omega t - \theta)}. \qquad (3)$$

The values of the coefficients A_n may be obtained by matching this solution with the inner, incompressible solution for $\lambda \gg r \gg a$. If use is made of the asymptotic form of the Hankel function, the acoustic potential is found to be

$$\phi(r,\theta,t) = -\frac{\pi^{1/2}\kappa_1\kappa_2(\kappa_1+\kappa_2)^{1/2}}{2c^{3/2}a}\times \\ \frac{1}{r^{1/2}}\exp[2\mathrm{i}(\omega t - \frac{\omega r}{c} - \theta + \frac{\pi}{8}]. \tag{4}$$

A different form of this solution was obtained by (Mitchell *et al*, 1992). Following (Gryanik, 1983) the energy flux to the acoustic field I_E, given by the integral of the intensity around a circle of large radius R, is found to be

$$I_E = \frac{\rho_0\pi^2}{c^4a^6}(\kappa_1\kappa_2)^2|\kappa_1+\kappa_2|^3. \tag{5}$$

The energy of the vortex system E, may be shown to be (Batchelor, 1967)

$$E = -2\rho_0\pi\kappa_1\kappa_2\ln a + \rho_0\pi(\kappa_1^2+\kappa_2^2)\ln(R/\epsilon). \tag{6}$$

In view of the infinite kinetic energy of the point vortices, the total kinetic energy is determined over a region that is bounded externally by a circle of radius R and internally by circles of small radius ϵ centered on each point vortex. Now, the rate of change of the energy of the vortex system must equal the energy flux to the acoustic field. Thus, an evolution equation for the vortex separation may be obtained by equating the time derivative of (6) to (5). This gives

$$a(t) = a_0\left(1+\frac{t}{\tau}\right)^{\frac{1}{6}}; \quad \tau = \frac{c^4a_0^6}{3\pi\kappa_1\kappa_2|\kappa_1+\kappa_2|^2} \tag{7}$$

where a_0 is the original separation distance of the vortices. It is clear that if the vortices have opposite, but unequal, circulations the separation distance will decrease and the vortices will collapse in a finite time τ.

Though this point vortex solution shows the back reaction of the sound radiation on the flow it is not suitable for numerical simulations. However, solutions with distributed vorticity can also exhibit similar effects. The present solution follows that of (Zeitlin, 1991). In the present analysis some minor corrections are made to his results and the solutions are cast in non-dimensional form with scales that are time-independent. The flow to be considered consists of an inner rotational region surrounded by an irrotational region. There are relatively few exact rotational solutions to the inviscid equations of motion. These include Gerstner waves and Kirchoff's elliptic vortex (Lamb, 1932). Recently (Abrashkin and Yakubovich, 1984) found a new class of vortex solutions. They found exact solutions to the Lagrangian equations of motion. They may be written

$$(x_a y_b - x_b y_a)_t = 0 \tag{8}$$

$$x_{tt}x_a + y_{tt}y_a + \frac{1}{\rho}p_a = 0 \tag{9}$$

$$x_{tt}x_b + y_{tt}y_b + \frac{1}{\rho}p_b = 0. \tag{10}$$

Here, x and y are the position of a fluid particle and a and b are the Lagrangian markers. If complex variables are introduced

$$z_i = x + \mathrm{i}y \quad \text{and} \quad \zeta = a + \mathrm{i}b \tag{11}$$

then an exact solution to the Lagrangian equations of motion is found to be given by

$$z_i(\zeta, \zeta^*) = G(\zeta)e^{\mathrm{i}\omega t} + F(\zeta^*)e^{\mathrm{i}\mu t} \tag{12}$$

where ω and μ are real, F and G are analytic functions, and an asterisk denotes the complex conjugate. The particle trajectories are seen to be hypocycloids. The complex velocity in the inner region is given by

$$V_i = dz_i/dt = \mathrm{i}\omega G(\zeta)e^{\mathrm{i}\omega t} + \mathrm{i}\mu F(\zeta^*)e^{\mathrm{i}\mu t} \tag{13}$$

and the vorticity, obtained by dividing the circulation of an element of the fluid by its area, is

$$\Omega = \frac{2(\lambda|G_\zeta|^2 - \mu|F_{\zeta*}|^2)}{|G_\zeta|^2 - |F_{\zeta*}|^2}. \tag{14}$$

(Zeitlin, 1991) considered the special case with $G(\zeta) = \alpha\zeta$ and $\mu = 0$. So that

$$z_i = \alpha\zeta e^{\mathrm{i}\omega t} + F(\zeta^*), \quad |\zeta| < 1. \tag{15}$$

An outer potential solution may be constructed for $|\zeta| > 1$ in the form

$$z_e = \alpha\zeta e^{\mathrm{i}\omega t} + F(\zeta^{-1}) \tag{16}$$

$$V_e = \mathrm{i}\omega\alpha\zeta^{*-1}e^{\mathrm{i}\omega t} \tag{17}$$

Since z_e is an analytic function of ζ and V_e is a function of ζ^* alone this represents a potential flow. At the boundary $|\zeta| = 1$ the velocity and pressure may be shown to be continuous. The only restriction on the definition of $F(\zeta)$ is that the transformation $z(\zeta, \zeta^*)$ be invertible. This insures that the boundary $|\zeta| = 1$ is not self–intersecting. Thus, the Jacobian of the transformation must satisfy

$$J = \frac{\partial(z, z^*)}{\partial(\zeta, \zeta^*)} \neq 0 \tag{18}$$

Where

$$J_i = |\alpha|^2 - |F'(\zeta^*)|^2 \neq 0 \tag{19}$$

$$J_e = \left| \alpha e^{i\omega t} - \frac{F'(\zeta^{-1})}{\zeta^2} \right|^2 \neq 0. \tag{20}$$

Now we let

$$F(\zeta) = \beta^* \zeta^N. \tag{21}$$

For $F = 0$ this yields a circular vortex of radius $|\alpha|$ with constant vorticity $\Omega = 2\omega$. For $N = 1$ we obtain an elliptic vortex with semi–major and minor axes given by $|\alpha| \pm |\beta|$ and a constant vorticity $\Omega = 2\omega|\alpha|^2/(|\alpha|^2 - |\beta|^2)$. For other values of N the boundary of the rotational region is a hypocycloid with $N + 1$ maxima. For example, Figure 1 shows a case for $N = 2$, $\alpha = 2$ and $\beta = 1/2$ at time $t = 0$. Lines of constant vorticity in the interior of the region are also shown.

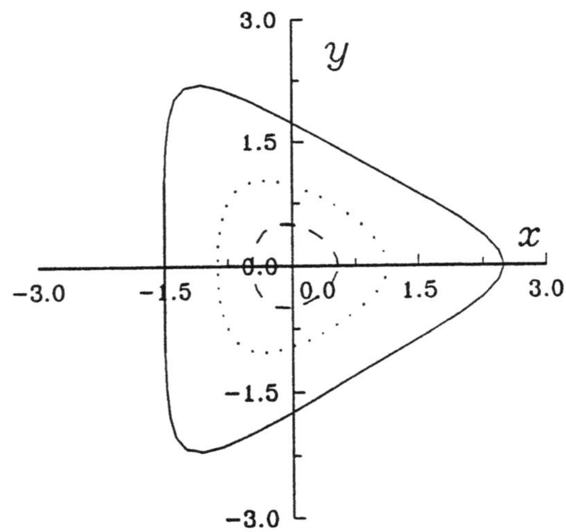

Fig. 1. Boundary of the rotational region and lines of constant vorticity. $N = 2$, $\alpha = 2$, $\beta = 0.5$, $t = 0$. ——, $|\zeta| = 1$.

As in the case of the vortex pair it is necessary next to obtain the form of the complex potential for $|z| \gg |z|_{|\zeta|=1}$. Following (Zeitlin, 1991) we first invert the function $z_e(\zeta)$ which gives

$$\zeta = \alpha^{-1} e^{i\omega t} z[1 - \beta\alpha^N e^{iN\omega t} z^{-(N+1)} + \cdots]. \tag{22}$$

Then the complex velocity is given by

$$V_e = -i\omega|\alpha|^2(z^{*-1} + \alpha^{*N}\beta^* e^{-iN\omega t} z^{*-(N+2)} + \cdots]. \tag{23}$$

If the complex conjugate of (23) is taken and the equation integrated with respect to z the complex potential for the irrotational inner flow may be obtained. The unsteady component of this complex potential for large $|z|$ is found to be

$$\Phi_s = \frac{i\omega|\alpha|^2\alpha^N\beta e^{iN\omega t} e^{-i(N+1)\theta}}{(N+1)r^{N+1}}. \tag{24}$$

Separable solutions of the wave equation for the acoustic potential may be obtained with a dependence on θ and t given by $\exp[iN\omega t - i(N+1)\theta]$. The amplitude of the acoustic potential is obtained by matching it in the limit as $r \to 0$ with (24). This gives

$$\begin{aligned} \phi = \; & \frac{\pi^{1/2} N^{N+1} \omega^{N+2} |\alpha|^2 \alpha^N \beta}{(N+1)!(2c)^{N+1}} \times \\ & H_{N+1}^{(2)}\left(\frac{N\omega}{c} r\right) e^{iN\omega t} e^{i(N+1)\theta}. \end{aligned} \tag{25}$$

The far field form of acoustic potential is obtained using the asymptotic form of the Hankel function. This leads to

$$\begin{aligned} \phi = \; & \frac{\pi^{1/2} N^{N+1/2} \omega^{N+3/2} |\alpha|^2 \alpha^N \beta}{(2c)^{N+1/2}(N+1)! r^{1/2}} \times \\ & \exp\left\{i\left[N\omega\left(t - \frac{r}{c}\right) - (N+1)\theta + \frac{2N+3}{4}\pi\right]\right\}. \end{aligned} \tag{26}$$

The evolution of the inner rotational region depends on the energy and angular momentum lost to the acoustic field. Once again the energy flux can be obtained by integration of the acoustic intensity over a circle of large radius R. If the acoustic velocity is given by

$$\mathbf{v}' = \nabla\Re(\phi)$$

and the acoustic pressure is denoted by $p' = -\rho_0 \partial[\Re(\phi)]/\partial t$ then the energy flux is given by

$$\begin{aligned} I_E &= \textstyle\int p'\mathbf{v}' \cdot d\mathbf{n} \\ &= \frac{\rho_0 \pi^2 N^{2N+3} \omega^{2N+5} |\alpha|^{2(N+2)} |\beta|^2}{2^{2N+1} c^{2(N+1)} [(N+1)!]^2} \end{aligned} \tag{27}$$

where $\mathbf{n}$ is the outward–pointing normal. Similarly, the angular momentum flux is given by

$$\begin{aligned} I_M &= \rho_0 \textstyle\int (\mathbf{r} \times \mathbf{v}') \cdot \mathbf{k}\mathbf{v}' d\mathbf{n} \\ &= \frac{\rho_0 \pi^2 (N+1) N^{2(N+1)} \omega^{2(N+2)} |\alpha|^{2(N+2)} |\beta|^2}{2^{2N+1} c^{2(N+1)} [(N+1)!]^2} \end{aligned} \tag{28}$$

where $\mathbf{k}$ is the unit vector in the direction normal to the z–plane. The total energy and angular momentum of the inner incompressible flow may also be determined by integration over a circular region of large radius R. In this case it is easier to perform the integration in the ζ–plane divided into two regions with $|\zeta| < 1$ and $|\zeta| > 1$. Further details are given by (Zeitlin, 1991). After some analysis the total energy and angular momentum of the incompressible flow are found to be

$$E = \rho_0 \pi \omega^2 |\alpha|^4 \left\{ \frac{1}{4} + \ln\left(\frac{R}{|\alpha|}\right) \right\} \tag{29}$$

and

$$M = \rho_0 \pi \left\{ |\alpha|^2 \omega R^2 - \frac{|\alpha|^4 \omega}{2} - \frac{|\alpha|^2 |\beta|^2 \omega}{N+1} \right\}. \tag{30}$$

Before we equate the rate of change of angular momentum and energy to their fluxes to the acoustic field it is important to determine what quantities are time–independent in (29) and (30). Otherwise the radius R, which should be arbitrary, could enter the final result. However, the circulation Γ will be constant. It is given by

$$\Gamma = \int dS \Omega(z, z^*) = 2\pi\omega|\alpha|^2. \tag{31}$$

Then if we set

$$\frac{dE}{dt} = -I_E \quad \text{and} \quad \frac{dM}{dt} = -I_M \tag{32}$$

it is readily shown that

$$\frac{d|\alpha|^2}{dt} = B_0^2 N |\alpha|^{-2(N+2)} |\beta|^2 \tag{33}$$

and

$$\frac{d|\beta|^2}{dt} = \frac{B_0^2}{2}(N+1)|\alpha|^{-2(N+2)}|\beta|^2 \tag{34}$$

where

$$B_0^2 = \frac{\pi N^{2(N+1)}(\Gamma/2\pi)^{2N+3}}{2^{2N} c^{2(N+1)}[(N+1)!]^2} = \text{ constant.} \tag{35}$$

From (33) and (34) it is clear that

$$\frac{(N+1)}{2}|\alpha|^2 - N|\beta|^2 = S^2 = \text{ constant} \tag{36}$$

(Zeitlin, 1991) notes that this conservation equation shows how the area of the elliptic vortex region is preserved when $N = 1$. However, since S has dimensions of length it can be used to nondimensionalize the results with scales that are time–independent. Thus, we introduce a length scale S and a velocity scale Γ/S. This gives the definition of a Mach number

$$M = \frac{\Gamma}{Sc}. \tag{37}$$

If an overbar is used to denote a nondimensional variable, the nondimensional form of the evolution equations for the inner incompressible flow may then be written

$$\frac{d|\overline{\alpha}|^2}{dt} = \overline{A}^2 N |\overline{\alpha}|^{-2(N+2)} |\overline{\beta}|^2 \tag{38}$$

$$\frac{d|\overline{\beta}|^2}{dt} = \frac{\overline{A}^2}{2}(N+1)|\overline{\alpha}|^{-2(N+2)}|\overline{\beta}|^2 \tag{39}$$

with

$$\overline{A} = \frac{N^{N+1} M^{N+1}}{\pi^{N+1} 2^{3(N+1)/2}(N+1)!} \tag{40}$$

and

$$\frac{(N+1)}{2}|\overline{\alpha}|^2 - N|\overline{\beta}|^2 = 1. \tag{41}$$

The nondimensional frequency of rotation of the inner rotational region is given by

$$\overline{\omega} = [2\pi|\overline{\alpha}|^2]^{-1}. \tag{42}$$

An example of the evolution of the scales $\overline{\alpha}$ and $\overline{\beta}$ is shown in Figure 2 for $N = 1$, the elliptic vortex. Also shown is the non-dimensional frequency. With increase in time, the length scales increase and the frequency decreases. The corresponding shape of the boundary of the rotational region is shown in Figure 3 at two times. As time increases so the aspect ratio of the elliptic region increases. However, these changes take place very slowly at low Mach numbers. In addition, the inner solution may not satisfy the constraint (18) for all values of N at all times. These issues are discussed in the next section.

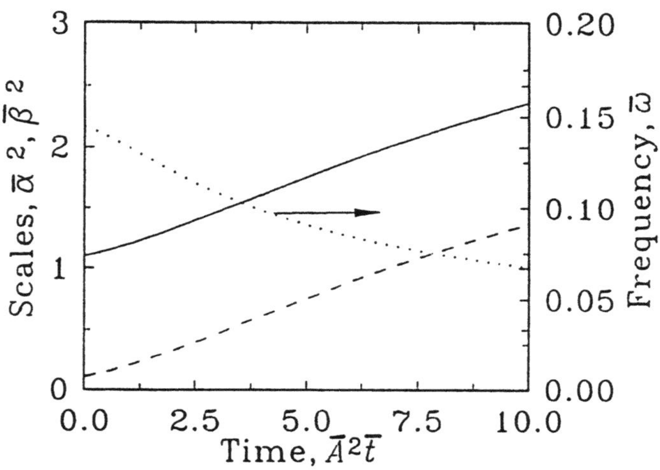

Fig. 2. Variation of length scales and frequency $\overline{\omega}$ with time. $N = 1$, $|\overline{\alpha}|_0^2 = 1.1$, $|\overline{\beta}|_0^2 = 0.1$. ——, $|\overline{\alpha}|^2$; – – –, $|\overline{\beta}|^2$; ·······, $\overline{\omega}$.

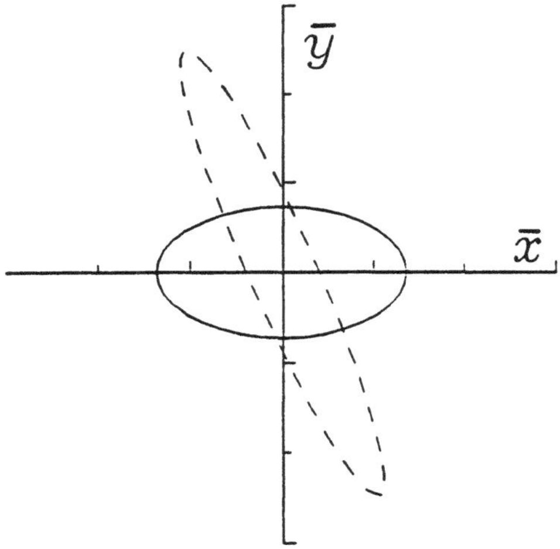

Fig. 3. Variation of the shape of the boundary of the rotational region with time. Same case as Figure 2. ——, $\overline{A}^2\overline{t} = 0.0$; – – –, $\overline{A}^2\overline{t} = 10.0$.

3. Discussion

In the previous section an exact solution of the inviscid equations of motion was constructed. It consisted of an inner rotational region surrounded by an irrotational flow. It was shown how the radiation of sound by this flow resulted in a change in the flow itself. Such a problem would appear to be an excellent candidate to validate computational aeroacoustic algorithms. Since the flow evolution is controlled by the energy flux to the acoustic field, any numerical dissipation or incorrect implementation of the boundary conditions would result in errors. However, several points about the solution need to be considered carefully. First, the solution is only exact in an asymptotic sense for very low Mach numbers. This permits the inner flow to be described by an incompressible set of equations. Also, the rate of change of the flow field due to the acoustic radiation, is very slow in comparison to the characteristic frequency of the flow. A time scale for the rate of change of the flow scales may be defined by

$$\tau_\alpha = \frac{1}{|\alpha|^2}\frac{d|\alpha|^2}{dt}. \tag{43}$$

In the present case an energy time scale $-I_E/E$ is somewhat arbitrary due to the unbounded nature of the flow energy. The ratio of the period of the flow rotation, $\tau_\omega = 2\pi/\omega$, to the flow time scale is given by

$$\tau_\omega/\tau_\alpha \sim C(N)M^{2(N+1)}. \tag{44}$$

Since $M \ll 1$, then $\tau_\omega \ll \tau_\alpha$. In addition, the factor C(N) is also very small being given by

$$C(N) \sim N^{2(N+1)}/(\pi^{2N}2^{3N+1}[(N+1)!]^2). \tag{45}$$

This decrease in the time scale for the rate of change of the flow with increase in N is due to the corresponding decrease in the acoustic intensity, see (27). (Zeitlin, 1991) notes, "the higher the symmetry of the vortex the weaker the radiation." Thus, for the changes to the flow to be evident in a numerical simulation the calculations must be performed for a long time by comparison to the period of the flow. However, in order to capture all the spectral components of the noise radiated by a turbulent flow relatively long calculation times are to be expected relative to the higher frequency fluctuations in the flow. Another difficulty that can occur with the flow solution is associated with self–intersections of the boundary between the rotational and irrotational flows. The condition for this not to occur is given by (18). In general self–intersections occur whenever

$$|\overline{\alpha}|^2 = \frac{2N}{N(N+1)-2}. \tag{46}$$

Clearly, this problem does not occur for the elliptic region but, since $|\overline{\alpha}|^2$ is an increasing function of time, self–intersections will occur for higher values of N. Finally, It should be noted that a stability analysis has not been performed for the incompressible flow. Clearly, if changes to the flow evolution due to flow instabilities occur more rapidly than changes due to the acoustic radiation the analysis would have to modified to include these effects.

Now we turn to the feasibility of the direct simulation of both the flow and the radiated noise. Some estimates can be made for typical calculations. For example, consider the calculations by (Mitchell,

Lele and Moin, 1992). They solved the two–dimensional Euler equations with a compact, high–order, spatial finite–difference scheme with a third–order Runge–Kutta method for the time step. The number of floating point operations per time step per node point may be estimated to be 750. For their grid with $(430)^2$ points with uneven spacing and the number of time steps calculated (42,000), this gives 5.8×10^{12} floating point operations. If we assume that the speed of the computer (a Cray YMP) is approximately 200 million floating point operations per second (mflops) this gives a computational time of 8 hours. However, their problem was a relatively simple vortex flow in two dimensions. We may also estimate the computational time for a three-dimensional Navier-Stokes calculation for a turbulent flow. (Reynolds, 1989) argued that the number of grid points should scale with the flow Reynolds number in order that all the scales of the turbulent motion may be captured. For a channel flow this gives the number of grid points as

$$N_{xyz} = 6 \times 10^{-4} Re_{\delta}^{2.7} \tag{47}$$

where the Reynolds number is based on the channel height and the centerline velocity. The time step may be estimated as the ratio of the grid spacing to the centerline velocity. For a three-dimensional Navier–Stokes solver we may estimate 1200 floating point operations per time step per node. For example, consider the turbulent channel flow simulation by (Kim, Moin and Moser, 1987). At a Reynolds number of 3300 the operations per time step are found to be approximately 2 billion. For a total simulation time of ten non-dimensional time units (channel height divided by the friction velocity), giving 3.3×10^4 time steps, approximately 210 hours of computer time is required on a 100mflop machine. If the acoustic field is to be determined as well then these estimates need to be multiplied by a factor of M^{-3}. At low Mach numbers this results in a significant further increase in computer resource requirements. To make such acoustic simulations viable at least three orders of magnitude increase over present computer power is required. It is likely that this will only be achieved through the use of massively parallel computers. Such computers, with teraflop (10^{12}flops) capabilities will be available in the near future. To take advantage of these machines it is important that the algorithms designed for the solution of computational aeroacoustic problems be chosen carefully with the parallel architecture in mind. For example, to reduce communications, compact, explicit, finite-difference schemes are likely to be more efficient than implicit or spectral schemes. It should also be noted that it may not be necessary to capture the details of all the scales of the turbulent flow in order to predict the radiated noise. This is particularly true for supersonic flows where the acoustic wavelengths are or the same order as the large scales in the turbulence. In this case a Large Eddy Simulation of the flow and its acoustic radiation would appear to be an appropriate model. It is not clear that the same is true in the subsonic case where the details of the random forcing of the large scales by the small scales could be important to the radiated noise.

Finally, it is useful to compare the methodologies and computation time requirements for the acoustic analogy and direct simulation approaches to computational aeroacoustics. In the former case it should be remembered that the evaluation of the pressure at a single far field point depends on a volume integral taken over the source region. If the pressure is required at all points on a spherical surface, with the same resolution as a direct simulation, this would require M^2 integrations. Care must also be taken that the important phase information contained in the sources be preserved. It is likely that this problem could be reduced by a calculation of the statistical properties of the turbulence which can be related directly to the statistical properties of the far field. However, if the near field pressure is desired then a direct simulation must be extended to include at least this region.

Thus the choice of methodology is dictated by the resolution and nature of the acoustic information required as well as the computational resources available. It not clear that the use of the acoustic analogy approach has any significant advantages over the direct simulation approach, especially for non–compact source regions and near field calculations. Thus, it may be argued that efforts should be directed towards the development of the direct approach in anticipation of the availability of adequate computer power in the near future.

References

Abrashkin, A. A. and Yakubovich, E. I., 1984. "Planar rotational flows of an ideal fluid," *Sov. Phys. Dokl.* **29**(5), p. 370.

Batchelor, G. K., 1967. *An Introduction to Fluid Dynamics*, Cambridge University Press, Cambridge (Eng.), p. 530.

Brentner, K. S., 1992. "A consideration of energy from the viewpoint of computational aeroacoustics," *Proc. of ICASE/NASA Workshop on Computational Aeroacoustics*, Hampton, VA, April 6–9.

Gryanik, V. M., 1983. "Emission of sound by linear vortical filaments," *Izvestiya Atmos. and Oceanic Phys.* **19**(2), p. 150.

Kambe, T., 1992. "Observed and computed waves of aerodynamic sound," *Proc. of ICASE/NASA Workshop on Computational Aeroacoustics*, Hampton, VA, April 6–9.

Kim, J., Moin, P. and Moser, R., 1987. "Turbulence statistics in fully developed channel flow at low Reynolds number," *J. Fluid Mechanics* **177**, p. 133.

Lamb, H., 1932. *Hydrodynamics*, 6th edition, Cambridge University Press, Cambridge (Eng.), p. 421.

Lilley, G. M., 1992. "On the noise radiated from aerodynamically unsteady (turbulent) shear flows. The high speed jet," *Proc. of ICASE/NASA Workshop on Computational Aeroacoustics*, Hampton, VA, April 6–9.

Milne-Thomson, L. M., 1968. *Theoretical Hydrodynamics*, 5th edition, Macmillan, New York, p. 358.

Mitchell, B. E., Lele, S. K. and Moin, P., 1992. "Direct computation of the sound from a compressible co–rotating vortex pair," *AIAA Paper* 92–0374, 30th Aerospace Sciences Meeting, Reno, January 6–9.

Reynolds, W. C., 1989. "The potential and limitations of direct and large eddy simulations," in *Whither Turbulence? Turbulence at the Crossroads*, ed. J. L. Lumley, Springer–Verlag, Berlin, p. 313.

Zeitlin, V., 1991. "On the backreaction of acoustic radiation for distributed two–dimensional vortex structures," *Phys. Fluids A.* **3**(6), p. 1677.

MODEL PROBLEMS ASSOCIATED WITH THE PREDICTION OF NOISE BY HIGH SPEED SHEAR LAYERS

John M. Seiner and T. R. S. Bhat[1]

NASA Langley Research Center
Hampton, Virginia 23681-0001

ABSTRACT

In addition to the need to develop innovative methods to control noise production by high speed shear layers, there is also the need to develop computational methods that accurately predict noise generated by these shear layers. With the successful application of Navier-Stokes codes to predict nozzle internal fluid mechanics, calculation of the noise emission from complex nozzle geometries is plausible if it were possible to compute the external noise source. This paper examines the prediction of mean flow structure associated with fully pressure balanced generic nozzle geometries using a parabolized Navier-Stokes solver with a $k-\varepsilon_d$ two equation turbulence model with compressible turbulence dissipation. The predicted mean flow structure is used to apply the linearized compressible Rayleigh equation to predict the principle components of noise radiation. Application of this method is made to predict the flow and acoustics associated with round and elliptic jets, each designed for an exhaust Mach number of 2. Where possible, comparison of predicted flow and sound is made to experimental data associated with round and elliptic supersonic nozzles.

1. Introduction

This paper examines the applicability of the compressible Rayleigh equation to predict noise from hot high Reynolds number supersonic jets with both round and elliptic nozzle geometry.

Supersonic nozzles with noise suppression geometry are extremely complex relative to the simple round supersonic jet. There is a great need to predict the noise associated with a jet nozzle suppressor and much less practical interest in prediction of noise from simple round

---

[1]NRC Fellow.

jets. However, the supersonic round jet offers significant opportunity to evaluate and develop computational strategies for the prediction of noise. This is due to the relative simplicity of the flow and availability of a large data base.

The computational strategy employed in this paper is based on a near term need to develop jet noise prediction methods for suppressor nozzle geometry. In its most useful state, the computational method would even assist in the nozzle design process to minimize noise production. This desirable feature presently excludes the use of strategies that compute noise using the Lighthill analogy along with direct numerical simulation of the flow or large eddy simulation with a sub-grid model. While algorithms have developed rapidly that show promise in minimizing wave reflection (Colonius, Lele, and Moin, 1992) and wave dispersion (Tam and Webb, 1992), these codes would require cumbersome computational power for flows much simpler than the single round jet. Previous studies (Seiner, 1992) show that the noise from high temperature supersonic jets is dominated by a Mach wave emission process, whose characteristics do not appear to depend on the detailed turbulent characteristics of the flow. This observation has a direct bearing on formulating an approach to compute high speed jet noise.

Ample scientific evidence exists that shows the dominant role of coherent large scale turbulence structure in the noise generation process of supersonic jets (Tam and Burton, 1984). Validation of the instability wave model of this structure primarily has been associated with low to moderate unheated round supersonic jets (Troutt and McLaughlin, 1982), but successful comparison of important flow and acoustic features has also been made to unheated round high Reynolds number supersonic jets by Seiner, McLaughlin, and Liu (1982) and by Lepicovsky, Ahuja, Brown, and Burrin (1985). More recently, the wave model has been applied to the hot supersonic jet by Tam, Chen, and Seiner (1992) and to elliptic jets by Morris and Bhat (1992).

This paper compares measured acoustic data from both round and elliptic supersonic nozzle geometries to that predicted by the compressible Rayleigh model. Both nozzles were designed to have exit Mach numbers of 2. The elliptic nozzle has an aspect ratio of 2. The round jet data was taken from a recent study by Seiner, Ponton, Jansen, and Lagen (1992) who obtained both aerodynamic and acoustic data from ambient temperature to 1530°K. The elliptic jet data was taken from a study conducted by Seiner and Ponton (1991).

Application of the compressible Rayleigh equation requires input of mean flow structure. In the present paper, the mean axial velocity profile is measured and mean density is inferred from Crocco's relation. For the round supersonic jet, the mean axial velocity is also input from numerical simulation using the SAIC TTJET code (Dash and Kenzakowski, 1992). TTJET is a parabolized Navier-Stokes code with a two equation turbulence model properly corrected for compressible turbulent dissipation (Sarkar, Erlebacher, Hussaini, and Kreiss, 1989). This paper therefore examines how well the numerically simulated jet can be used with the Rayleigh equation to predict noise without reference to measured flow field data.

Only the first two fundamental modes are considered, i.e. axisymmetric and helical for the round jet. These contain most of the energy at the flow field temperatures investigated (i.e. 755°K). Solutions for supersonic instability waves (Tam and Hu, 1989) are not considered, due to their relatively low contribution to noise at temperatures of only 755°K. After discussing the method of analysis used to predict the growth of large scale structure and noise, the paper examines the fundamental differences that exist between the wave growth, phase speed, directivity and relative amplitude for the fundamental modes of each nozzle studied.

2. Analysis

The analytical method used to compute noise from a round supersonic jet is considered first. The nozzle is assumed to be operating fully pressure balanced at its design point. This means the jet plume is shock free. Both round and elliptic nozzles were designed for exhaust Mach numbers of 2. For convenience, a cylindrical-polar coordinate system, (r, θ, x), is chosen with the jet axis aligned in the x-direction.

The development of an instability wave of fixed real frequency, ω, can be described by the inviscid compressible, linearized equations of motion. The equation for the fluctuating pressure associated with an instability wave is obtained by taking the divergence of the momentum equation and using the equation of continuity and is given by:

$$\frac{\partial^2\hat{p}}{\partial r^2}+\frac{\partial\hat{p}}{\partial r}\left\{\frac{2\omega}{\Omega}\frac{\partial\overline{u}}{\partial r}+\frac{1}{r}-\frac{1}{\overline{\rho}}\frac{\partial\overline{\rho}}{\partial r}\right\}+\left\{\overline{\rho}M_j^2\Omega^2-\frac{n^2}{r^2}-\alpha^2\right\}\hat{p}=0. \quad (1)$$

The boundary conditions are:

$$\hat{p}\rightarrow 0 \quad \text{as} \quad |r|\rightarrow\infty \qquad (2)$$

and

$$\hat{p} \quad \text{is finite as} \quad r\rightarrow 0 \qquad (3)$$

where $\Omega = \omega - \alpha\overline{u}$, α is the axial wavenumber, n the azimuthal mode number, ω is the instability wave frequency, and $\overline{u}$ and $\overline{\rho}$ are respectively the mean velocity and density. The exit Mach number is M_j. All variables are non-dimensionalized with respect to flow conditions at the jet exit and with the jet exit radius. This is the compressible Rayleigh's equation, where the fluctuating pressure is assumed to take the form:

$$p(r,\theta,x,t)=A(x)\hat{p}(r)\exp\{i(\alpha x+n\theta-\omega t)\} \qquad (4)$$

where $A(x)$ is the amplitude function. In Equation (1), it is assumed that the mean velocity is in the axial direction only and is only a function of the radial coordinate, r. This is the locally parallel flow approximation to the slowly diverging flow. The solutions for the Rayleigh's equation in the inner and outer regions of the mixing layer are developed next.

2.1. Inner and outer solution

The development of the inner, outer, and far-field solutions are very similar to that of Tam and Burton (1984). Thus only a brief outline is given of the method. The only significant difference is that Tam and Burton have included the effects of slowly diverging flow. This results in the need for a multiple scales approach. In the present study, the parallel-flow approximation is used.

In the region just outside the jet flow, it is assumed that the ambient fluid is static and the mean flow properties are constant. The Rayleigh's equation in this region reduces to

$$\frac{\partial^2\hat{p}}{\partial r^2}+\frac{1}{r}\frac{\partial\hat{p}}{\partial r}+\left\{\overline{\rho}_\infty M_j^2\omega^2-\frac{n^2}{r^2}-\alpha^2\right\}\hat{p}=0 \qquad (5)$$

where $\overline{\rho}_\infty$ is the ambient density. The solution to Equation (5), satisfying the boundary condition (2), is

$$\hat{p}(r) = H_n^{(1)}(i\lambda_0 r) \tag{6}$$

where, $H_n^{(1)}$ is the nth order Hankel function of the first kind and

$$\lambda_0(\alpha) = (\alpha^2 - \overline{\rho}_\infty M_j^2 \omega^2)^{1/2}. \tag{7}$$

A similar solution may be obtained for $\hat{p}(r)$ close to the jet centerline. This can be determined by solving Equation (1) using the method of Frobenius series. It can also be shown that the solution satisfying the boundedness condition at $r = 0$ is given by

$$\hat{p}(r) = J_n(i\lambda_i r) \tag{8}$$

where J_n is the nth order Bessel function and

$$\lambda_i(\alpha) = [\alpha^2 - M_j^2(\omega - \alpha\overline{u})^2]^{1/2}. \tag{9}$$

The solutions (6) and (8) serve as the boundary conditions for the numerical solution of Equation (1) in the jet mixing layer.

In the outer region, the mean velocity is zero and the pressure fluctuation satisfies the equation

$$\frac{\partial^2 \tilde{\phi}}{\partial x^2} + \frac{\partial^2 \tilde{\phi}}{\partial r^2} + \frac{1}{r}\frac{\partial \tilde{\phi}}{\partial r} + \left\{ \overline{\rho}_\infty M_j^2 \omega^2 - \frac{n^2}{r^2} \right\} \tilde{\phi} = 0. \tag{10}$$

Define the Fourier transform pair as

$$P(k) = \frac{1}{2\pi} \int_{-\infty}^{\infty} \tilde{\phi}(k) e^{-ikx} dx \tag{11}$$

and,

$$\tilde{\phi}(x) = \int_{-\infty}^{\infty} P(k) e^{ikx} dk \tag{12}$$

where k is the Fourier transform variable. The equation for $P(k)$ is found to be,

$$\frac{\partial^2 P}{\partial r^2} + \frac{1}{r}\frac{\partial P}{\partial r} + \left\{ \overline{\rho}_\infty M_j^2 \omega^2 - \frac{n^2}{r^2} - k^2 \right\} P = 0. \tag{13}$$

It can be seen that equations (5) and (13) are similar with α replaced by k. This implies that the form of the solution to (13) should be the same as that given by (6). Thus the outer solution may be written as,

$$P(r;k) = C_n H_n^{(1)}(i\lambda_k r) \tag{14}$$

$$\lambda_k = (k^2 - \overline{\rho}_\infty M_j^2 \omega^2)^{1/2}. \tag{15}$$

The unknown coefficient C_n can be obtained by matching this outer solution to the inner solution. The pressure outside the jet flow can then be written as

$$p(r,\theta,x,t) = \int_{-\infty}^{\infty} P(k) e^{ikx} e^{in\theta} e^{-i\omega t} dk. \tag{16}$$

2.2. Directivity of far field sound

Following the procedure of Tam and Burton (1984) of matching the inner and outer solutions, the near pressure field in the region outside the jet flow is given by

$$p(r,\theta,x,t) = \int_{-\infty}^{\infty} g(k) H_n^{(1)}(i\lambda_k r) e^{i(kx+n\theta-\omega t)} dk \tag{17}$$

$$g(k) = \frac{1}{2\pi}\int_{-\infty}^{\infty} A(x) e^{-ikx} dx \tag{18}$$

where $\lambda_k = (k^2 - \overline{\rho}_\infty M_j^2 \omega^2)^{1/2}$, $A(x)$ is the streamwise variation in amplitude and phase of the instability wave and is given by

$$A(x) = A_0(x_0) \exp\left[i \int_{x_0}^{x} \alpha dx \right] \tag{19}$$

where α is the complex wavenumber and A_0 is the initial amplitude of the wave at the jet exit.

Not all the near-field pressure fluctuations are radiated into the far-field as sound. To determine the far-field sound pressure, it is convenient to use the spherical coordinate system, (R, χ, θ), where

$$x = R\cos\chi \quad \text{and} \quad r = R\sin\chi. \tag{20}$$

The far-field pressure can then be written as,

$$p(R,\chi,\theta,t) \underset{R\to\infty}{=} \frac{2}{R} g(\overline{\rho}_\infty^{1/2} M_j \omega \cos\chi)$$

$$\exp\left[i(\overline{\rho}_\infty^{1/2} M_j \omega R + n\theta - \omega t) - \frac{i}{2}(n+1)\lambda \right]. \tag{21}$$

The far field directivity function, $D(\chi)$, defined as the sound power radiated in a direction per unit solid angle by an instability wave of frequency ω, is given by,

$$D(\chi) = \lim_{R\to\infty} \frac{1}{2} R^2 |p|^2 = 2|g(\overline{\rho}_\infty^{1/2} M_j \omega \cos\chi|. \tag{22}$$

2.3. Elliptic supersonic jet

The analysis for the elliptic jet is done along similar lines. For convenience, the problem is recast in terms of elliptic-cylindrical coordinates. The Rayleigh equation is solved numerically in this frame of reference with Mathieu and modified Mathieu's functions as the starting solutions, instead of Bessel and Hankel functions as in the axisymmetric problem. Finally, an expression for the far-field directivity using a method similar to the circular jet is obtained. The complete details of this analysis, along with derivation of the various expressions, are given by Morris and Bhat (1992).

3. Numerical Analysis

This section presents the procedure used to calculate the local eigenvalue (or wavenumber), the axial distribution of the amplitude function, the corresponding wavenumber spectrum, and the far field directivity. To enable numerical solution of the Rayleigh equation, both the mean density and velocity fields need to be defined. The mean velocity profile can be closely approximated by a half-Gaussian function. This choice is, however, not unique. Here the mean velocity in the potential core region will be approximated by,

$$\overline{u}(x) = \begin{cases} 1 & \text{for } r \le h \\ \exp[-ln2((r-h)/b)^2] & \text{for } r > h \end{cases} \tag{23}$$

and in the fully developed region by

$$\overline{u}(x) = u_c(x)\exp[-ln2(r/b)^2]. \tag{24}$$

In these equations, $h(x)$ is the radius of the potential core, $b(x)$ is the half-width of the shear layer, and $u_c(x)$ is the centerline velocity. For the present calculations, the values of $h(x)$, $b(x)$, and $u_c(x)$ are obtained from either the experimental data of Seiner, Ponton,

Jansen, and Lagen (1992) or the numerical data of the TTJET code of Dash and Kenzakowski (1992). Once the mean velocity is known, the mean density is determined from Crocco's relationship. The experimental data of Seiner and Ponton (1991) is used for analysis of the elliptic jet.

To compute the growth and decay of the amplitude associated with an instability wave and the spatial evolution of the perturbation pressure first requires calculation of the eigenvalue or the axial wavenumber α. This eigenvalue is determined in an iterative fashion. An initial guess of the eigenvalue is used to set up solutions (Equations 6 and 8) along the inner and outer edges of the mixing layer. The Rayleigh Equation (1) is then integrated numerically using a variable step-size Runge-Kutta scheme. The integration is carried out from the two edges towards an intermediate matching point. At this point, the matching point condition, i.e. the continuity of pressure and its derivative, leads to a set of homogeneous equations. For a non-trivial solution, the determinant of the coefficient matrix should be zero. The axial wavenumber is then obtained by finding the zeroes of the determinant of the matrix using a Newton-Raphson scheme.

In order to determine the axial evolution, the above procedure has to be repeated to calculate the eigenvalue as a function of downstream distance. The instability wave grows until some point downstream and then starts to decay. Eventually the wave is damped. In the damped regions, the integration path must be deformed around the critical point r_c, where $\omega - \alpha\overline{u}(r_c) = 0$. Secondly, for those waves whose phase velocity is greater than the ambient sound speed, i.e. supersonic damped waves, an appropriate value of $\lambda_0(\alpha)$ in the starting solution has to be used (see Tam and Burton, 1984).

After obtaining the axial wavenumber, the variation of the amplitude function is determined next. This can be done using Equation (19) and by assuming an initial amplitude. This is the only unknown multiplicative constant in the present analysis. The wavenumber spectrum of the pressure fluctuation, Equation (18), is then calculated by the method of fast Fourier transform. Of all the wave components, only those with supersonic phase velocity relative to the ambient sound speed radiate into the far-field. Finally the far-field directivity is calculated using Equation (22). The procedure for the elliptic jet is similar to the circular jet. Details of the method are given in Morris and Bhat (1992).

4. Numerical Results

This section compares experimentally determined axial mean flow jet properties to those predicted by the SAIC TTJET code. Also presented are the numerical results associated with axial variation of the phase velocity and growth rate for the instability waves. These are thought to be important parameters governing both the amplitude and direction for noise radiation. Finally the acoustic far-field directivity of circular and elliptic jets is presented for characteristic Strouhal numbers.

4.1. Prediction of mean velocity and jet spread rate

Figure 1, taken from Seiner, Ponton, Jansen, and Lagen (1992) compares the measured mean centerline velocity, V_c, to that predicted by the SAIC TTJET PNS code with the compressibility corrected $k - \varepsilon_d$. Here, the mean velocity and axial coordinates are respectively non-dimensionalized by the jet exit velocity and nozzle exit radius. The experimental and numerical predictions are relative to a round Mach 2 jet operating fully pressure balanced with jet total temperatures of $755°K$ and $1370°K$.

From Figure 1, one sees that the jet potential core length is a weak function of jet temperature. The core extends to between 18 and 19 jet nozzle exit radii. The numerical predictions only slightly overpredict the length of the core.

Beyond the potential core, both the experimental and numerical simulation show that the centerline velocity decays at a faster rate at higher jet temperature. The predicted decay, however, is much greater than that measured. This faster decay can be better understood by the jet spread rate comparisons of Figure 2. Figure 2 presents the predicted and measured axial variation of the jet half velocity width, $R_{.5}$, and the half width of the shear layer, b. This data is normalized by the jet exit radius.

The data of Figure 2 shows that in the potential core region, also known as the linear region, TTJET predicts a similar growth rate for the velocity half width $R_{.5}(x)$. However, a small difference in overall shear layer growth rate exists, as exhibited by the predicted and measured differences in $b(x)$. These differences can be primarily attributed to a thicker initial boundary thickness assumed in the TTJET calculations at the nozzle exit. Downstream of the potential core, the numerically simulated jet spreads at a much more significant rate than measured.

The cause of the over-prediction of shear layer growth rate downstream of the potential core is still under investigation. The over-prediction, however, prompts the following consideration. How well does one need to simulate the flow field to enable sufficiently accurate prediction of dominant noise for supersonic jets using the compressible Rayleigh equation model? In the following, this will be addressed.

4.2. Prediction of phase speed and wave growth rate

Instability wave calculations have been performed for both the axisymmetric and helical modes at different Strouhal numbers, $S_t = \omega r_j/\pi V_j$. Again these calculations have been carried out for the round jet operating at its Mach 2 design point and with jet total temperatures of $313°K$ (cold jet) and $755°K$ (hot jet). In all the results presented in this section, the axial distance is normalized with respect to the jet exit radius, R_j, and the phase velocity is referenced to the ambient speed of sound, C_a.

Figures 3 and 4 show the axial variation of the predicted phase velocity and growth rate of the axisymetric mode ($n = 0$) for three Strouhal numbers with the jet operating at $755°K$. These Strouhal numbers are $S_t = 0.11$, 0.22, and 0.43, which are representative of important high amplitude noise as shown previously in Seiner, Ponton, Jansen, and Lagen (1992). As can be seen, the phase velocity is supersonic over the entire downstream distance considered. Thus all these waves would be expected to radiate noise. The growth rate curves of Figure 4 show that the higher frequency wave decays rapidly initially and becomes damped closer to the jet exit. The lower frequency wave remains unstable over a greater distance.

The corresponding predictions for the helical mode ($n = 1$) phase speeds and growth rates are shown respectively in Figures 5 and 6. The results are similar to those obtained for the axisymmetric mode. The most significant differences are that the helical modes have lower phase speeds with higher growth rates. This latter behavior suggests that the helical mode is the more significant noise producer.

In all of the above calculations, the shear layer parameters are obtained from experimental data. The results obtained for the helical mode, using the TTJET code predicted shear layer parameters, are shown respectively in Figures 7 and 8 for the phase speed and wave growth rate. The difference between these results and those contained in Figures 5 and 6 is very small.

Two cases are considered for the elliptic jet. The first is that of an isothermal (jet static = ambient temperature) elliptic jet of aspect ratio 2, operating fully pressure balanced and shock free at its design Mach number of 1.52. The second case is an aspect ratio 3 elliptic nozzle operating fully pressure balanced at its design Mach number of 2.0 at a temperature of $313°K$.

The results presented for the elliptic jet are normalized by the area equivalent radius. The results for both cases are only for a $S_t =$ 0.2 and the varicose mode. Previous results obtained by Morris and Bhat (1991 and 1992) show that the varicose mode is the dominant mode for these elliptic jets. Figure 9 shows the predicted results for phase velocity for both cases and Figure 10 the corresponding growth rates. The shear layer parameters for these calculations were obtained from Seiner and Ponton (1991). Comparison of these results to the axisymmetric mode results for the round jet show significantly lower phase speed with a similar growth rate. The predicted phase speed for each case shows anomalous behavior. The phase speed both decreases and increases in the linear region of the jet, a result not observed in the round jet calculations. In addition, the growth rate for the $AR = 3.0$, Mach 2 case begins to increase near 30 R_j. This suggests unresolved numerical problems.

4.3. Prediction of far-field directivity

Figures 11 and 12 show respectively the predicted far-field directivity for the axisymmetric and helical modes of a circular jet at $755°K$. The initial amplitude of the instability wave of various modes and at different Strouhal numbers are taken to be equal. The predicted data is normalized by the maximum amplitude calculated for the two modes and three Strouhal numbers. The maximum value is found to be associated with the helical mode and for $S_t = 0.11$. This peak value defines the normalizing amplitude P_0 used in Figures 11 and 12.

The corresponding experimental data is shown in Figure 13. The measured noise levels are normalized using the same procedure used in the predicted data. This procedure is followed to eliminate the arbitrariness associated with selection of the initial amplitude in the Rayleigh equation formulation. The measured peak amplitude level is also associated with the $S_t = 0.11$ component.

The helical mode is seen to dominate the noise field. The axisymmetric mode does not show any significant shift of angle corresponding to peak noise level with frequency. This is a result of all three S_t components having a nearly constant phase speed in the region of maximum wave growth (see Figures 3 and 4). The helical mode shows significant shift in emission direction with S_t, which is consistent with the helical mode's predicted phase speed in Figure 5.

The numerically simulated jet data has also been used to compute the far-field noise levels for different Strouhal numbers. The results obtained are very similar to those obtained using experimentally generated data. The table summarizes the agreement obtained between measured and simulated flow field data for both axisymmetric and helical modes and two jet temperatures. The numerical jet data results are shown in parenthesis. From the table, it is clear that the helical mode amplitude and emission direction are in good agreement with experimental data. Secondly, the TTJET calculations appear adequate for calculating the principal Strouhal numbers for noise radiation.

In Figures 14 and 15, a comparison is made between measured and predicted noise data for the two elliptic nozzles. The experimental data is from Seiner and Ponton (1991). The comparisons are made of the directivity along the major and minor axes of the jet. For the Mach 1.5 aspect ratio 2 elliptic jet the peak angle for noise emission is for the $S_t = 0.2$ component is correctly predicted. However, the maximum noise is observed along the minor axis direction, $\phi = 90°$. The predicted result shows maximum noise in the major axis direction, $\phi = 0°$. For the Mach 2 aspect ratio 3 elliptic jet the principle nozzle axis for maximum noise emission is correctly predicted, but the angle of emission is substantially different than observed. Further work is required in applying the Rayleigh equation to compute noise associated with large scale structure for the elliptic jet. In the present calculations it has been assumed that the problem can be cast in the elliptical cylindrical coordinate system, an approximation that becomes less valid with significant three dimensional distortion of the jet shear layer.

5. Conclusions

Several important issues arise based on comparisons between solutions of the compressible Rayleigh equation and experimental data. First, it is seen that the wave solutions to the compressible Rayleigh equation provide a reasonable account for the relative amplitude and direction of the fundamental noise components emitted from round and elliptic jets. It appears that this is true for temperatures to at least $755°K$. Also simulation of jet flow parameters using a PNS code with a compressibility corrected two equation turbulence model is adequate for representing the flow field for purposes of noise calculations. This is consistent with the observation that the most significant noise radiated from the supersonic jet is produced in the linear region of jet flow development. The helical mode is shown to dominate noise radiated by jets heated to $755°K$. Improvements are required to apply the Rayleigh equation to elliptic jets. In the present formulation, elliptic cylindrical coordinates were assumed to apply to the whole jet. The measured experimental data supports this assumption for the Mach 1.52, $AR = 2.0$ nozzle, but significant measured distortion of the jet shear layer in the Mach 2.0, $AR = 3.0$ case raises concerns for this assumption. Substantial improvement to the formulation is required to handle highly complex three dimensional jet flows.

References

Colonius, T., Lele, S. K., and Moin, P., 1992. "Boundary conditions for direct computation of aerodynamic sound generation," 14th Aeroacoustics Conf., Aachen, Germany, DGLR/AIAA 92-02-075.

Dash, S. M. and Kenzakowski, D. C., 1992. Private Communication.

Lepicovsky, J., Ahuja, K. K., Brown, W. H., and Burrin, R. H., 1985. "Coherent large-scale structures in high Reynolds number supersonic jets," NASA CR-3952.

Morris, P. J. and Bhat, T. R. S., 1991. "The prediction of noise radiated from supersonic elliptic jets," AGARD 78th Specialist Meeting on Combat Aircraft Noise, Bonn, Germany, Paper No. 20.

Morris, P. J. and Bhat, T. R. S., 1992. "The noise from supersonic elliptical jets," 14th Aeroacoustic Conf., Aachen, Germany, DGLR/AIAA 92-02-042.

Sarkar, S., Erlebacher, G., Hussaini, M. Y., and Kreiss, H. O., 1989. "The analysis and modeling of dilatational terms in compressible turbulence," ICASE Report No. 89-79.

Seiner, J. M., 1992. "Fluid dynamics and noise emission associated with supersonic jets," Studies In Turbulence, Springer-Verlag, NY, pp. 297-323.

Seiner, J. M., McLaughlin, D. K., Liu, C. H., 1982. "Supersonic jet noise generated by large scale instabilities," NASA TP-2072.

Seiner, J. M. and Ponton, M. K., 1991. "Supersonic acoustic sources for free jets of various geometries," AGARD 78th Specialist Meeting on Combat Aircraft Noise, Bonn, Germany, Paper 16.

Seiner, J. M., Ponton, M. K., Jansen, B. J., and Lagen, N. T., 1992. "The effects of temperature on supersonic jet noise emission," 14th Aeroacoustics Conf., DGLR/AIAA 92-02-046.

Tam, C. K. W. and Burton, D. E., 1984. "Sound generated by instability waves of supersonic flows, Part 2. Axisymmetric Jets," *J. Fluid Mech.* **138**, pp. 273-295.

Tam, C. K. W., Chen, P., and Seiner, J. M., 1992. "Relationship between instability waves and noise of high speed jets," *AIAA J.* **30**(7), pp. 1747-1752.

Tam, C. K. W. and Hu, F. Q., 1989. "On the three families of instability waves of high speed jets," *J. Fluid Mech.* **201**, pp. 447-483.

Tam, C. K. W. and Webb, J. C., 1992. "Dispersion-relation-preserving schemes for computational acoustics," 14th Aeroacoustics Conf., Aachen, Germany, DGLR/AIAA 92-02-033.

Troutt, T. R. and McLaughlin, D. K., 1982. "Experiment on the flow and acoustic properties of a moderate Reynolds number supersonic jet," *J. Fluid Mech.* **116**, pp. 123-156.

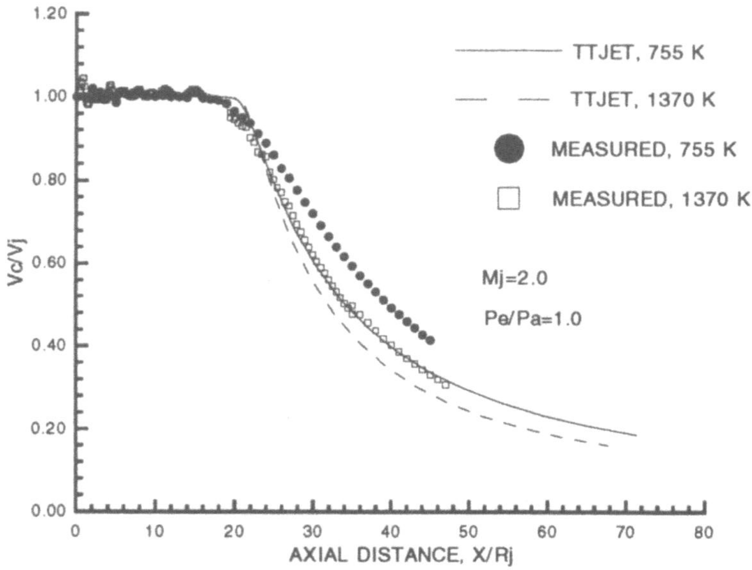

Figure 1. Comparison of jet centerline velocities.

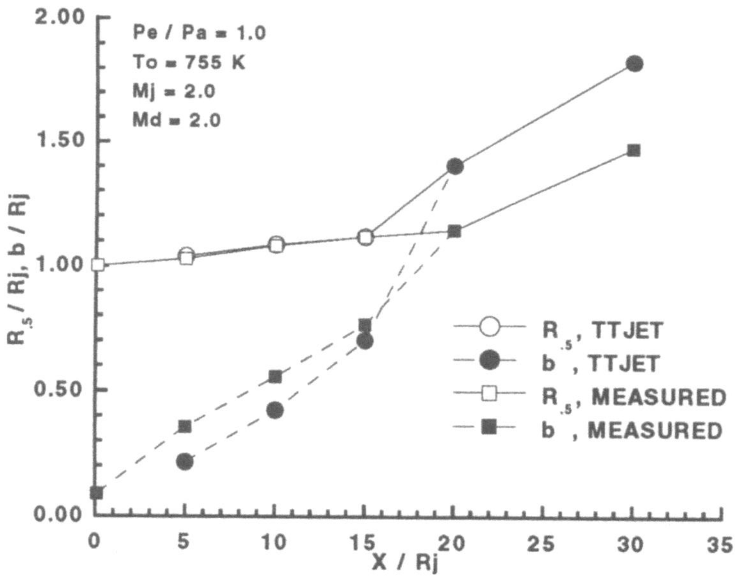

Figure 2. Comparison of jet spread parameters.

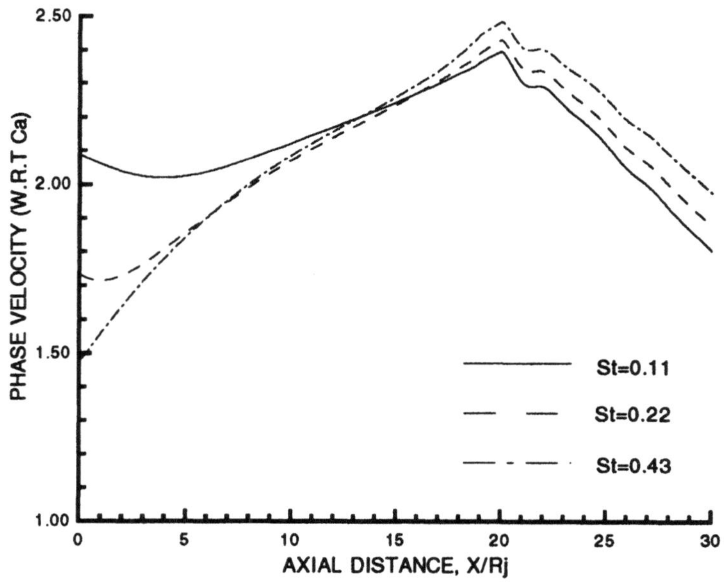

Figure 3. Axial variation of phase velocity. $Mj = 2.0$, $To = 900$ deg. F, $n = 0$ (based on experimental data).

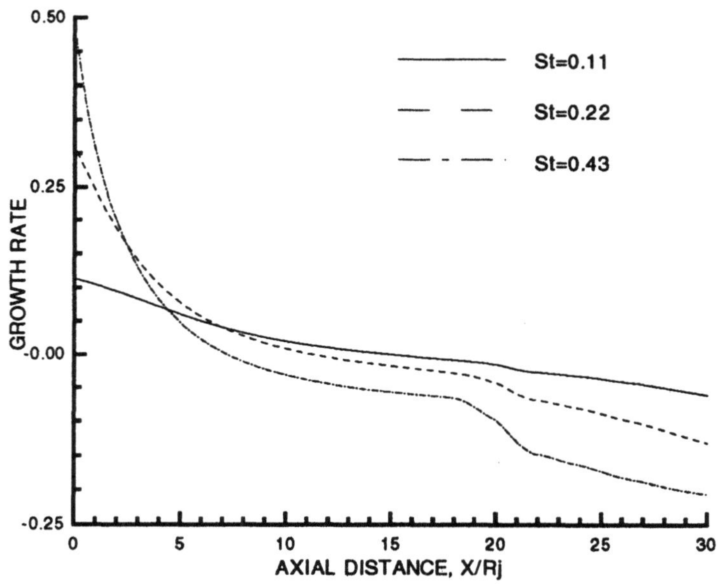

Figure 4. Axial variation of growth rate. $Mj = 2.0$, $To = 900$ deg. F, $n = 0$ (based on experimental data).

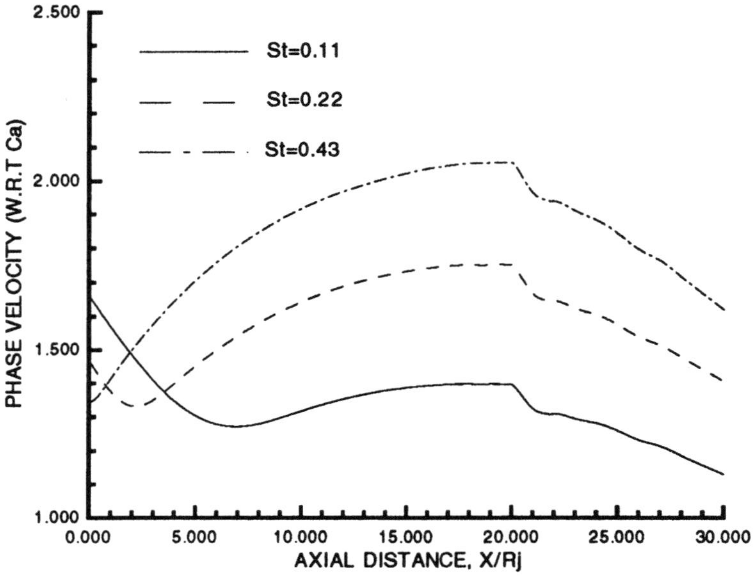

Figure 5. Axial variation of phase velocity. $Mj = 2.0$, $To = 900$ deg. F, $n = 1$ (based on experimental data).

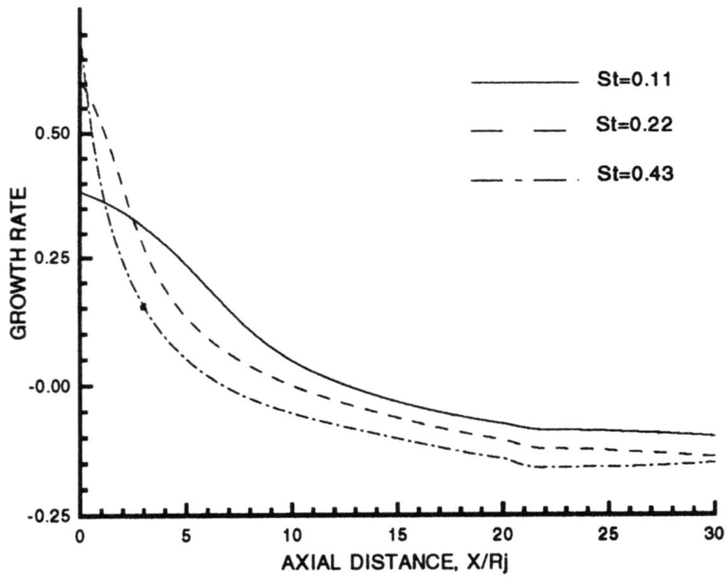

Figure 6. Axial variation of growth rate. $Mj = 2.0$, $To = 900$ deg. F, $n = 1$ (based on experimental data).

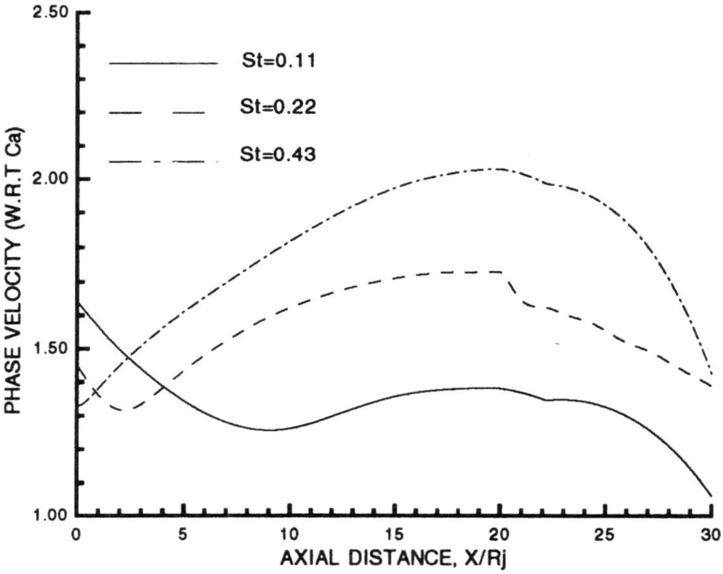

Figure 7. Axial variation of phase velocity. $Mj = 2.0$, $To = 900$ deg. F, $n = 1$ (based on TTJET code data).

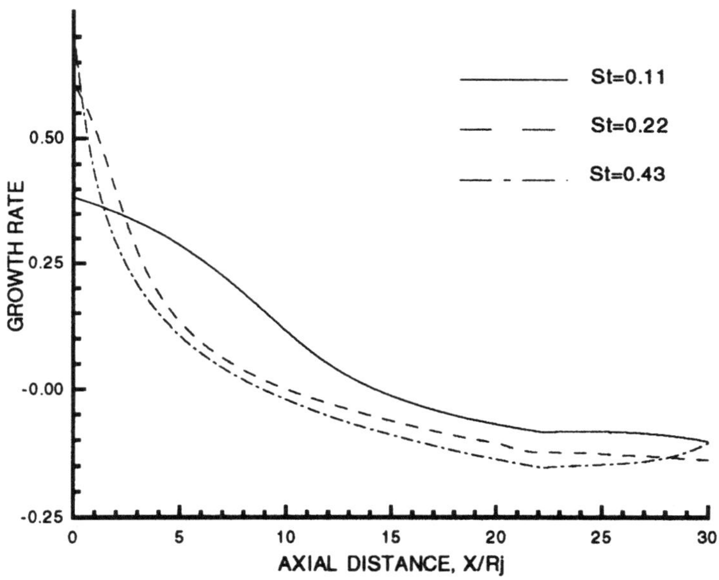

Figure 8. Axial variation of growth rate. $Mj = 2.0$, $To = 900$ deg. F, $n = 1$ (based on TTJET code data).

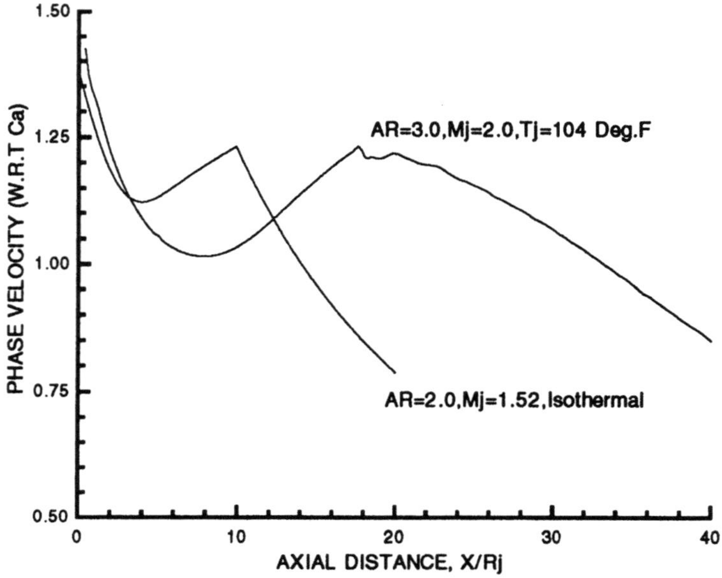

Figure 9. Axial variation of phase velocity. Elliptic jet, $St = 0.2$, varicose mode.

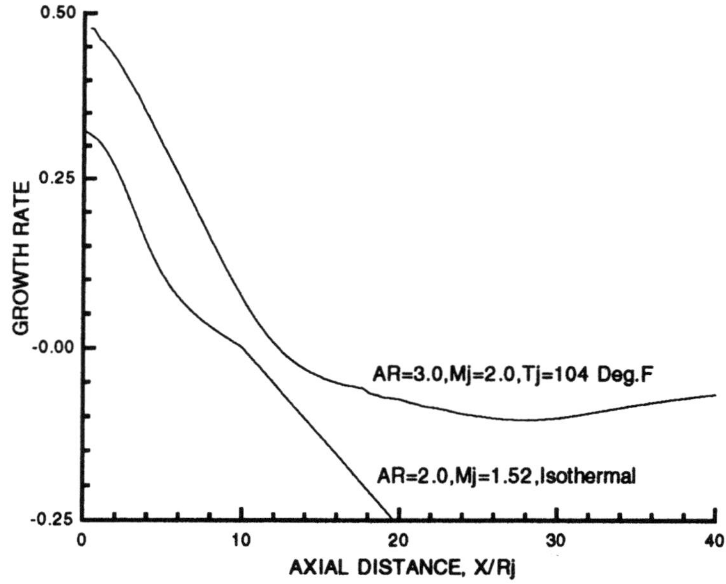

Figure 10. Axial variation of growth rate. Elliptic jet, $St = 0.2$, varicose mode.

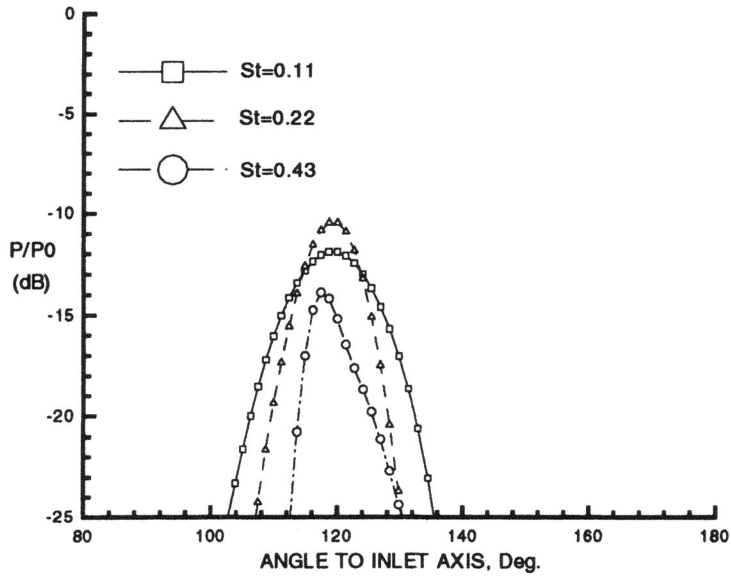

Figure 11. Strouhal dependence with jet angle. $Mj = 2.0$, $To = 900$ deg. F, $n = 0$.

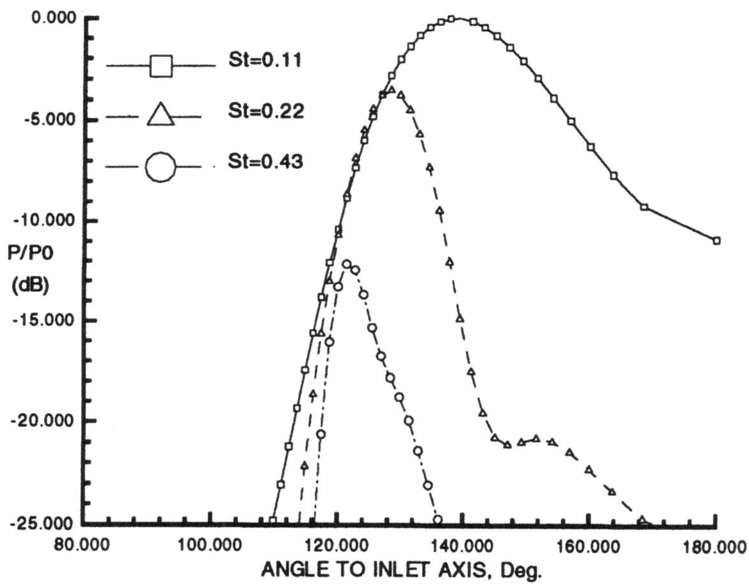

Figure 12. Strouhal dependence with jet angle. $Mj = 2.0$, $To = 900$ deg. F, $n = 1$.

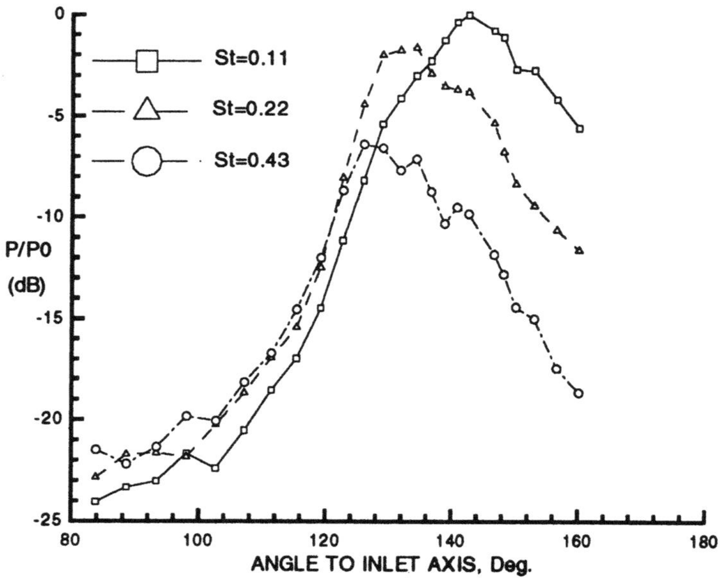

Figure 13. Strouhal dependence with jet angle. $Mj = 2.0$, $To = 900$ deg. F, experimental data.

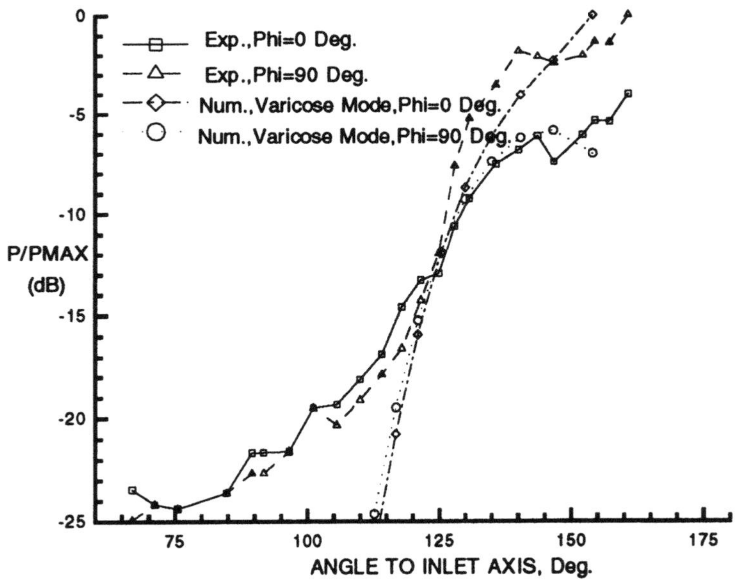

Figure 14. Far-field directivity, $AR = 2.0$, $Mj = 1.52$, $St = 0.2$, isothermal jet.

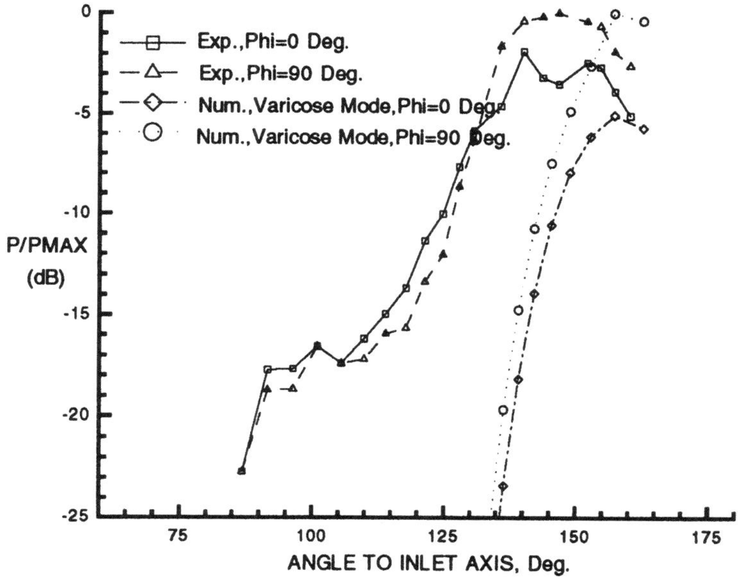

Figure 15. Far-field directivity, $AR = 3.0$, $Mj = 2.0$, $St = 0.2$, $To =$ 104 deg. F.

DIRECT SIMULATION

DIRECT SIMULATION: REVIEW AND COMMENTS

M. K. Myers

The George Washington University, JIAFS
Mail Stop 269, NASA Langley Research Center
Hampton, Virginia 23681-0001

1. Introduction

The session entitled "Direct Simulation" comprised six invited papers, each illustrating timely and interesting applications of CFD methodology to aeroacoustics. Probably the most important aspect of the presentations, however, was the fact that each pointed up issues that are and will be of major concern as computational aeroacoustics (CAA) evolves into a working discipline. In this review, an attempt is made to provide a brief summary of the thrust of each of the contributions to the session as well as, where appropriate, some highlights of computational issues raised explicitly or implicitly by their authors. Further, the review contains some observations on these issues. To some extent, these represent personal views of the current author, but they are also meant to reflect the outcome of discussions that occurred during the panel session held at the conclusion of the ICASE/NASA Workshop.

In order that no misunderstanding arises relative to the title of the session, it is noted that the term "direct simulation" has come to be used somewhat differently in CAA than it has been heretofore in CFD. Three of the papers presented dealt with aspects of direct simulation of turbulence and its use in predicting acoustic propagation in turbulent flows. The remaining three papers represent the broader definition of the term that has lately been used in CAA to denote any direct computational description of sound generation, transmission and radiation utilizing techniques of CFD to solve the appropriate fundamental system of governing equations. Although it is no doubt true that the interaction of sound and turbulent flow is a problem area significantly more complex than those of a more classical nature, it is certainly also true that all computational efforts in aeroacoustics are subject to a common set of numerical pitfalls. One can thus argue that their consideration together, as in the present grouping, makes good sense given the current state of development of CAA. In fact, one recurring theme of the Workshop has been to emphasize the fundamental value of benchmark problems, usually of the more classical type, as the means by which computational issues can be understood and resolved prior to numerical attacks on problems for which physical and analytical descriptions are limited.

2. Review of Session Contents

The session commenced with presentation of the paper by C. H. Berman, G. Gordon, S. A. Orszag and G. E. Karniadakis in which computations of jet noise are described. In this work, a complete near field is obtained using a three-dimensional spectral finite-element technique that includes a direct simulation of turbulent effects. This is coupled to an acoustic model based on solution of Phillips' equation to yield the sound field at points sufficiently far from the jet boundary. The paper illustrates the potential utility of incorporating CFD methods along with acoustic analogy representations provided that, as the authors emphasize, proper attention is paid to the formulation and accurate numerical implementation of the source model.

In the next paper, by K. S. Brentner, the two-dimensional flow around a cylinder accelerating in a stationary, inviscid fluid is computed with a finite-volume algorithm and acoustic portions of the disturbance in the entire field. An exact disturbance energy conservation equation is employed to effect a partition of the numerically determined energy into local aerodynamic, acoustic and entropy-associated components. The work indicates that energy considerations can be of considerable value in conjunction with CAA methods in that they provide means by which numerical error can be detected in the absence of available analytical solutions. In this paper it is found that the entropy component of energy, which is spurious so long as no shocks develop in the field, arises largely on account of numerical error associated with the computational grid. Examples are given showing that a very fine grid is necessary to eliminate it completely.

The third contribution to the session was that of S. A. Orszag, who laid out in detail various fundamental aspects of compressible turbulence simulation according to renormalization group theory. Particular attention here focuses on the determination of the proper definition of eddy viscosity for high Mach number flows. This theory is the basis for the simulations employed in the work discussed in the first paper above.

The work of S. Sarkar and M. Y. Hussaini represents another example in which an acoustic analogy is used in conjunction with a direct numerical simulation. Here the objective is to obtain the acoustic output of a bounded region of turbulent flow. Three simple homogeneous flows are considered for which numerical solutions of the compressible Navier-Stokes equations are obtained. The instantaneous fluid state in a bounded region is used to compute the Lighthill stress tensor components and thus the radiated acoustic field. The authors discuss relevant issues associated with the process, including the effect of shear on the predicted acoustic fluctuations and the degree of resolution required in the flow field to obtain sufficiently accurate acoustic results.

In the fifth paper of the session, W. R. Watson reported results of computations of periodic steady-state acoustic disturbances in subsonic internal flows using a MacCormack scheme. Attention is concentrated on solutions in a fully bounded domain for which inflow/outflow boundary conditions become of critical concern. The author explores a numerical treatment of the equations governing the sound alone and contrasts this procedure with that of obtaining the mean flow and acoustic perturbations together. It is found that the latter approach is inferior to the former, primarily because mean flow computational errors can easily be of the same magnitude as the desired acoustic quantities. In addition, it is simple to neglect direct viscous influence on sound, as is often justifiable in practice, if the perturbations are computed separately. Doing so, however, leads to difficulties because undesired transients persist in the inviscid disturbance field unless totally non-reflecting boundary conditions are physically appropriate. Even when appropriate, however, such conditions are not known in multi-dimensional cases. In the paper, a method of filtering the numerical results to eliminate transients is discussed and shown to be effective.

The session concluded with the paper by S. K. Lele, in which direct computations of far-field sound radiated by known unsteady flows are discussed. Several model problems are considered, and the direct simulations are compared with corresponding predictions of the acoustic analogy using numerically determined source terms. It is found that the incompressible form of the source yields good agreement with the full numerical simulation at sufficiently low Mach numbers. In general, however, they deviate as the Mach number increases, and the author finds that compressibility effects in the near-field flow must be carefully taken into account in the source terms of the analogies to improve agreement with the directly computed results.

3. Comments on Issues in CAA

Consideration of the discussions that followed each of the separate presentations in sessions throughout the Workshop and those of the panel session on the final day indicates that there are clearly two points of view taken by those concerned with CAA. One seems to presume that computational capability is largely in place to deal with acoustics and seeks to explore its application to unsolved problems that are likely to yield only to numerical treatment and that promise significant practical benefit. Attention here is naturally drawn to the general area of flow generated noise and turbulent effects on sound. The other point of view, much less sanguine with regard to computing, puts CAA in the category of a new discipline for which many of the tools have yet to be invented or at least have yet to be shown to be sufficiently reliable to be applied with any degree of confidence. Those who identify with this view tend to focus on problems about which much is already known, using them as vehicles by which to validate numerical schemes and then employing the numerics to carry out extensions and generalizations as appropriate.

As a result, a wide range of issues attracted attention at the Workshop. To a certain extent, however, discussion at the panel session reflected emphasis more heavily placed on the first point of view above. This is not without justification if it is kept in mind that the ultimate goal of CAA is, after all, to understand sound, not to understand computational methods. Nevertheless, those who lean to the second view, including the current author, could argue that such emphasis tended to limit some of the interchange between acousticians and numerical analysts on purely computational questions that would have been desirable. Detailed dialogue between them is essential, because it is clear that the acoustics community in general lacks sufficient familiarity with the state of affairs in CFD, just as many in the CFD community are not fully aware of the extent to which aeroacoustics presents new demands on computation. Enormous progress has been made in CFD over the past two decades or so in the direction of understanding the ranges of applicability of algorithms and in utilizing them successfully in a wide variety of fluid dynamic applications. Even a cursory review, however, leads one to the conclusion that the successes have usually been achieved only after considerable fine-tuning of general numerical schemes to match special mathematical or physical characteristics of individual problems. The result is that there has grown in CFD a coherent body of computational knowledge about what works and what does not, and why. The only rational approach to CAA must be one that proceeds as rapidly as possible to assimilate this existing expertise, but it seems far from obvious just how much of it is applicable to aeroacoustics. One point that cannot be overemphasized in this context is that, because much of the thrust of CFD has been toward steady flow problems, it has evolved in many respects in directions that are antithetical to successful treatment of sound.

Readers of this volume are likely to arrive at differing interpretations of the issues bearing on CAA and at differing opinions as to which are most fundamental. Taken as a whole, however, the papers in the volume and the activities at the Workshop itself indicate that many have yet to be resolved. In the remainder of this section a few of these are discussed from the current author's perspective with a view toward summarizing and recording certain conclusions one can draw from the proceedings.

3.1. Algorithms

Reservations have been expressed by many recently about the extent to which existing CFD methodology is adequate to be applied directly to aeroacoustics. Such concerns were echoed repeatedly at the Workshop, and readers of this volume will find a number of papers specifically addressing numerical difficulties inherent in particular classes of algorithms. On the whole, however, there is a tendency thus far for researchers in CAA to adopt numerical schemes based more on considerations of individual familiarity and simplicity than on those of suitability for calculating sound. This is a logical initial approach, of course, and it may prove that many well-known schemes are indeed suitable. Nevertheless, there is a need for a substantial coordinated effort to explore the dissipative and dispersive character of existing schemes to determine which can be used on a routine basis in CAA and to what extent development of new algorithms may be necessary.

A number of distinguishing factors impact the numerical simulation of acoustic wave motion. They include the very small magnitude of acoustic quantities relative to their ambient values, the fact that frequencies of practical interest range up to 20 kHz or so, the need for spectral data and therefore for accurate computations over many cycles of the fluctuations, and, correspondingly, the necessity to resolve waves accurately over many spatial wavelengths. Algorithms useful for aeroacoustics must deal with all of these and must do so while requiring as few grid points per cycle in time and space as possible (many current schemes appear to need of the order of twenty, whereas five or six is perhaps the least that can be hoped for).

These and other factors constitute severe tests for existing numerical schemes, many of which have been designed explicitly to eliminate wave motion generated in the process of driving solutions to steady conditions. In fact, a number of participants in discussions at the Workshop suggested that the degree of accuracy demanded of acoustic computations will be achieved only by routine use of high-order numerical methods. Whatever the case, it seems clear at present that the question of adequacy of algorithms remains to a great extent an open one, and that much more must be learned before CAA will be on secure ground.

3.2. Complete flow vs. perturbations

An issue closely related to the above arises from the fact that there are two approaches that can be followed in computing an aeroacoustic field propagating through a moving fluid. One is to consider only the fundamental set of equations governing the complete fluid motion, solving once for the ambient flow and then obtaining the disturbance quantities by subtraction. The alternate approach consists of solving for the ambient state and then computing the sound field from the derived set of equations describing the disturbance quantities alone. Examples of each approach appear in this volume. They are, of course, analytically equivalent, and one would expect no significant numerical differences between them if the ambient flow is known exactly. Nevertheless, difficulties arise when the flow must be calculated numerically. Although it is not possible at present to judge either approach consistently superior, each has certain numerical advantages and disadvantages in the context of specific problems.

The first strategy yields the complete field, flow plus sound, using the same algorithm and the same numerical boundary conditions. This may be advantageous in some cases, but it may well be that acoustic fluctuations are better obtained with a scheme different from that used for the underlying flow. One clear advantage of the approach is that it allows the governing equations to be utilized in conservative form so that shocks that occur in either the flow or the acoustic field are, in principle, straightforward to accommodate. On the other hand, a principal disadvantage of the combined field approach is the fact that numerical errors present in the ambient flow quantities, unless driven to extraordinarily small residual values, can easily be large enough to destroy the simulation of the very small acoustic quantities being sought.

The second procedure has the obvious advantage of compatibility with any algorithm and boundary conditions that are appropriate for predicting the disturbances. Also it can be expected to be more forgiving to errors in the ambient quantities because these are fixed once they are brought into the coefficients of the system governing the perturbations. Moreover, the strategy provides a simple means by which direct viscous effects on the acoustic field can be ignored while retaining them in the calculation of the ambient flow. When this is an acceptable idealization, an additional benefit accrues from the fact that the disturbance equations become a hyperbolic system for which formulation of boundary conditions is greatly simplified. This comes at some cost, however, because the absence of viscosity can cause transients to persist in the solution domain in many cases, thus making it more difficult to extract periodic acoustic solutions when these are of interest. A definite disadvantage of the approach is that the non-conservative form of the perturbation equations makes it unsuitable for dealing with shocked acoustic waves.

As noted before, neither of the two approaches for simulating wave propagation can be considered superior in all circumstances. It does appear, however, on the basis of the remarks in the preceding paragraph, that the perturbation approach may be the more effective one so long as the sound field remains continuous.

3.3. Boundary conditions

The development and numerical implementation of appropriate boundary conditions remains of substantial and continuing concern to CAA. What is appropriate, of course, depends strongly on the problem being considered and on the specific numerical scheme being employed. Thus, the topic has been given considerable explicit attention in presentations at the Workshop and in numerous papers in this volume. In general, these tend to support a belief that computation of sound requires a level of accuracy in specifying boundary data that is significantly greater than that previously considered adequate in CFD. For example, if periodic steady-state sound is being sought then small errors introduced by improper modeling of the interaction of the field and the boundary of the solution domain can easily give rise to spurious reflections that completely destroy convergence to the desired results. Readers are referred to the individual papers in these Proceedings for more detailed discussions of numerical difficulties associated with boundary conditions, and only a few general remarks are included here.

First, for free boundaries (inflow/outflow), emphasis in CFD has been on formulating so-called non-reflecting conditions, usually through application of characteristic theory. Conditions of this sort appear to work well in simulating sound propagation into infinite space across artificial "far-field" boundaries. However, it is important to keep in mind that for many acoustic problems on finite domains free boundaries are actually reflectors of sound at which annihilation of waves incoming to the domain is physically incorrect. For this type of problem, it appears that research is still needed to determine the most effective means of simulating boundary behavior.

A second point concerns the imposition of proper physical conditions at solid walls. It is common in CFD to treat these by extrapolation of values from the interior of the domain when steady flow solutions are of interest. When an acoustic field coexists with the flow, however, it can easily be found that extrapolation introduces spurious vorticity which can then spread throughout the solution domain. Again, it appears that this is an area in which more research is required.

Finally, the author would emphasize that there is a lack of understanding as to how best to model acoustically absorbing solid boundaries in a fully unsteady computational description. Wall absorption is an area of intense interest in many acoustic situations that will need to be addressed by CAA in the future, and there is an obvious need here for research closely coordinated with the experimental acoustics community.

3.4. Incorporation of classical theory

Progress in computational aeroacoustics is in its infancy compared to that in other areas of fluid mechanics primarily because computing capacity has been unavailable previously to deal with it. While this situation is beginning to change, the fact remains that practical acoustic fields demand very large computational efforts. This makes it of great value for CAA to incorporate whenever possible as working computational tools the existing integral representations of sound in the form of the acoustic analogy, various Kirchhoff formulations, and so on. Using CFD to obtain source terms for an acoustic analogy approach to flow generated noise is an obvious example of this process, and, with due attention paid to the source model, is demonstrated to be effective in several papers in these Proceedings. The power of integral representations can be applied more widely, however, including using them as the basis for boundary integral techniques to be combined with finite difference solutions as a means to reduce computational size. It would appear that there is significant potential here for innovative research to produce results of major impact on CAA.

4. Concluding Remarks

An attempt has been made in this article to highlight a few of the issues emerging from research reported at the ICASE/NASA Workshop that appear to have significant impact upon the evolution of CAA. Although many of these cannot be said to be resolved, the Workshop provided what is actually the first forum of its kind at which they could be debated in some depth by both the computational and the aeroacoustics communities, and there can be little doubt that it represents a major milestone in this respect. It is to be hoped that the process of exchange of ideas represented by the papers in this volume is one that will continue in similar forms in the future.

TIME-DEPENDENT JET NOISE COMPUTATION TECHNIQUES

Charles H. Berman and Gary Gordon[1]

AeroChem Research Laboratories, Inc.
P.O. Box 12
Princeton, New Jersey 08542

Eric Jackson, George E. Karniadakis, and Steven A. Orszag

Princeton University
Princeton, New Jersey 08544

ABSTRACT

The coupling of a three-dimensional, time-dependent, incompressible flow code to an acoustic analysis based on numerical solutions of the Phillips convected wave equation is discussed. A major challenge is finding techniques to accurately evaluate the noise source strength, which is expressed as a product of velocity derivatives in the Phillips equation. It is found to be advantageous to reexpress the Phillips source in a double space divergence form similar to that of Lighthill when the flow is incompressible. The validity of this approach is tested against both numerical solutions based on the more conventional time derivative source form in Lighthill's equation and exact analytical solutions. Work in progress on developing a compressible turbulent flow code holds out the possibility of eventually using Lighthill's integral equation directly to compute noise. Although the motivation for the current studies is to predict the noise of high speed jet flows, it should be noted that the incompressible flow code provides the capability of computing the noise of very low Mach number flows directly using Lighthill's equation.

1. Introduction

The least well understood aspect of jet noise prediction is the specification of the time-dependent flow field. The ability to de-

[1]Past support on this project was provided by the National Aeronautics and Space Administration, Langley Research Center under Contract No. NAS1-18849 and current support by the NASA Lewis Research Center under Contract No. NAS3-26247.

scribe the turbulent fluctuations and extract those properties responsible for jet noise has been a dream for decades. Until recently our knowledge of turbulence has come primarily from experimental measurements. Modeling of this flow information has been used with time-averaged turbulent flow computations to develop a predictive capability for jet noise (Berman, 1979). This is a practical approach when the jet geometry and operating conditions are similar to those for which the experimental data was obtained.

The present study is directed toward the development of numerical techniques for predicting the noise due to suppressors for high speed jets which may produce a turbulence structure quite different from conventional jet flows. Indeed an objective might be to ask what changes in the turbulence structure result in reduced noise for a given thrust. We believe that only a time-dependent three-dimensional solution of the Navier-Stokes equations can provide the necessary information. The details of such solutions and how the sound is to be extracted are discussed next.

2. General Formulation

Treating high Reynolds number flows using a direct numerical simulation is still impractical given today's computer resources. However, large eddy simulations, LES, which properly account for the effect of the small scale turbulence on the larger scale turbulent structure that is most likely to be important for noise generation, are within the realm of today's computer resources.

The particular subgrid scale model used here is based on renormalization group theory, RNG (Yakhot and Orszag, 1986). Comparisons of computed results with standard turbulence flow data have been quite good for channel flows (Yakhot et al., 1989). Qualitative agreement has been found for jet flows (Berman et al., 1991), and work is continuing on obtaining better agreement. Specific issues are inflow conditions and grid resolution.

The most general approach would be to numerically solve the compressible, time-dependent, three-dimensional Navier-Stokes equations and substitute the variable pressure, density, and velocity into the integrand of Lighthill's integral solution (Lighthill, 1952), keeping proper track of the time history. This would be less prone to error and also easier than following the sound waves in the region exterior to the jet.

An equivalent procedure is to formulate Lighthill's integral in the frequency domain with the integrand given as a Fourier transform of the quadrupole stress tensor T_{ij}. The Fourier transform is easier to work with in terms of keeping track of retarded time differences, and its statistical reliability is increased as it is averaged over more data records. Memory can be minimized by performing a standard finite Fourier transform, rather than an FFT, at each point of interest in the jet. The standard Fourier transform can be formulated as a summation which is updated after each time step until the desired time data record length is obtained, whereas all the time data must be stored to perform an FFT. Neither a uniform grid spacing nor a constant time increment is needed for performing the integration.

At present a compressible code is being developed, and it is difficult to estimate its computer requirements at this time. An available incompressible LES code can be operated in either a uniform or a nonuniform density mode. For this reason our work on noise computation is based on using the incompressible codes, which obviously lack important information for a high Mach number calculation using Lighthill's approach. Thus we have broken the problem up into two parts. The first is to solve the time-dependent flow equations for the instantaneous turbulent velocity field and the second is to solve a convected wave equation, specifically the Phillips equation (Phillips, 1960) in the jet and within its near radiation field. This pressure is then extrapolated from an imaginary plane to the far field using a Green's function technique.

Mach number effects on propagation through the incompressible jet and the radiation field are dependent on the jet velocity and the speed of sound in the wave equation, which introduces compressibility. Velocity and speed of sound in the Phillips equation wave operator are fully time-dependent.

Our formulation is directed at treating realistic Mach number jet flows. In this case numerical techniques for evaluating noise generation and propagation are practical because the wavelengths encountered are not excessively large. For example, the ratio of wavelength, λ, to jet nozzle diameter, D is

$$\frac{\lambda}{D} = \frac{1}{St\ u_j/c_o} \tag{1}$$

where u_j is the jet exit velocity, c_o is the ambient speed of sound, and St is the jet Strouhal number. For $u_j = c_o$ and $St = 0.1$, a full octave below the jet noise peak frequency, $\lambda/D = 10$. Numerical simulations performed for dipole and quadrupole sources show that the near field ends, i.e., inverse distance scaling holds, at distances on the order of $\lambda/4$ from the source. For the above example the acoustic field is reached at a distance of 2.5 D, so that a long distance wave computation capability is not required.

A more detailed description of the overall approach has been presented elsewhere (Berman et al., 1990; Berman et al., 1991).

3. Source Terms

Dropping entropy and viscous effects the Phillips equation is written as

$$\frac{D^2\sigma}{Dt^2} - \frac{\partial}{\partial x_i}\left[c^2\frac{\partial\sigma}{\partial x_i}\right] = \gamma\frac{\partial u_i}{\partial x_j}\frac{\partial u_j}{\partial x_i}$$

$$\sigma = \ln\frac{p}{p_o}. \tag{2}$$

Since the flow is considered to be incompressible, the source term is also equal to a double divergence of a velocity tensor product, similar in form to Lighthill's turbulence tensor

$$\frac{\partial u_i}{\partial x_j}\frac{\partial u_j}{\partial x_i} = \frac{\partial^2 u_i u_j}{\partial x_i \partial x_j} + \textit{compressible terms}. \tag{3}$$

We next sought the best way to represent the source term, either as a product of derivatives or as a double divergence. The latter has the property that its integral over all space at zero retarded time is zero so that the sound is zero at zero frequency according to theory. We at first thought that for higher frequencies the double divergence form would be numerically less accurate. However, we were surprised to find that although the Phillips product derivative form gave a better point by point representation of the source, the double divergence form gave more accurate sound values at both high and low frequencies.

One strong point of the double divergence form is that the standard second order accurate finite difference second derivative is identically conservative in that its integral over all space is zero, just as in the analytical representation, if the integrand is zero at large distances.

We display next how the double divergence finite difference form can provide some insight on the structure of quadrupole sources.

Consider the finite difference longitudinal quadrupole composed of -2 units at grid point $N = 0, 1$ unit each at $N = \pm 2$, and zero strength at all other locations. Obviously, as shown in Figure 1, the integral of the quadrupole sources is identically zero at zero retarded time. The distribution of T_{11} producing that quadrupole when the second-order differencing is applied is illustrated in the same figure. Here T_{11} is a simple finite difference representation of a bell shaped curve whose curvature changes three times. These changes in curvature are the origin of the sound and their values are given by the structure of the quadrupole. Note that for a distribution of quadrupoles, positive and negative regions need not cancel out locally, but only when integrated over the entire domain of the turbulent region at zero retarded time.

When the quadrupoles are produced by turbulent velocity fluctuations, a longitudinal quadrupole's strength is, as a first approximation, proportional to the square of the first derivative of the velocity component along the quadrupole axis. This can be verified by expanding the velocity as a Taylor series about the point where the space derivative is to be obtained, squaring the velocity, and taking the second derivative of this product. For cross-derivative quadrupoles the strength is, to lowest order, similarly proportional to the shear velocity derivatives corresponding to the two quadrupole axes. Thus, the values of the Lighthill quadrupoles are products of velocity derivatives, precisely as specified by the Phillips equation. However, the direct Phillips form is numerically less accurate because it does not identically satisfy the double divergence zero integral condition at zero retarded time. Thus, although the Lighthill form we plan to use involves second-order derivatives in space, the accuracy requirements are primarily on first derivatives of velocity and not on second derivatives of velocity.

The accuracy of the double divergence form needs to be determined since the standard Lighthill method involves a second time-derivative of the source strength. In experiments, time and frequency information is more readily available and reliable than spatial information. However, when performing computer flow simulations there is a wealth of spatial information.

One, two, and three-dimensional cases were studied to answer the basic question of whether the double divergence space derivative form can provide sufficiently accurate results. We believe that this can be done.

The one-dimensional test case considered that T_{ij} had the form of a Gaussian modulated by a traveling cosine wave

$$T_{ij} = T_{11} = \exp[-(x/b)^2]\cos[\omega(t - x/u_c)]. \quad (4)$$

This can be viewed as a turbulent eddy traveling at the speed u_c with an amplitude that rises and then falls in space. A similar model has been studied to explain jet noise (Michalke, 1977), but our main interest in it is that an exact analytical formula can be obtained with which to compare the numerical results. By using the time-derivative form of the source it is then straightforward to obtain the amplitude of the sound field as

$$\rho = (\omega/c_o)^2\sqrt{\pi}b\exp[-(\omega/c_o)^2(1 - c_o/u_c)^2]. \quad (5)$$

Note that the maximum sound level occurs when $u_c = c_o$, and that the pressure varies as ω^2 so that there is no sound at zero frequency.

The finite difference double space derivative form was next integrated over x and the far field sound determined by including the proper retarded time in the cosine factor using the distances between points in the source region and the far field location.

In Figure 2 the comparison between the theoretical result and the space derivative source computations is good even for a relatively small number of grid points per acoustic wavelength and per Gaussian half-width of source. The results were unaffected by the value of u_c in the range $1 \leq u_c/c_o < \infty$. The results become markedly better for increased resolution. The space derivative method is extremely accurate for longer wavelengths since the grid points per wavelength increase. For example, significant differences are observed only for wavelengths greater than 1000 b. The time derivative computations are so close to the theoretical results that they cannot be distinguished from the latter on the scale shown.

The next case studied was that of a two-dimensional vortex with a Gaussian distribution of vorticity. The amplitude of the velocities changes uniformly sinusoidally in time. This could be viewed as one Fourier frequency component of a decaying vortex. We have no analytical solution in this case, but we can compare the double divergence form to the time derivative result, which will serve as our standard for comparison, and this in turn to the Phillips derivative product form. Both the time and double space divergence results agree quite well over a range of wavelengths. For example, excellent results are obtained for $\lambda/r_c = 100$ where r_c is the vortex core radius. The Phillips result becomes inaccurate at smaller values of λ/r_c in the low frequency range and is not consistently good at higher frequencies.

For the three-dimensional case a spherically symmetric distribution of source strength whose magnitude fell exponentially with distance from the center and varied sinusoidally in time was chosen. This could be viewed as a model for sound due to fluctuations in fluid density in Lighthill's source term.

In this case the following theoretical result for amplitude can be obtained by analytically using the double time derivative approach where b is the $1/e$ point of the source

$$\rho = \frac{2(\omega b/c_o)^2 b}{|x|[1 + (\omega b/c_o)^2]^2}. \tag{6}$$

In Figure 3a the double space and double time derivative computations based on finite difference techniques are displayed along a line 128 grid points in length through the center of the source at one time in comparison to the theoretical envelope of the sinusoidal fluctuations. There are 12.8 points per wavelength, and the source reaches its $1/e$ point at two grid points from its center. The other two domain dimensions are 40 points in width. The three-dimensional results shown differ from the other results in that the wave equation was solved here to account for propagation time effects whereas in the one and two dimensional cases an analytic formula was used for the retarded time. Figure 3b displays results along a line that is halfway between the centerline and a sideline domain boundary.

The observed deviations from the theoretical envelope are due principally to reflections with the computational domain boundaries since only a simple normal impedance boundary condition is used. Differences between the two source model results are manifested mainly as vertical displacements, i.e., if the peak of one curve is higher than the peak of the other curve, then its closest minimum is also higher by about the same amount.

Additional work is in progress to better define the grid spacing required to achieve accurate results for particular source parameters.

The double divergence source form was applied to the turbulent velocity field of an incompressible, uniform density jet issuing from a round nozzle simulated by a short length of a very thin cylinder. The noise source term is shown in Figure 4 at one instant of time on a plane through the jet centerline.

4. Numerical Considerations

At present a second-order accurate in space method of solving the wave equation is being utilized. We believe that this is adequate for the current problem rather than going to a fourth-order technique since the propagation distances are rather short. Although the characteristic scale of the wave operator is λ and that of the flow is an eddy length l, the grid spacing for the wave equation must be controlled by the smallest scale, which is generally l. The grid must resolve the turbulence sufficiently well to obtain accurate values for the source derivatives. The generally accepted requirement that acoustic grid spacing be constant to avoid mesh reflections means that we must be prepared to set up acoustic problems to have large numbers of grid points. This is not so much a problem of computation time but one of memory on a computer. In general, it is the turbulent flow computations, not the acoustic computations, that use the most computer time. A fourth-order method, which needs fewer mesh points, would not help since the finer grid is needed to resolve the turbulence anyway. The problem is a tractable one for the Mach number range of our current interest but becomes worse at lower Mach numbers.

However, at very low Mach numbers the convective term in the Phillips equation is unimportant and the integral form of Lighthill's equation, using the time derivative formulation, is once again practical. In that case using a Fourier frequency transform as discussed above for the compressible case is an attractive approach. Thus, we conclude that techniques for the numerical solution of very low Mach number turbulence flows are currently available. Low Mach number turbulent flows often involve an interaction with solid boundaries due to the long wavelengths encountered, and the sound produced by these interactions can be computed using Curle's theory (Curle, 1955).

5. Conclusions

Currently available LES codes have the potential to facilitate the computation of time-dependent three-dimensional jet noise. For transonic and low supersonic Mach numbers the assumption is that the turbulence structure is relatively independent of Mach number. In this case an acoustic analogy based on the Phillips convected wave equation can be used with the turbulent source information generated by the LES code. We see no fundamental problems in obtaining acoustic results in these cases.

At higher Mach numbers, compressibility is important; a compressible LES code is currently under development. In this case it should be possible to apply Lighthill's integral solution method directly. Thus, while the solution of a compressible flow problem is expected to be much more costly than an incompressible one, the acoustic solution may actually be easier to implement, primarily due to reduced memory requirements.

The same direct Lighthill approach can also be applied to the very low Mach number case using the incompressible LES code, thus avoiding the need to consider differential acoustic wave equation solutions over extremely long distances.

It is then the intermediate Mach number range that produces the most difficulty. Here wave equation solutions would be needed over long distances if incompressible flow were used, and compressible flow solutions would need to be extremely accurate to account for propagation effects in a direct Lighthill approach.

References

Berman, C. H., 1979. "The generation and propagation of sound in turbulent jets," AIAA Paper No. 79-0573.

Berman, C. H., Ramos, J. I., Karniadakis, G. E., and Orszag, S. A., 1990. "Time-dependent jet flow and noise computations," AIAA Paper No. 90-3961.

Berman, C. H., Gordon, G., Karniadakis, G. E., and Orszag, S. A., 1991. "Three-dimensional jet noise computations," AIAA Paper No. 91-2256.

Curle, N., 1955. "The influence of solid boundaries upon aerodynamic sound," *Proc. Roy. Soc.* **231A**, p. 505.

Lighthill, M. J., 1952. "On sound generated aerodynamically. I. General theory," *Proc. Roy. Soc.* **211A**, p. 564.

Michalke, A., 1977. "On the effect of spatial source coherence on the radiation of noise," *J. Sound Vibration* **55**, p. 377.

Phillips, O. M., 1960. "On the generation of sound by supersonic turbulent shear layers," *J. Fluid Mech.* **9**, p. 1.

Yakhot, V. and Orszag, S. A., 1986. "Renormalization group analysis of turbulence. I. Basic theory," *J. Sci. Comput.* **1**, p. 3.

Yakhot, A., Orszag, S. A., Yakhot, V., and Israeli, M., 1989. "Renormalization group formulation of large-eddy simulations," *J. Sci. Comput.* **4**, p. 139.

$$\partial^2 T_{11}/\partial x_1^2 = \frac{T_{11}(n+1) - 2T_{11}(n) + T_{11}(n-1)}{(\Delta x_1)^2}$$

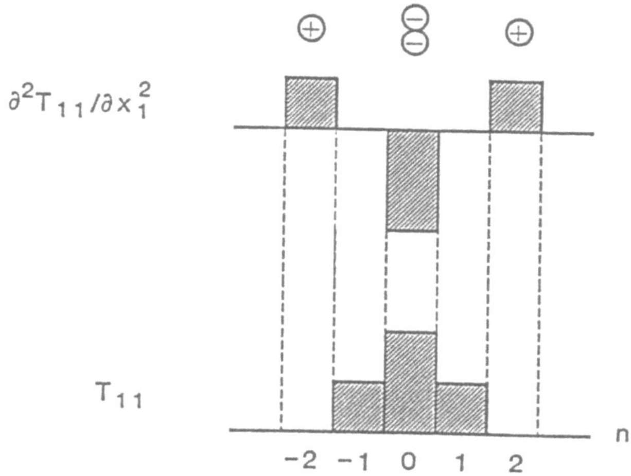

FIGURE 1. DOUBLE DIVERGENCE FORM APPLIED TO LIGHTHILL LONGITUDINAL QUADRUPOLE

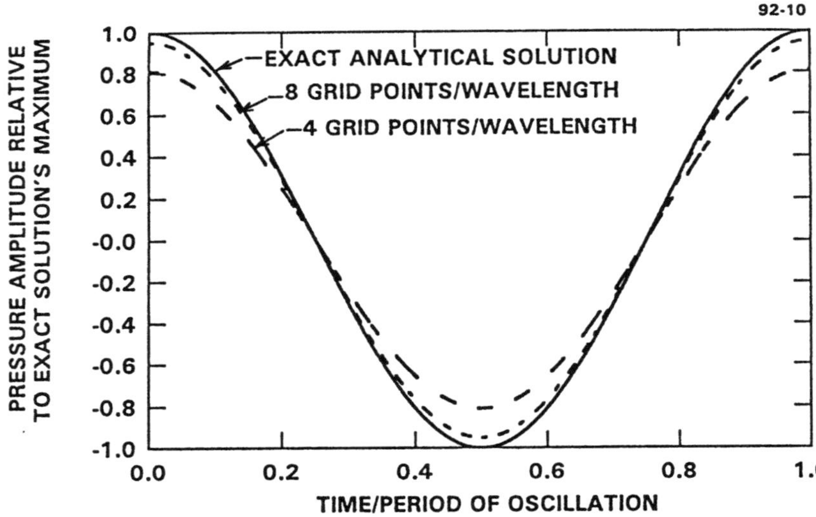

FIGURE 2. EFFECT OF GRID SPACING ON LIGHTHILL SPACE DERIVATIVE SOURCE MODEL

One-dimensional Gaussian space dependence
Wavelength = Gaussian half-width

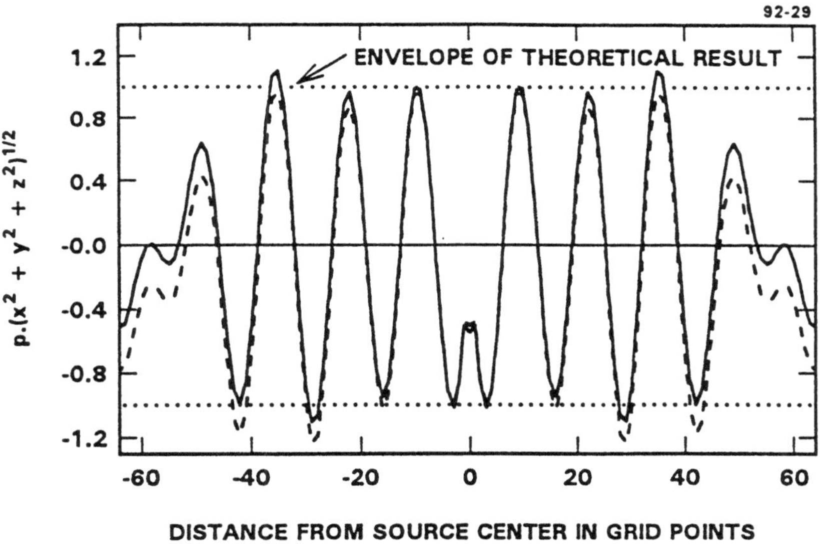

a. Pressure along centerline

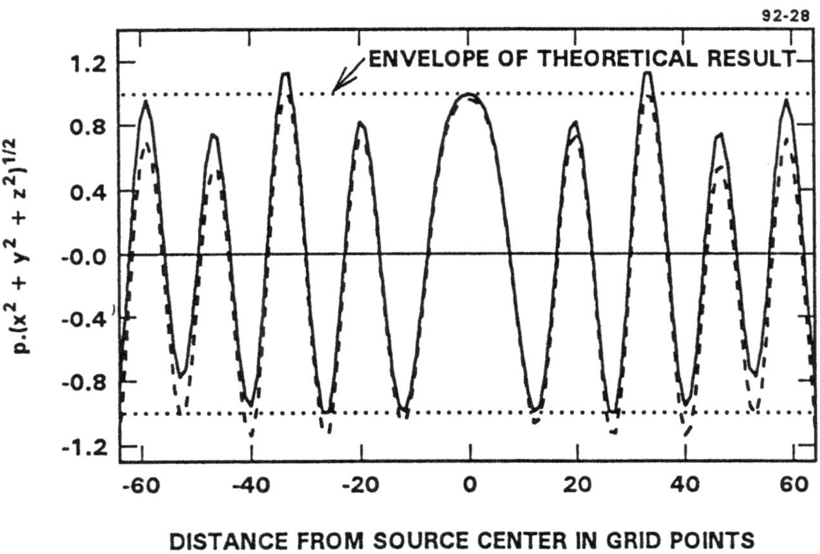

b. Pressure along line midway between centerline and domain boundary

FIGURE 3 PRESSURE DUE TO SPHERICAL VOLUME QUADRUPOLE SOURCE

Wavelength = 12.8 grid points

——— Double space derivative form of source

- - - - - - Double time derivative form of source

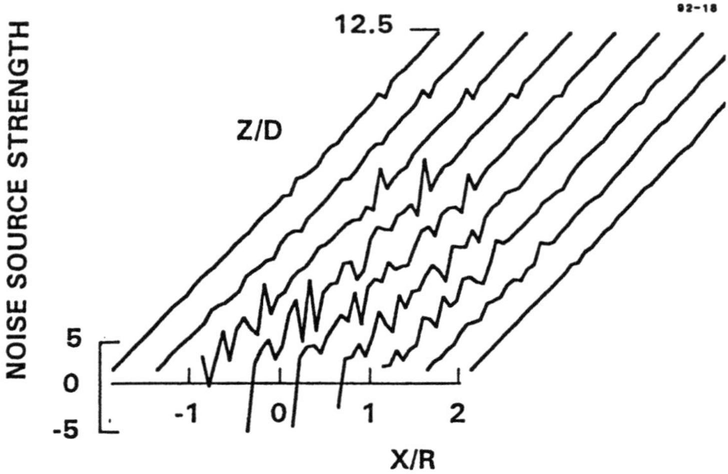

FIGURE 4. QUADRUPOLE NOISE SOURCE STRENGTH ON PLANE THROUGH JET CENTERLINE

Nondimensionalized by jet velocity and radius

A CONSIDERATION OF ENERGY FROM THE VIEWPOINT OF COMPUTATIONAL AEROACOUSTICS

Kenneth S. Brentner

NASA Langley Research Center
Hampton, Virginia 23681-0001

ABSTRACT

The propagation of acoustic energy from a sound source to the far field is a fundamental problem of acoustics. In this paper, a finite volume, multistage time stepping, Euler code is used to investigate the use of CFD algorithms for the direct calculation of the acoustic field. The two dimensional, compressible, inviscid flow about an accelerating circular cylinder is used as a model problem. The time evolution of the energy transfer from the cylinder surface to the fluid, as the cylinder is moved from rest to some nonnegligible velocity, is shown. Energy is the quantity of interest in the calculations since various components of energy have physical meaning. By examining the temporal and spatial characteristics of the numerical solution, a distinction can be made between the propagating acoustic energy, the convecting energy associated with the entropy change in the fluid, and the energy following the body. In the calculations, entropy generation is due to a combination of physical mechanisms and numerical error. In the case of propagating acoustic waves, entropy generation seems to be a measure of numerical damping associated with the discrete flow solver.

1. Introduction

The process of rapidly accelerating a body can be a potent noise source. It is natural then to consider the transient behavior of the compressible flow field in this situation. Several model problems have been studied analytically for various limiting conditions. Ffowcs Williams and Lovely (1977) considered the case of a sphere that is suddenly brought into a low Mach number translation in an inviscid, compressible fluid. G. I. Taylor (1942) also considered a sphere which, after an impulsive start, decelerated due to the drag induced by the transient pressure field. When the sphere is impulsively started to low Mach number, both of these studies found an equipartitioning of the energy between the kinetic energy following the body and the radiated acoustic energy. Longhorn (1952) found that the work required to start a sphere impulsively was twice the amount needed if the sphere was started slowly. The additional work required to rapidly accelerate the sphere supplies the energy which radiates as sound for this low Mach number problem.

Although low Mach number problems have been treated, transonic flows are difficult to handle analytically with any generality. In this paper, the Euler equations, together with the continuity and energy equations, are solved numerically for the two dimensional model problem of a circular cylinder accelerating from rest. The Euler equations were chosen so that the artificially viscous nature of the discrete numerical algorithm may be considered independently from real viscous effects. The Euler equations are also useful because they allow strong shock waves and the transport of vorticity.

2. The Accelerating Cylinder

Inspired by the accelerating sphere problem, an investigation of the transient flow field around a circular cylinder which has impulsively started is now given. Leftward translation at Mach 0.4, the critical Mach number for the cylinder, was chosen so that the flow around the cylinder would be transonic without having shock waves in the ultimate steady flow. The flow field is assumed to be symmetric: therefore the computational grid used for these calculations is a polar grid which models half of the flow field and moves with the cylinder. The azimuthal direction is divided into 95 cells ($\Delta\theta = 0.0331$ radians). The radial dimension of the grid cells increases linearly up to ten cylinder radii from the center of the cylinder, after which the radial increment remains constant out to forty cylinder radii (the outer boundary of the grid). Figure 1 shows the inner part of the computational domain. Notice the very fine grid resolution, especially near the cylinder surface. The numerical method is based on Jameson's finite volume, multistage time stepping scheme (Jameson et al., 1981; Schmidt and Jameson, 1982) and is nominally second order accurate in space and time. More details of the Euler code used for these calculations and a wider range of cases are given by Brentner (1990).

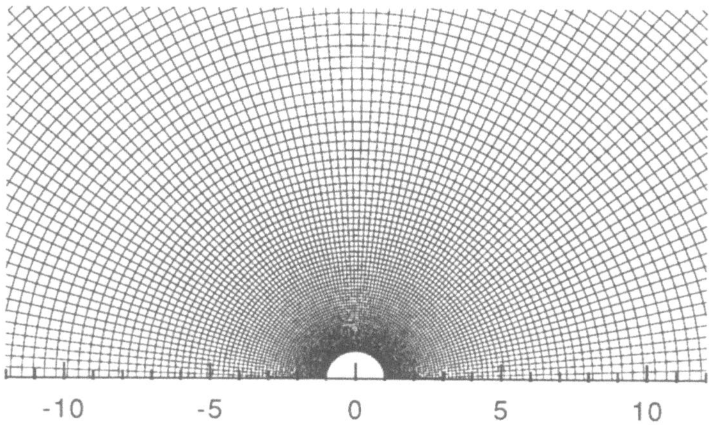

Figure 1: The inner part of the computational grid used for the numerical calculations.

2.1. Description of the starting process

As the cylinder impulsively starts moving leftward, an expansion wave propagates to the right, behind the cylinder, while a compression wave moves to the left, ahead of the cylinder. This is shown in the time sequence of density perturbation contours plotted in Figure 2. The density perturbation, ρ', is scaled by $r_o^{1/2}$ to account for the effect of cylindrical spreading. Here r_o is the distance from the point in the field to the position where the cylinder center was located a time $t = 0$. Notice that the expansion moves away from the cylinder more quickly than the compression. This is due to the relative motion of the cylinder. The steady flow field near the cylinder surface is established by the nondimensional time t=20. Note that in all the contour plots, the range of contours is limited: therefore the extreme values in the contour plot may be greater than the maximum or less than the minimum shown in the legend.

Immediately after the impulsive start, the pressure on the surface corresponds to $p = p_o + \rho_o c_o v_n$, which is expected for an impulsive motion. (Here the subscript o refers to the undisturbed fluid state and v_n is the normal component of the local surface velocity.) Subsequently a shock forms on the downstream side of the cylinder, strengthens, and moves forward. As the shock reaches the top of the cylinder, it begins to weaken as it continues to move forward until it

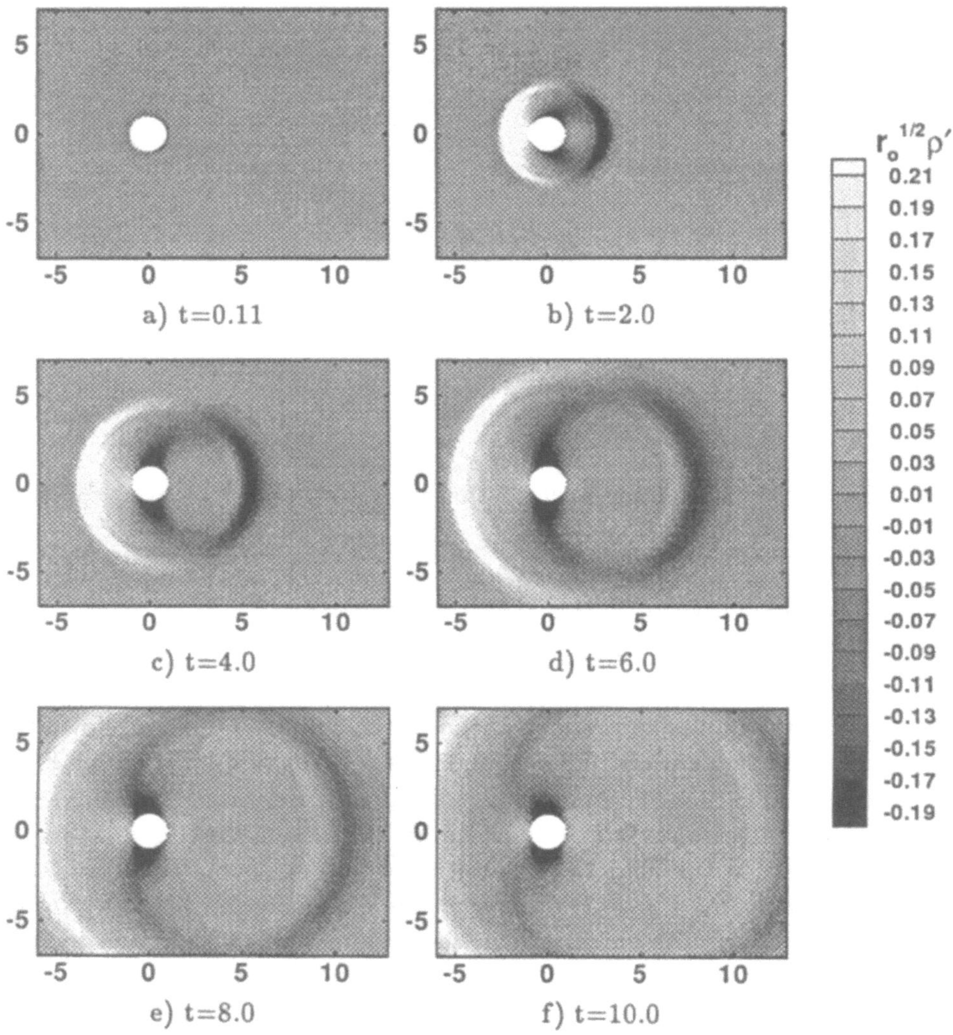

Figure 2: The time sequence of density perturbation ρ' contours for the cylinder impulsively started to Mach 0.4. The density perturbation has been scaled by $r_o^{1/2}$ to account for cylindrical spreading.

finally leaves the cylinder surface. Shortly after the shock leaves the cylinder region, the flow field around the cylinder approaches that expected for a steady potential flow. The steady state solution for a potential flow is unique. Hence it does not depend upon the time history, but the transient flow field and the work required to maintain the cylinder velocity during the transient phase vary greatly with the manner in which the steady state flow is achieved. This will be discussed in more detail later.

2.2. Energy partition

With a qualitative understanding of the physics, it is useful to discuss the problem more quantitatively. Longhorn has done this using the total work required to accelerate a sphere to low Mach number translation as a quantitative measure. The work required to accelerate the cylinder gives a measure of energy input into the system. Once in the fluid, the energy can be separated into its various components. Energy flux out of a control volume enclosing the cylinder can be used to determine the propagation and convection of energy to the far field. The form in which energy leaves the control volume can be determined by the time required to exit the volume, since the acoustic propagation and fluid convection speeds are disparate.

Myers (1991) has recently developed an exact energy corollary which is well suited for consideration of energy transport in flow fields which may include shock waves and vorticity. Myers' result is a generalization of the concept of acoustic energy, which he recognized from a perturbation expansion of the general energy equation of fluid mechanics. Since this corollary is exact, it is uniquely appropriate for the nonlinear problems under current consideration. For the case of perturbation about an undisturbed inviscid fluid, Myers' corollary can be written as

$$\frac{\partial \varepsilon}{\partial t} + \nabla \cdot \vec{\mathcal{F}} = 0. \tag{1}$$

where

$$\varepsilon = \rho(h - h_o + \frac{u^2}{2} - T_o(s - s_o)) - (p - p_o) \tag{2}$$

and

$$\vec{\mathcal{F}} = \rho(h - h_o + \frac{u^2}{2} - T_o(s - s_o))\vec{q}. \tag{3}$$

In these equations, h is the specific enthalpy, $e + p/\rho$, and s is the specific entropy. The subscript o refers to the value of the quantity in the undisturbed medium. This is a substantially simplified version of Myers' result since the effect of mean flow, viscosity, and heat conduction have been neglected.

The energy density given in Equation 2 contains three components,

$$\varepsilon_k = \frac{\rho u^2}{2} \quad (4)$$
$$\varepsilon_p = \rho(\Delta h - T_o\Delta s) - \Delta p \quad (5)$$
$$\varepsilon_s = \rho T_o \Delta s \quad (6)$$

which are the kinetic, potential, and entropy energy densities, respectively. The kinetic energy has its normal physical meaning and is the only component of energy in the inviscid, incompressible limit. The potential energy is related to the compression of the fluid, and the entropy energy is energy corresponding to the increase of entropy in the fluid. The statement of energy conservation, written in terms of these components, is

$$\frac{\partial}{\partial t}[\varepsilon_k + \varepsilon_p + \varepsilon_s] + \nabla \cdot [(\varepsilon_k + \varepsilon_p + \varepsilon_s - p_o)(\vec{q} - \vec{v}) + p\vec{q}] = 0 \quad (7)$$

for a moving control volume. Upon integration over a control volume and using the divergence theorem, the global conservation of energy statement can be written in component form as

$$\frac{\partial}{\partial t}(E_k + E_p + E_s) + \dot{F}_a + \dot{F}_s = \dot{W}. \quad (8)$$

Here E_k, E_p, and E_s refer to the total amount of kinetic, potential, and entropy energy in the volume at a particular time. F_a and F_s are the time integrated flux of acoustic and entropy energy out of the volume (i.e. through the outer boundary of the volume) since time $t = -\infty$, and W is the total work done by the body on the fluid since $t = -\infty$. The dot over F_a, F_s and W in Equation 8 represents time differentiation and is used only on boundary integrals. These global measures of energy are also useful for quantitatively examining the transfer of energy from the cylinder surface to the fluid and ultimately the far field. A more precise definition of these quantities is given by Brentner (1990).

2.3. Energy distribution

The components of energy discussed in the previous section are now used to develop an understanding of the energy transport from the cylinder surface to the near and far fields after a Mach .4 impulsive start. In figures 3, 4, and 5, the kinetic, potential, and entropy energy density contours are plotted in a time sequence. Note that the energy density is scaled by r_o to account for cylindrical spreading of the energy as it propagates away from the cylinder surface. Figure 3 shows that the kinetic energy is concentrated both near the cylinder surface and in the acoustic wavefronts. The kinetic energy following the cylinder, which is part of the local aerodynamic field, distributes itself uniformly around the cylinder. The kinetic energy component of the acoustic energy radiates mainly to the left and right of the cylinder, with the highest intensity forward of the cylinder in the direction of the motion. The shock can be seen in the local kinetic energy distribution as a discontinuity in the contours near the cylinder, from $t = 4.0$ to $t = 8.0$, in figures 3c-e.

As in the case of the kinetic energy, the potential energy is found primarily in the wave fronts and near the cylinder, but unlike kinetic energy, the potential energy following the cylinder is found predominantly in the region of maximum fluid velocity. The shock is much more apparent in the potential energy density contour sequence shown in Figure 4. This is to be expected since the potential energy in the fluid identifies the presence of compressibility. For a low Mach number flow, the potential energy contribution to the local field following the body, i.e. the aerodynamic field, would be negligible compared with the kinetic energy. Hence the potential energy would be primarily of interest in the acoustic waves. A comparison of figures 3 and 4 reveals that the energy in the acoustic waves, both the compression and expansion, is nearly equally divided between kinetic and potential energy components.

The contours of entropy energy density, shown in Figure 5, reveal the unexpected relationship between the acoustic waves and entropy in the numerical solution. This is unexpected since sound propagation is an inviscid, *isentropic* process. Upon closer examination of Figure 5, it is evident that the area of the highest level of entropy energy remains near the surface of the cylinder for the short time sequence shown in the figure. This concentrated region of entropy highlights a vortex generated by the strong transient shock which

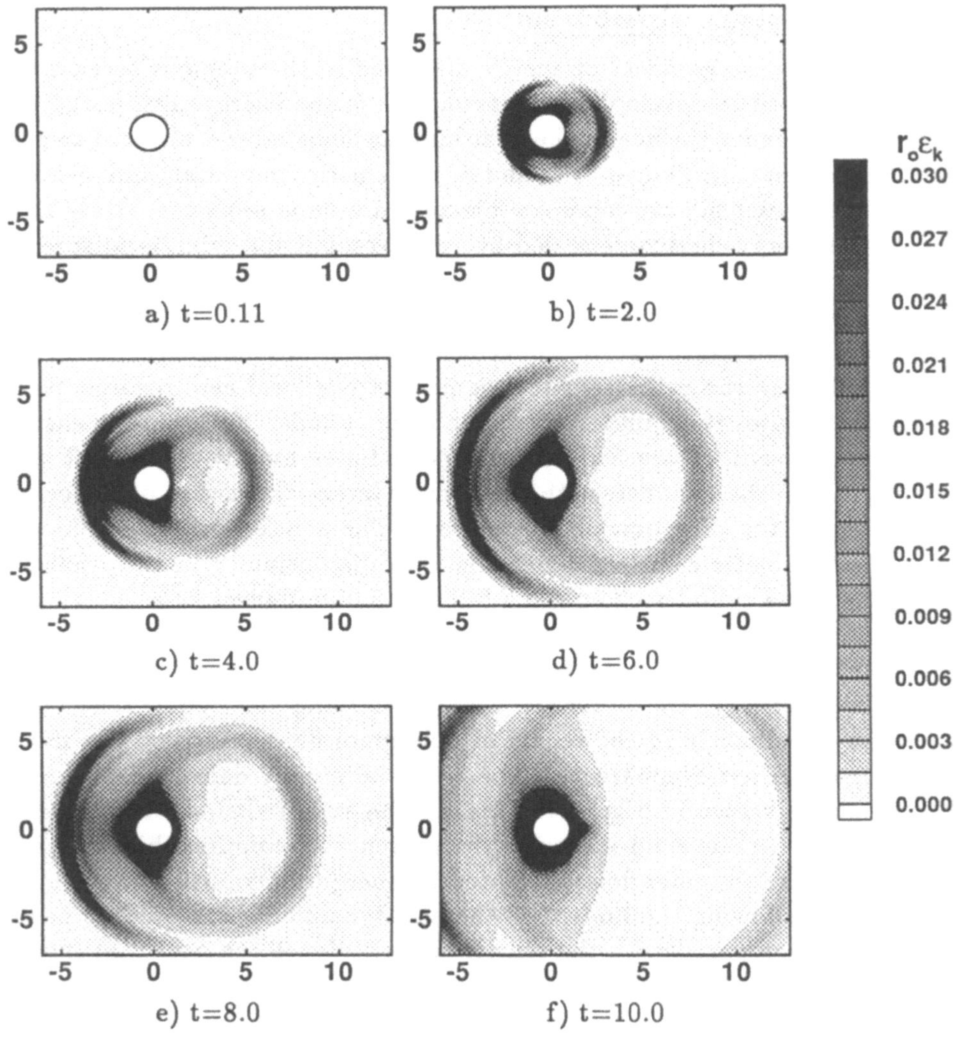

Figure 3: The time history of the kinetic energy field ε_k is shown in these figures for a cylinder which has been impulsively started to Mach 0.4 translation. Energy has been scaled by r_o to account for cylindrical spreading.

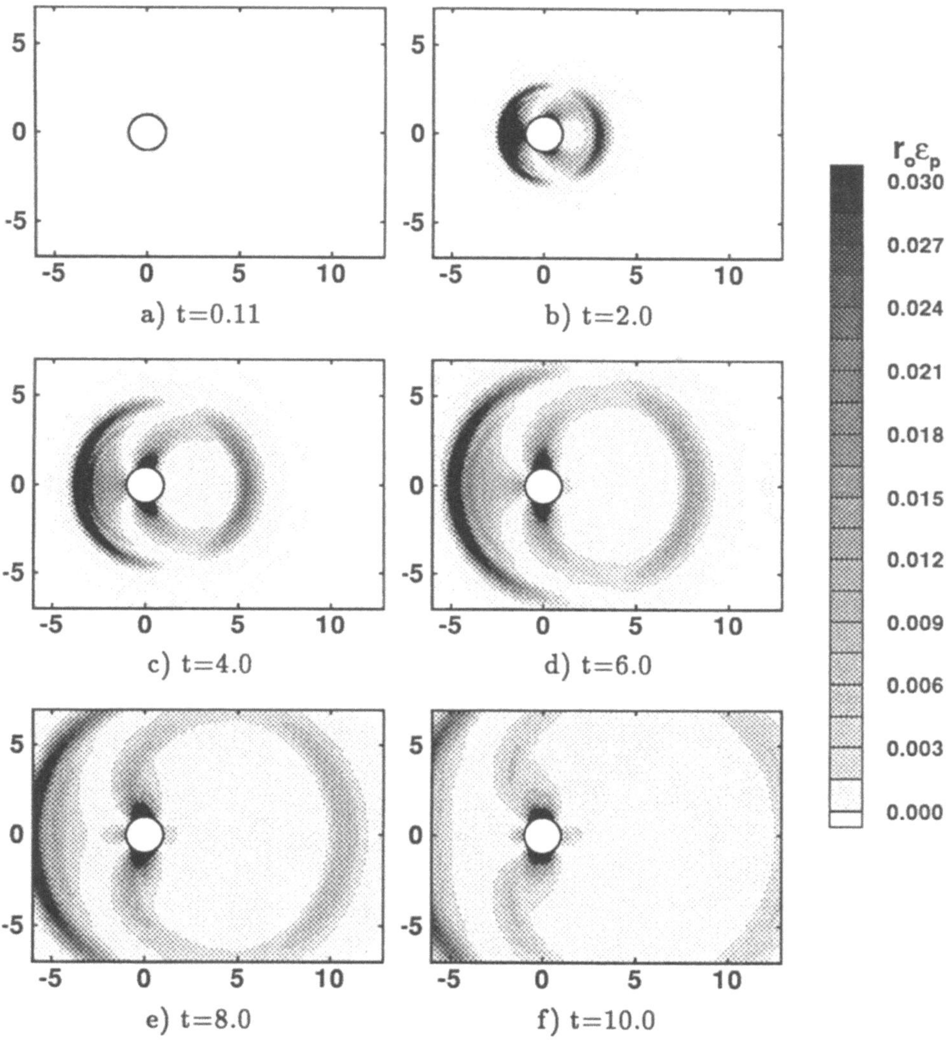

Figure 4: The time history of the potential energy field ε_p is shown in these figures for a cylinder which has been impulsively started to Mach 0.4 translation. Energy has been scaled by r_o to account for cylindrical spreading.

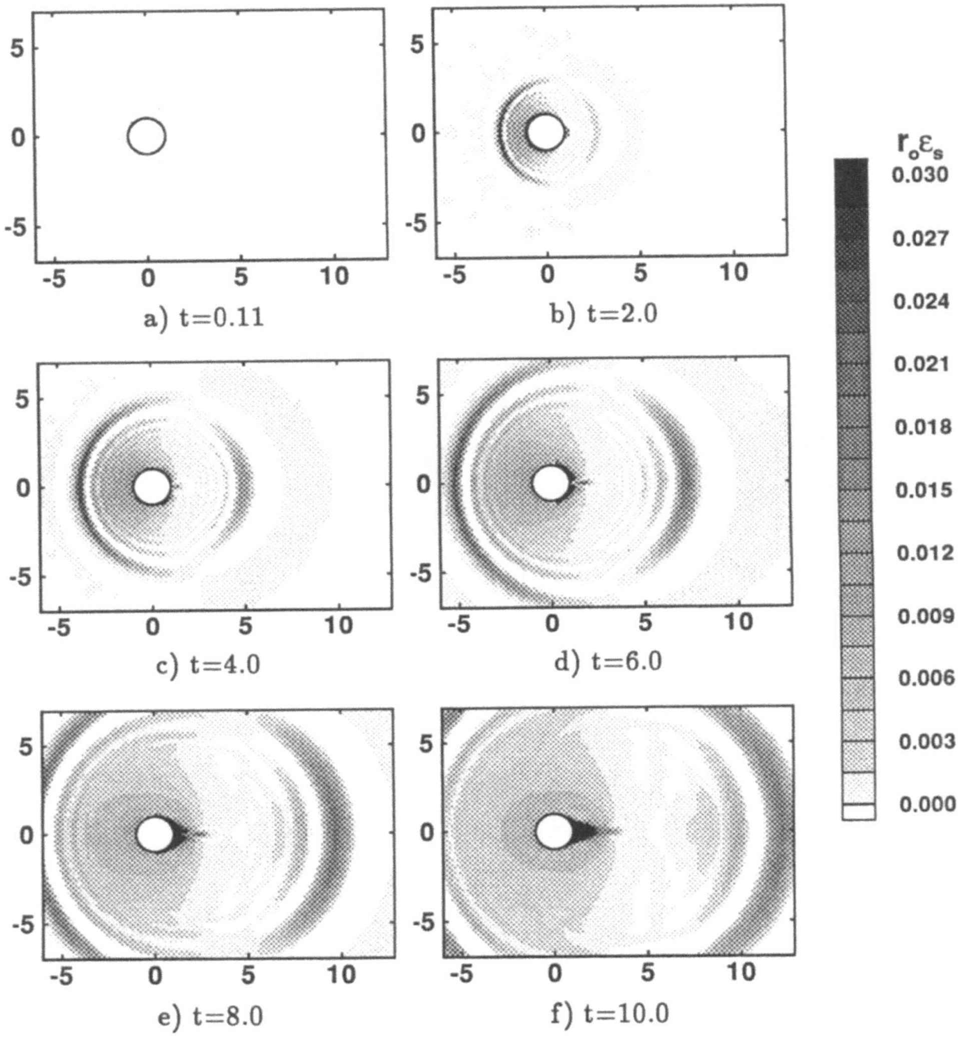

Figure 5: The time history of the entropy energy field ε_s is shown in these figures for a cylinder which has been impulsively started to Mach 0.4 translation. Energy has been scaled by r_o to account for cylindrical spreading.

exists shortly after the start of motion. Even so, the most striking feature is the undeniable entropy generation by the propagating acoustic waves - a phenomenon which must have numerical rather than physical origins. This result suggests that the numerically induced errors or damping are identified by the entropy energy term. The significance of this finding is that entropy energy gives a quantitative local measure of the numerical error which can be used to judge the acceptability of the acoustic solution. This idea is central to this paper and shall be considered in more detail in what follows.

2.4. Global energy balance

With a better understanding of how the components of energy are spatially and temporarily distributed, it is useful to consider the global energy balance as a function of time for a control volume surrounding the cylinder. In the following calculations, a circular control surface, ten cylinder radii in diameter and concentric with the cylinder, defines a volume around the cylinder. The total kinetic, potential, and entropy energy in the volume along with the acoustic and entropy energy flux out of the volume, indicated in Equation 8, are computed as a function of time for various starting scenarios.

In Figure 6, the energy components are added such that the envelope of the curve represents the total work input into the fluid by the cylinder and each of the components are nondimensionalized by the total kinetic energy of the incompressible case, $\rho\pi a^2 v^2/2$. Notice that kinetic energy, potential energy and acoustic energy flux are the first three components. They are plotted in this order because acoustic energy is counted as kinetic and potential energy while inside the control volume and acoustic energy flux as it leaves the control volume. Likewise entropy energy and entropy energy flux are plotted together since they correspond to the same component of energy either inside or outside of the specified volume. When the cylinder is impulsively started to Mach 0.4, most of the work goes initially into kinetic and potential energy modes and eventually a significant entropy energy component before $t = 10.0$ at which time acoustic energy begins to leave the control volume. This is shown in Figure 6a. By time $t = 40.0$, all of the acoustic energy has left and the entropy energy begins to leave. The entropy energy convects with the fluid velocity rather than the sound speed, thus accounting for the delay in leaving the control volume. Acoustic waves have completely left the control volume when the acoustic flux contribution reaches its final constant value. Notice in Figure 6 that there is no increase in entropy after the acoustic waves have left the control volume. This is another indication that acoustic energy is being dissipated in the discrete numerical calculation.

Figure 6a also shows that total energy input into the fluid after an impulsive start is equally divided between the kinetic and potential energy which follow the body and the acoustic and entropy energy which are transported to the far field. This finding is consistent with the low Mach number theory for the sphere (Taylor, 1971; Longhorn, 1952; Ffowcs Williams and Lovely, 1977) and is an extension of previous results since the numerical computations have no restriction to low Mach number. In fact the transient flow field is transonic, vorticity is generated, and there is significant potential energy even in the ultimate steady aerodynamic field. Based on Figure 6a, the equipartition of energy is no longer simply a balance between the kinetic energy in the local field and the acoustic energy propagating to the far field as in the low Mach number case. After the impulsive start, half of the input energy remains near the cylinder and the other half goes to the far field. The local energy field is comprised of both kinetic and potential energy while both acoustic propagation and entropy convection account for the energy transport to the far field in the more general case of compressible flow.

In the remainder of Figure 6, parts b-f show the energy transfer for a cylinder accelerated at various nonimpulsive, constant nondimensional acceleration rates. An examination of Figure 6a-f reveals that the total work input at steady state (the sum of all the energy components at large time) declines significantly with reduced acceleration even though the steady state flow field is the Mach .4 potential flow solution for each case. As the rate of acceleration is decreased, the production of entropy and acoustic energies decline until for the case of very low acceleration in Figure 6f, the energy input into the fluid nearly corresponds to the energy in the local field. Thus a very small amount of acoustic energy leaves the control volume. Also note in the figure that the entropy generated is greatly reduced when the acceleration rate is decreased.

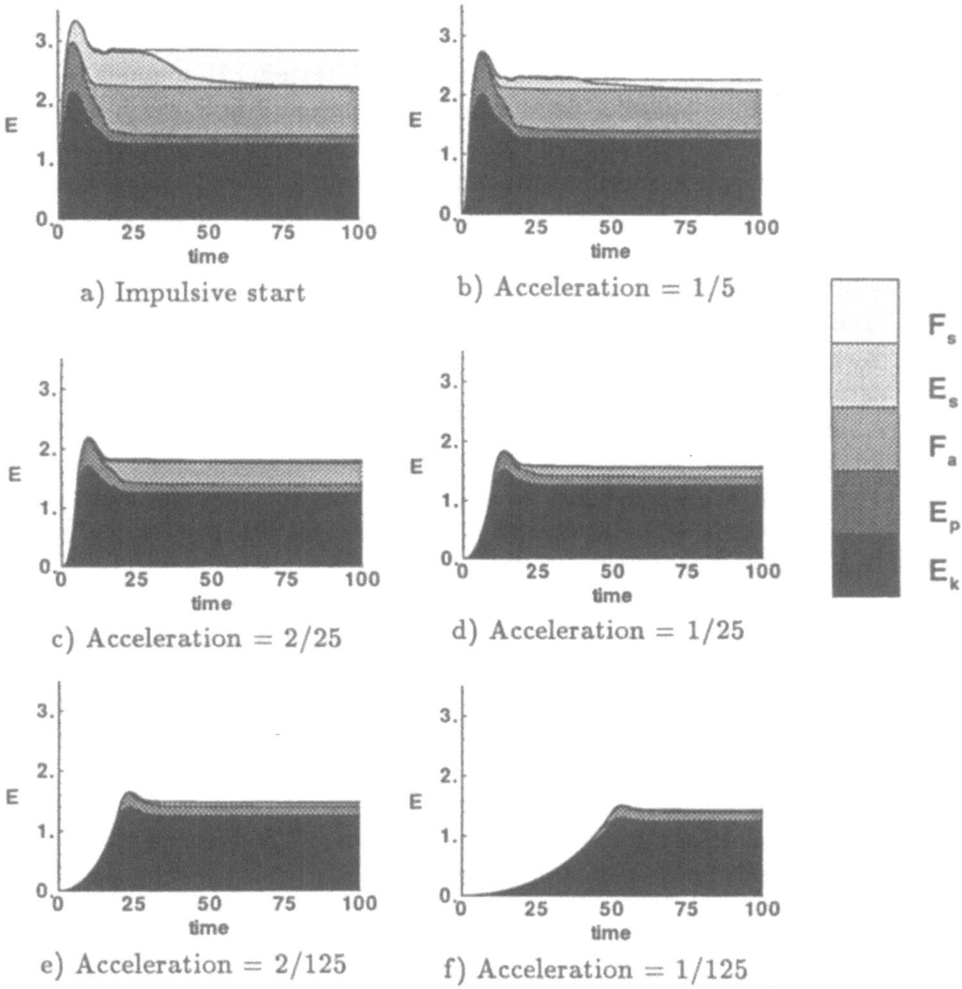

Figure 6: Global energy balance time history for a control volume with radius $R = 10$. The total kinetic energy E_k, potential energy E_p, acoustic energy flux F_a, entropy energy E_s, and entropy energy flux F_s are nondimensionalized by the total incompressible kinetic energy, $\rho\pi R^2 v^2/2$, where v is the steady state velocity Mach 0.4.

3. Sources of Entropy

One nagging question must still be addressed – How much of the entropy energy results from numerical error and how much is real? For the impulsively started cylinder, both physical and numerical entropy generation mechanisms must be present. The physical source of entropy in this inviscid calculation is the strong shock generated during the transient startup period. The exact mechanism responsible for the numerical dissipation of the acoustic waves is not fully understood. This is clearly a serious problem since the desired numerical calculation should mimic the isentropic nature of the physical situation. There is some hope that this will not be an insurmountable problem, however, since for lower acceleration rates there is very little energy in the entropy component even though there is still a significant acoustic energy component. (See Figure 6c for example.) Hence even if acoustic energy is dissipated, it may be at negligible levels. In the remainder of this paper, an investigation will be made to determine the source of the nonphysical origins of the entropy. Specifically, the effect of the explicitly added artificial viscosity and the role of grid resolution will be studied separately.

3.1. Effect of artificial viscosity

The artificial viscosity added for numerical stability seems to be an obvious generator of unwanted entropy. The form of the artificial dissipation used in the computations is that given by Jameson (Jameson et al., 1981; Schmidt and Jameson, 1982). It is composed of a combination of second and fourth order terms, with the fourth order terms providing background dissipation and the second order terms adaptively turned on in regions of sufficiently high gradients. The purpose of the second order dissipation is to eliminate oscillations around shock waves, but the present form of the scheme does not distinguish between shock waves and acoustic waves. To examine the amount of entropy generation by the artificial viscosity terms, the magnitude of the dissipation terms were increased independently. In Figure 7a, the effect of changing the damping coefficients on the entropy energy is shown. Even when the second order term is tripled or the fourth order term is increased by an order of magnitude, the effect on entropy generation is minimal.

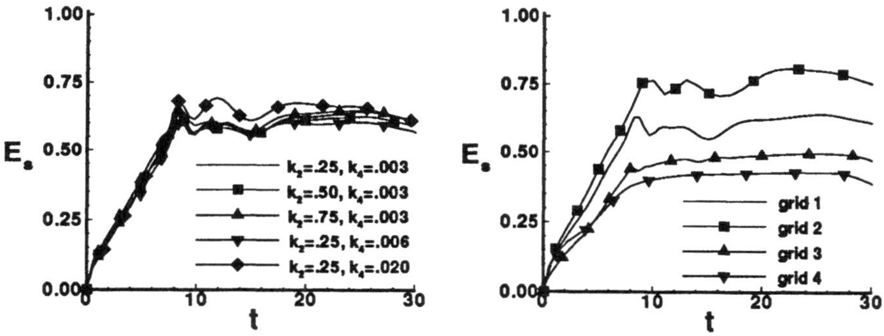

a) Effect of artificial viscosity b) Effect of grid resolution

Figure 7: A comparison of the entropy energy in the control volume for various levels of the artificial viscosity and different grids. The nominal values of the second order and fourth order viscosity coefficients are $k_2 = .25$ and $k_4 = .003$, respectively. Grid 1 is the nominal grid.

3.2. Effect of grid resolution

The grid resolution is certainly related to the numerical dissipation observed in the calculations. In fact, the grid probably does not sufficiently resolve the high frequency component of the acoustic waves generated during the impulsive start. The high level of entropy energy observed in the impulsive case may be an indication of this. To study this effect, three additional computational grids are considered. The grid used in the previous calculations and shown in Figure 1 is designated here as Grid 1. Grid 2 has half the resolution of Grid 1 in both r and θ directions while Grids 3 and 4 both have significantly finer resolution in the r direction. The constant radial spacing of Grid 4 starts at a much smaller r than the other grids, giving it the finest radial resolution. Grid 3 has twice the azimuthal resolution of Grid 1 and Grid 4. The CFL number was held constant for all grids by changing the time step to account for changes in the grid size.

The grid resolution seems directly related to the entropy level, as shown in Figure 7b. The coarser grid has a high level of entropy corresponding to a loss in acoustic energy while the entropy of the finer grids is significantly reduced. Upon closer examination of the solution, the amount of entropy computed using the fine grids seems to be near that which was physically generated by the shock immediately

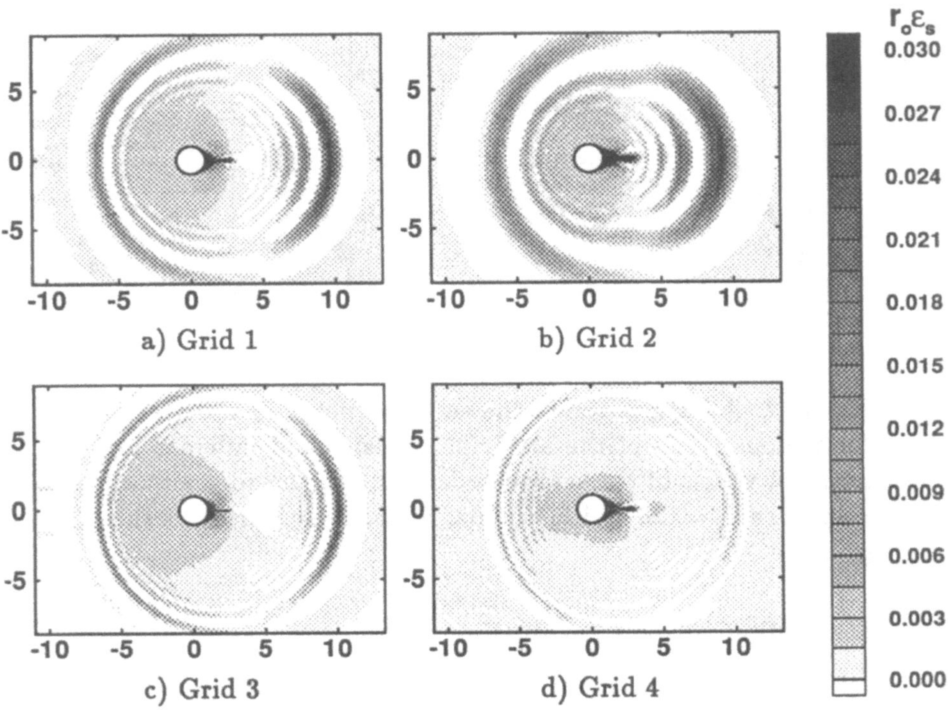

Figure 8: A comparison of the entropy energy ε_s distribution for various grids at $t = 8.0$. The cylinder was impulsively started in Mach 0.4 translational motion.

after startup. Figure 8 compares the entropy energy distribution for the impulsively started cylinder at time $t = 8.0$. Notice that both Grids 3 and 4 have significantly less entropy energy associated with the acoustic waves than the original grid. These calculations agree with the prevailing view that very high accuracy is required for the prediction of acoustics. Thus grid resolution and the order of accuracy of the numerical algorithm are of primary importance in direct computation of acoustics with CFD methods.

Even though grid refinement improved the calculations, a polar grid is not be the best choice since the azimuthal arc length of the cells grows linearly with r. Other types of grids which maintain the grid cell size uniformly throughout the computational domain are more suitable for acoustics since they minimize anisotropy effects. Perhaps adaptive, unstructured grids can be used to actually increase the grid density only where waves exit in the field. Higher order accuracy algorithms are also needed to reduce the demand for extremely fine grids. Finally, the local energy field seems to be much less sensitive to the grid resolution away from the body – an observation which explains why stretched grids don't seem to cause problems in aerodynamic calculations.

4. Conclusions

The aim of this paper is twofold: 1) to understand the nonlinear, transient energy transfer from the surface of an accelerating cylinder to the far field; and 2) to better understand the capabilities and limitations of present day CFD methodology as applied to acoustic problems. Both of these aims have been addressed through the numerical study of the circular cylinder model problem.

The transition from rest to a Mach .4 translation is somewhat more complicated than in the low Mach number problems studied previously. When the cylinder is accelerated rapidly, a shock forms in the compressible fluid generating entropy and vorticity in the early stages of the motion. As the shock disappears, the vorticity convects away from the cylinder and the steady flow around the cylinder becomes essentially potential. It is somewhat surprising then that an equipartition of energy exists for this compressible problem in the same way as a low Mach number case. Nevertheless, the present calculations show that half of the energy input during an impulsive start follows the cylinder and the other half is transported to the far field.

Finally, the separation of energy into kinetic, potential, and entropy energy components is useful in understanding both the physics of the problem and the effect of the numerical damping. The entropy term is especially useful because it gives a quantitative measure for comparing the effect of algorithms and grids on the unsteady solution. This use of energy is apparently new. For the second order accurate, finite volume, Euler solutions computed for this paper, the discretization errors related to the grid resolution appear to be the primary source of numerically generated entropy. If the grid is too coarse, acoustic energy is transferred to entropy energy as it propagates. The role of the explicitly added artificial viscosity has a small effect on the time dependent energy solution.

Acknowledgements

This research was carried out under the supervision of Professor J. E. Ffowcs Williams as part of the author's doctoral work.

References

Brentner, K. S., 1990. "The sound of moving bodies," Ph.D. dissertation, University of Cambridge.

Ffowcs Williams, J. E., and Lovely, D. J., 1977. "An approximate method for evaluating the sound of impulsively accelerated bodies," *Journal of Sound and Vibration*, 50(3), pp. 333-343.

Jameson, A., Schmidt, W., and Turkel, E., 1981. "Numerical solution of the euler equations by finite volume methods using runge-kutta time-stepping schemes," AIAA paper 81-1259.

Longhorn, A. L., 1952. "The unsteady, subsonic motion of a sphere in a compressible inviscid fluid," *Quarterly Journal of Mechanics and Applied Mathematics*, V(1), pp. 64-81.

Myers, M. K., 1991. "Transport of energy by disturbances in arbitrary steady flows," *Journal of Fluid Mechanics*, **226**, pp. 383-400.

Schmidt, W. and Jameson, A., 1982. "Recent developments in finite-volume time-dependent techniques for two and three dimensional transonic flows," Von Karman Lecture Series 1982-04.

Taylor, G. I., 1942. "The motion of a body in water when subjected to a sudden impulse," in *Scientific Papers of G. I. Taylor*, III, Aerodynamics and the Mechanics of Projectiles and Explosions, (G. K. Batchelor, editor), Cambridge University Press, 1971, pp. 306-308.

DIRECT COMPUTATION OF AERODYNAMIC NOISE

Sanjiva K. Lele, Parviz Moin, Tim Colonius, and Brian Mitchell

Department of Mechanical Engineering
Stanford University
Stanford, California 94305

In his pioneering work Lighthill formulated the problem of predicting the noise radiated by an unsteady flow in terms of an acoustic analogy. However, in order to predict the far-field noise the acoustic source terms of such acoustic analogies must be known accurately. Such detailed information is usually not available and empirical models of the acoustic sources are utilized to arrive at an empirical prediction of the far-field noise. Recent advances in direct and large eddy simulations of turbulent flows provide an alternative to this empiricism. The far-field noise can be obtained from first principles by using the 'exact' source terms from DNS/LES in an acoustic analogy. Developing the methods to make this possible and studying the nature of particular flow processes responsible for the most intense noise are the principal objectives of the research underway at Stanford.

The mathematical derivation of acoustic analogies is based on a rearrangement of the conservation equations describing the mass, momentum and energy in the flow field. Unsteady flows which possess a compact vorticity field generate compact acoustic sources when the flow Mach number is small. Under these conditions the acoustic analogies can be derived by matched asymptotic expansions (Kambe, 1986) and their error can also be assessed by computing the higher order terms of the expansion. For flows of practical interest such as those involving jets, mixing layers, wakes and boundary layers the flow field in not compact and the rearrangement of the equations leading to the differentiation of acoustic sources from wave propagation and refraction/scattering effects is ambiguous. The numerical task of 'inverting' a suitable wave propagation operator to obtain the far-field sound is far from trivial even when a rigorously derived source term is available. To address these issues it becomes necessary to have access to the 'exact' far-field sound and the associated unsteady flow fields. It does not seem practical to achieve this by experimental measurements. Our approach relies on direct computations of sound from unsteady flows which can be compared with acoustic analogy based predictions.

In a direct computation of sound radiated by an unsteady flow the far-field sound is obtained directly by solving the unsteady compressible flow equations. Such a calculation must overcome a number of difficulties before it can be regraded as reliable. Typical acoustic fluctuations have energy levels that are 8-10 orders of magnitude smaller than the near-field hydrodynamic fluctuations. For a numerical calculation to accurately represent both the flow and the sound the discretization schemes must maintain the amplitude and phase errors to a minimum over the full range of scales of interest. Typically, the wavelength of acoustic waves is many times the characteristic hydrodynamic length scale in the flow. This ratio being M^{-1} for small M, where M the flow Mach number. To calculate the sound directly the computational domain must extend to the acoustic far-field and maintain a proper resolution of both the sound waves and the unsteady flow features. This makes the use of stretched meshes necessary and care must be taken to avoid spurious numerical reflection errors that may occur if the mesh is stretched too rapidly. It has been suggested (Crighton, 1988) that the very act of numerical discretization may introduce spurious low-order acoustic sources which may, due to their higher acoustic efficiency, overwhelm any flow generated sound. To further add to the numerical challenges the numerical reflections, which may occur at the computational boundaries, must be kept to a minimum.

At the outset it may appear that the numerical challenges to be overcome in a direct computation are far too overwhelming. Our research, which has evolved through a series of idealized unsteady flow problems, has shown that these difficulties can, indeed, be overcome with proper attention paid to the numerical issues. The simplest unsteady flow, which is also technically interesting for its acoustics, is the evolution of a temporally developing compressible mixing layer. Our early success (Lele and Ho, 1988 and Lele and Ho, 1992) in accurately capturing the acoustic radiation from such flows over a Mach number range of 0.05 to 0.6 established that the numerical discretization errors (when using a compact Pade scheme for the spatial derivatives as described in Lele, 1990) were small enough and the spurious monopole and dipole contributions of the discretization errors were too small to be observed. In Figure 1 the acoustic pressure measured in the simulations is compared with an acoustic analogy based prediction. The sources for the acoustic analogy are taken from the numerical simulation. A prediction using the sources from an incompressible flow simulation is also plotted on the figure. The maximum deviation, which is obtained for the highest Mach number case plotted in the figure, is 2.5%. As the convective Mach number $M_c = \frac{U_1 - U_2}{C_1 + C_2}$ is increased significant departures from the incompressible aeroacoustic theory arise. To display these departures the peak radiated acoustic power normalized by M_c^6 is plotted against M_c in Figure 2. If the incompressible theory were to apply no change with M_c is expected. Most of the departure from the incompressible theory can, however, be accounted for by using the sources from a compressible simulation. In Figure 3 a comparison at $M_c = 0.4$ is made between the 'measured' and predicted sound. The deviations are about 10% while they would be larger than 100% if the incompressible source terms were used.

In the next set of test problems computations of acoustic wave scattering by compressible vortices (Colonius, Lele, and Moin, 1991) were performed. The computations were compared with acoustic analogy predictions using the sources from the simulations and were found to agree very well. This established that the boundary conditions were sufficiently accurate for near planar waves. Additional tests on non-planar waves, such as those generated by a quadrupole in shear flow which become nonplanar due to refraction effects, showed the deficiency of the 'nonreflecting' boundary conditions. Stable higher order boundary conditions were developed and were found to be adequate for wave propagation problems.

The problem of acoustic radiation from a corotating compressible vortex pair was undertaken (Mitchell, Lele, and Moin, 1992) to assess how well the numerical simulations performed when a large length scale disparity (a factor of 50 in the case studied) between the acoustic waves and the unsteady vortical flow was present. To validate the computations a two-dimensional version of an acoustic analogy due to Mohring (1977) was developed by following the method given by Kambe (1986). This form of the acoustic analogy is particularly suited for flows with compact vorticity distributions and expresses the far-field sound in terms of the third time derivatives of the second order moments of the vorticity distribution. The comparison between the direct computation and the acoustic analogy shown in Figure 4 established that the numerical difficulties alluded to earlier could indeed be overcome. The study also showed that when the

Mach number in the source region was increased to be around 0.5 the acoustic analogy based on small Mach number expansion of the flow became significantly inaccurate (Figure 5). Retaining higher order terms which account for the acoustic time delay within the source region did not significantly improve the prediction of the acoustic analogy. A similar effect of compressibility on the unsteady flow had also been observed in the study of Lele and Ho on radiation from a temporal mixing layer. Only when the compressibility effect on the source terms was taken into account the acoustic analogy prediction became accurate.

Since the problems of practical interest involve extended flow fields with inflow and outflow it is necessary to use boundary conditions which at the inflow allow an accurate nonreflection of the upstream propagating acoustic waves (in a subsonic flow) while providing control over the inflowing vortical flow. Similarly, the outflow boundary conditions must allow a smooth passage to both vortices and acoustic waves. In addition the boundary conditions must also allow the flow to relax to a laminar steady state in a convectively unstable system. Our work (Colonius, Lele, and Moin, 1992) shows that the higher order boundary conditions which perform well on wave propagation problems are inadequate for direct computations of sound from a spatially evolving flow. The difficulty arises from significant spurious acoustic reflections from vortices propagating through the outflow boundary. Once again the difficulties can be overcome by a suitable modification of the boundary treatment. Figure 6 shows the improvement achieved with the new formulation. In the work under progress these modified boundary treatments are being used in direct computations of sound from spatially evolving mixing layers and jets.

Acknowledgements

The work reported here is conducted with support from the Office of Naval Research. The computations were carried out on the computers facilities of NAS and NASA-Ames.

References

Colonius, T., Lele, S. K., and Moin, P., 1991. "Scattering of Sound Waves by a Compressible Vortex," *AIAA Paper*, AIAA-91-0494, Reno, 1991.

Colonius, T., Lele, S. K., and Moin, P., 1992. "Boundary Conditions for Direct Computations of Aerodynamic Sound Generation," *AIAA-DGLR Aeroacoustics Conference*, AIAA-92-02-075, Aachen, Germany, 1992.

Crighton, D. G., 1988. "Goals for Computational Aeroacoustics," *Computational Acoustics*, Elsevier, IMACS.

Kambe, T., 1986. "Acoustic Emission by Vortex Motions," *J. Fluid Mech.* **173**, pp. 643–666.

Lele, S. K. and Ho, C. M., 1988. "Acoustic Radiation From Temporally Evolving Free Shear Layers," *Bull. Amer. Physical Soc.* **33**(10), p. 2255, and for details see Lele, S. K. and Ho, C. M., 1992, submitted to *J. Fluid Mech.*

Lele, S. K., 1990. "Compact Finite Difference Schemes with Spectral-Like Resolution," *CTR Manuscript 107*, submitted to *J. Comput. Physics.*

Mitchell, B., Lele, S. K. and Moin, P., 1992. "Direct Computation of the Sound from a Compressible Co-Rotating Vortex Pair," *AIAA Paper*, AIAA-92-0374, Reno, 1992.

Mohring, W., 1977. "On Vortex Sound at Low Mach Number," *J. Fluid Mech.* **85**, pp. 685–691.

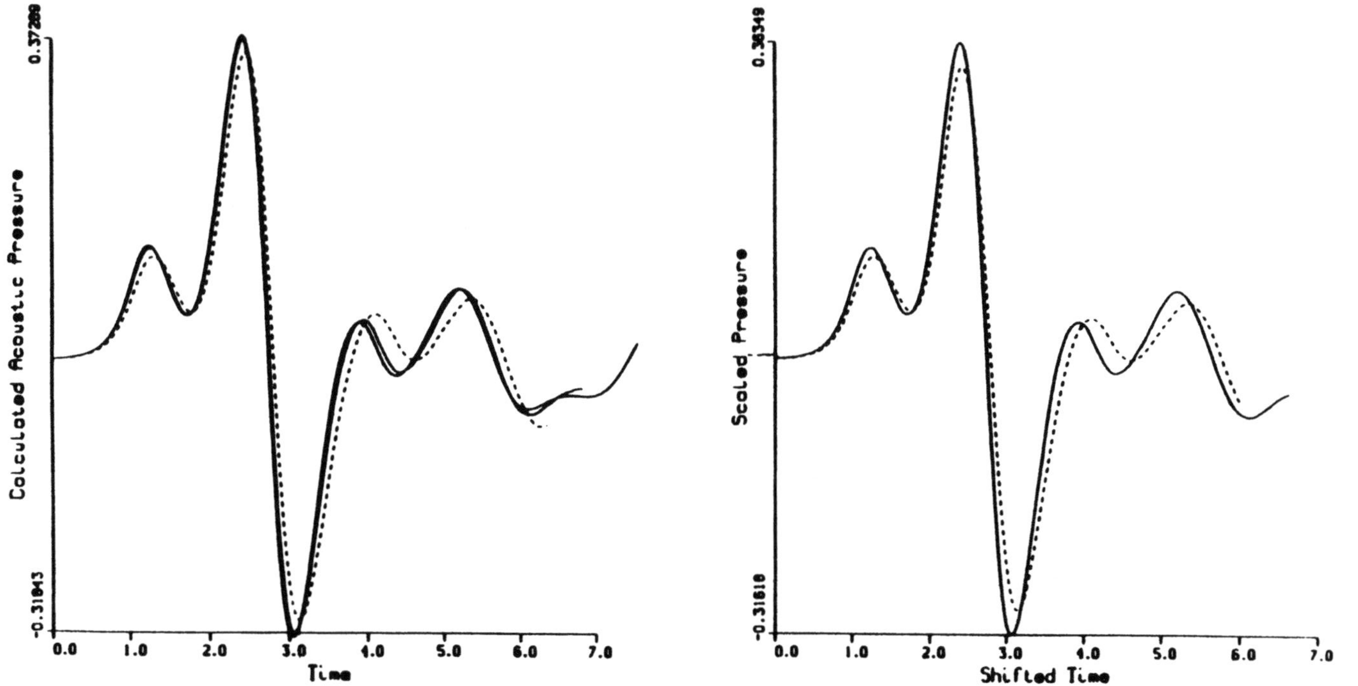

Figure 1. Calculated and measured acoustic pressure signals for vortex roll up and pairing at different M_c; a) Calculated pressure, solid line $M_c = 0$ and $M_c = 0.1$, dashed line $M_c = 0.2$, b) Measured pressure, solid line $M_c = 0.1$, dashed line $M_c = 0.2$.

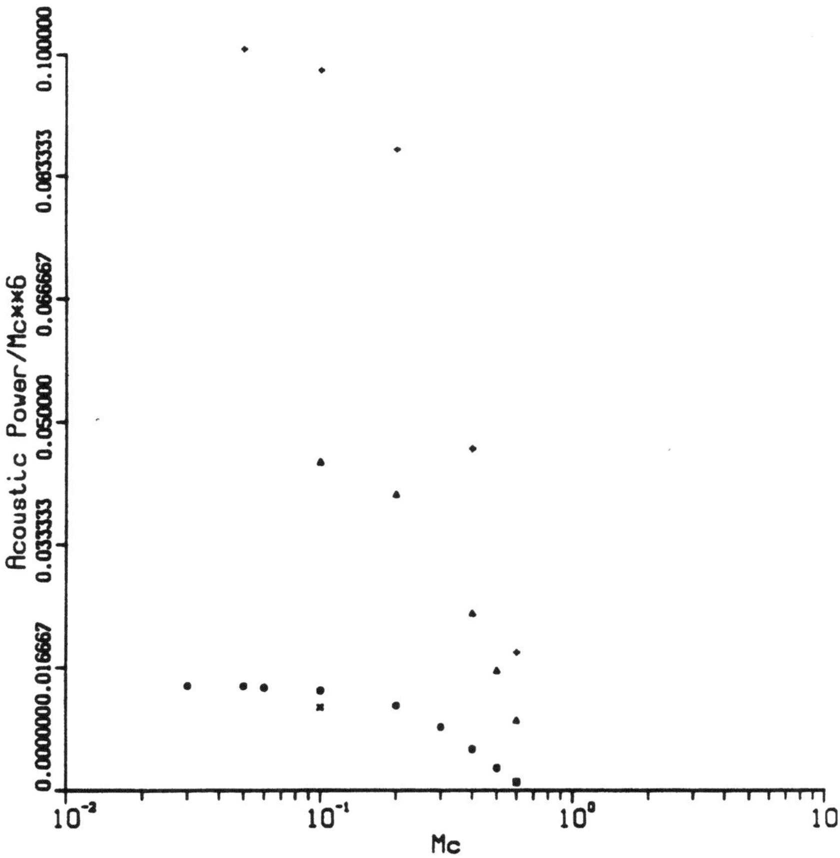

Figure 2. Normalized peak acoustic flux as a function of Mach number M_C; fundamental roll up (circles), subharmonic roll up (triangles), vortex pairing (pluses), and vortex shredding/tearing (cross).

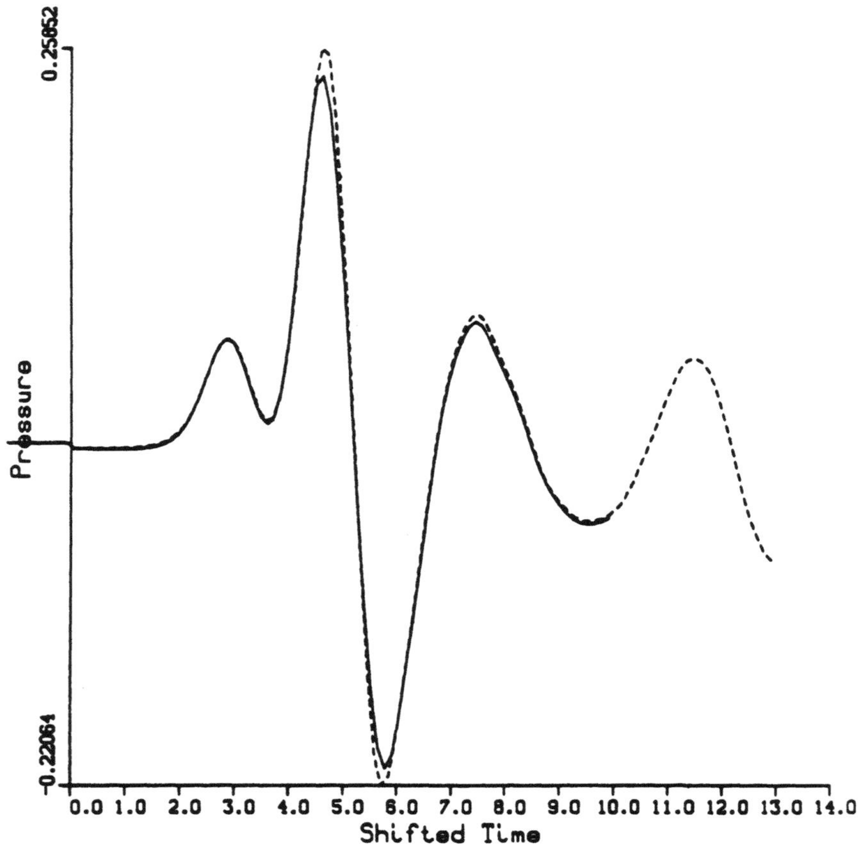

Figure 3. Comparison of the calculated acoustic pressure from the two term approximation with the measured acoustic pressure for roll up and pairing at $M_c = 0.4$; solid line is the measurement and the dashed line is the acoustic analogy prediction.

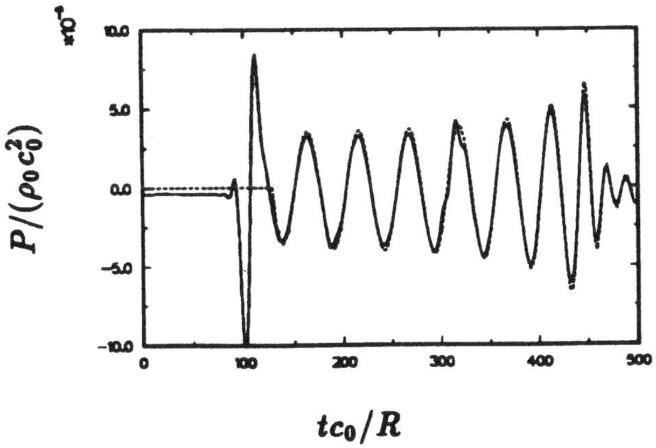

Figure 4. The far-field pressure at $r/\lambda = 2$ for a low Mach number, acoustically compact co-rotating vortex pair. Shown are the results of the simulation, ——, and the predictions of a two-dimensional form of Möhring's equation, – – ––.

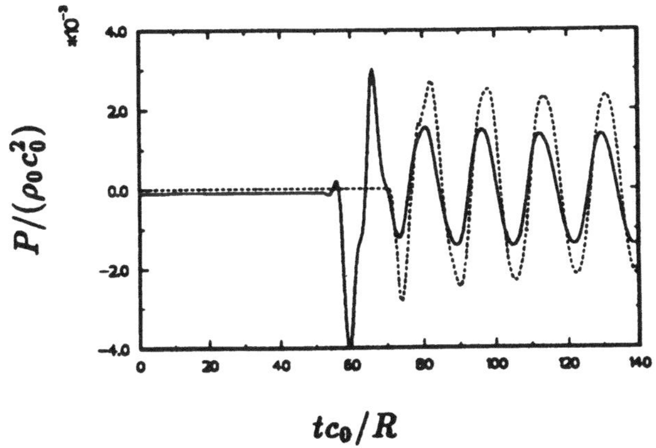

Figure 5. Far-field pressure traces at $r/\lambda = 3.6$ for a higher Mach number, co-rotating vortex pair. Curves as defined in Figure 4. The over-prediction is due to increased compressibility in the near-field.

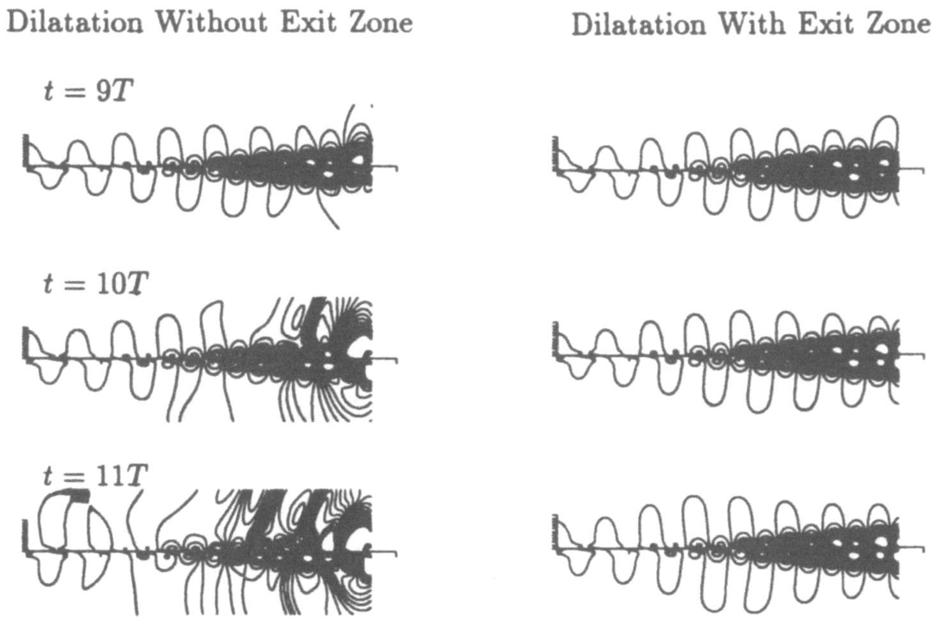

Figure 6. Comparison of outflow boundary condition with and without exit zone for compressible mixing layer forced at its most unstable frequency, $f = 1/T$. Contours of dilatation are plotted at various times. The startup vortex passes through the outflow boundary between $t = 8T$ and $t = 9T$. Large amplitude spurious acoustic waves are reflected from the outflow boundary and propagate upstream when no exit zone is used. $Re = (U_1 - U_2)\delta_\omega/\nu_1 = 500$, $T_2/T_1 = 1$, $(U_1 - U_2)/U_1 = .5$. Box length is $55\delta_\omega$. Coutour Levels: Min $= -3. \times 10^{-4}$, Max $= 3. \times 10^{-4}$, Inc $= 3. \times 10^{-5}$.

COMPUTATION OF THE ACOUSTIC RADIATION FROM BOUNDED HOMOGENEOUS FLOWS

S. Sarkar and M. Y. Hussaini

Institute for Computer Applications in Science and Engineering
NASA Langley Research Center
Hampton, Virginia 23681-0001

ABSTRACT

A numerical study of the problem of sound radiation from isotropic turbulence is carried out. The hybrid DNS approach which combines direct numerical simulation (DNS) of turbulence with an acoustic analogy for the far-field sound is used as the computational tool. The feasibility of the hybrid DNS approach for the computation of the instantaneous acoustic pressure is demonstrated. Preliminary results indicate that the sound is predominantly associated with the energy-containing scales of motion. Furthermore, it appears that the acoustic output of isotropic turbulence is less than that of subsonic jets.

1. Introduction

In 1952, Lighthill posed the problem of estimating the sound radiated by "a fluctuating fluid flow occupying a limited part of a very large volume of fluid of which the remainder is at rest - the fluid flow of course is either ordered, with coherent distributions of vorticity, or disordered and turbulent". More than four decades later, it is quite pertinent to reconsider the same problem given the advances in computational fluid dynamics and computer technology. This may be considered as a fundamental problem in computational aeroacoustics.

In the present work, we consider a simple model of this problem, i.e., the sound radiated by isotropic turbulence bounded by a quiescent fluid. Since homogeneous turbulence is usually assumed to be equivalent to turbulence in an infinite domain, at first sight the conditions of boundedness and homogeneity seem to be contradictory. We interpret homogeneous turbulence in a bounded domain as a model wherein the turbulence is uninfluenced by the boundary layer surrounding the turbulence, and the mean flow (linear in the incompressible case) is such that the statistical moments are spatially invariant. Homogeneous grid turbulence in the laboratory is a similar realization, in that the turbulence is enclosed by the wind tunnel walls and the wall boundary layers are assumed not to affect the homogeneity of the flow in the central region. The isotropic turbulence is represented in our numerical study by a periodic box which contains the fluid motion and is surrounded by a layer in which the velocities decrease to zero. It is assumed that the sound radiated by the periodic box is unaffected by the surrounding thin layer. This may be viewed as one of the "benchmark problems" (to validate methods of computational aeroacoustics) which come within the scope of Lighthill's statement, and are amenable to analytical (e.g., Proudman (1952)), computational and even possibly experimental study. This simple model problem in three dimensions may still not be amenable to a direct numerical simulation (DNS) in which both the aerodynamic and acoustic fluctuations are computed consistently in a single scheme. The alternate choice is a hybrid direct numerical simulation where the aerodynamic fluctuations are directly computed whereas the acoustics are obtained by Lighthill's acoustic analogy or any other similar analogy.

The acoustic analogy approach which was first proposed by Lighthill (1952, 1954) reduces the aeroacoustics problem to an inhomogeneous wave equation. In his analogy, Lighthill chose the homogeneous part to be a linear wave equation for the density and obtained the acoustic far-field as an integral of the equivalent acoustic sources in a uniform medium at rest which replace the fluid motion inside a bounded volume. Furthermore, by considering the statistics of the integrand, Lighthill deduced the eighth-power law for the net acoustic power radiated from a turbulent region.

Lighthill's analogy was followed by other analogies (Phillips (1960), Powell (1964), Ribner (1964), Lilley (1974), and Howe (1975)). In the analogies of Phillips, Lilley and Howe, the homogeneous part of the equation for the acoustic variable is a nonlinear equation rather than the linear wave equation in Lighthill's analogy, and as discussed by Goldstein (1974), Crighton (1975) and FfowcsWilliams (1977) has the advantage of explicit consideration of the convection and refraction of sound by turbulence. Powell and Howe rewrite the term $\mathbf{u}\cdot\nabla\mathbf{u}$ as $\boldsymbol{\omega}\times\mathbf{u}+\nabla\mathbf{u}^2/2$ in their analogies which leads to an explicit vorticity-dependent term in the inhomogeneous part of the wave equation. The explicit appearance of the vorticity in the forcing term is convenient in unsteady flows such as the Von Karman vortex shedding in the laminar wake of a cylinder, but perhaps not so important in high Reynolds number turbulent flows where the flow is less coherent and cannot be modeled as a simple collection of idealized vortices. It is also more convenient in the context of vortex methods (Leonard (1985)).

With the DNS approach, the instantaneous, compressible fluid motion can be accurately calculated for moderate Reynolds number. The important question is whether the instantaneous velocity field can be used to obtain the *instantaneous* far-field acoustic pressure by a *hybrid DNS* which combines DNS of turbulence with the acoustic analogy approach. The present study which employs this approach to study noise radiated by isotropic turbulence provides the answer in the positive. Then the total record of the instantaneous acoustic pressure with time thus obtained not only suffices for the determination of the sound intensity, rms acoustic pressure, acoustic power, and frequency spectrum, but also provides maximum sound levels and acoustic loading on nearby structures. It is perhaps worth mentioning that Proudman (1952) considered the same problem using a statistical model of isotropic turbulence and Lighthill's acoustic analogy. Lighthill's analogy has also been used previously to compute the sound from the large-scale instability waves in a shear layer by Gatski (1979), but we are not aware of any previous attempt to use the analogy within the framework of direct numerical simulation of three-dimensional turbulence without any turbulence models.

The relative advantages and disadvantages of the two approaches – i) the DNS of sound, and ii) the hybrid DNS of sound – are discussed in Section 2. The numerical implementation of the hybrid DNS approach is discussed in Section 3. The simulation method is briefly explained in Section 4, and Sections 5 and 6 complete the paper with results and conclusions.

2. The Role of DNS of Turbulence in the Computation of Sound

We first discuss DNS of sound wherein the turbulence and sound are computed together in a sufficiently large domain, and then consider the alternate approach of hybrid DNS. For the following discussion, it is sufficient to assume quasi-incompressible turbulence. Let L be the characteristic length of the computational domain and η be the Kolmogorov length scale of the turbulence. Then the number of spatial grid points required for a three-dimensional simulation scales as

$$N \simeq (\frac{L}{\eta})^3$$

In order to obtain statistics of the energy-containing eddies, L must be sufficiently large compared to the integral length scale of the turbulence. The choice of $L = 6l$ where l is the integral length scale of the turbulence is typical of DNS calculations of homogeneous turbulence. Then the number of points required to resolve the flow is

$$N_1 = 216\alpha Re_t^{9/4} \tag{1}$$

where α is a $O(1)$ coefficient, Re_t is the Reynolds number based on integral length scale, and we have used the result $l/\eta = O(Re_t^{3/4})$.

The dominant wavelength λ_s of the radiated sound can be estimated as l/M_t where l is the integral length scale and M_t is the turbulent Mach number. In the DNS of sound approach, the computational domain must include both the sound-generating vortical motion and the far-field sound. Let us estimate the dimension L of the computational domain which is large enough to include far-field sound as $L = 4\lambda_s = 4l/M_t$. The required number of spatial grid points on a uniform grid to resolve the flow and acoustic far-field is then

$$N_2 = 64M_t^{-3}\alpha Re_t^{9/4} \tag{2}$$

Eq. (1) relates the number of points N_1 required for simulation of turbulence to the Reynolds number Re_t, and Eq. (2) estimates the number of points N_2 required for simulation of both turbulence and far-field acoustics. Let the number of grid points N be a constant determined by available memory and CPU resources; therefore $N_1 = N_2 = N$. Then it follows from Eqs. (1) and (2) that, for $M_t = 0.01$ which is representative of low-speed jets and boundary layers, resolution of both flow and acoustics rather than only the flow leads to a decrease in Re_t by a factor of 0.004! Eqs. (1) and (2) assume that the acoustic far-field is three-dimensional. However, in circular jets, the acoustic far-field can be assumed to be axisymmetric. After the modification of Eqs. (1) and (2) to account for the dependence on two spatial dimensions rather than three, it follows that Re_t decreases by a factor of 0.02 for a jet with $M_t = 0.01$. In the case of an inhomogeneous free shear layer, the Re_t that can be achieved with present computing resources in a simulation which resolves only the flow is barely enough to sustain fully developed turbulence. Thus, it is impossible to resolve both the fluctuating vortical motion and the far-field acoustics in low-speed jets. However, for high-speed jets with $M_t = 0.3$, the reduction in Re_t with the direct approach is by a factor of only 0.3. Thus, the direct approach may be feasible for the computation of sound from highly compressible turbulent flows. Although the direct approach is perhaps not feasible at the present time for most three-dimensional turbulent flows, simple model problems in two dimensions can be computed in order to investigate aeroacoustic physics. The direct approach can also be compared against acoustic analogies to verify mutual consistency.

Aside from stringent memory resources, there is of course the question of accuracy required for computing the fluid motion in the far-field which, in the direct approach, is the acoustic radiation. Acoustical pressures are several orders of magnitude lower than typical aerodynamic pressures within the flow field and therefore the accuracy requirement is more severe than in the case of traditional computational fluid dynamics problems.

We now consider the second approach to the computation of sound – the hybrid DNS method. The hybrid DNS approach is relatively straightforward to implement in principle. However, there are open questions regarding the most appropriate form of the analogy, the accuracy required in the spatial and temporal discretization of the fluid motion, the accuracy required in capturing retarded time effects, and algorithms with efficient use of memory. Section 4 addresses some of the numerical issues in the implementation of an acoustic analogy.

Compressibility can significantly modify the characteristics of turbulent flow. For example, it is well known from experiments that high-speed shear layers show a relative decrease in growth rate; at a convective Mach number of 1.5, the growth rate parameter decreases to about one-third of the incompressible value. Recent DNS of compressible turbulence by Sarkar, Erlebacher and Hussaini (1991b) for the case of shear flow and by Sarkar, Erlebacher, Hussaini and Kreiss (1991a) in decaying isotropic turbulence has clearly shown that sheared turbulence grows slower and isotropic turbulence decays faster than in corresponding incompressible cases. The DNS and analysis of Sarkar et al. (1991a) and Sarkar (1992) indicate that compressible dissipation and pressure-dilatation are important terms which contribute to the mentioned compressibility effects. Acoustic analogies have been criticized because they do not completely account for compressibility. In principle, computations with the Lighthill analogy can account for compressibility effects on the generation of sound and acoustic feedback, if the compressible Navier-Stokes equations are used for the flow simulation and the instantaneous density ρ and the instantaneous velocity $\mathbf{u}$ are used for computing the nonlinear forcing T_{ij}. However, it is unclear at present, if such a procedure is a sound practice. According to Lilley (private communication), it is essential to include the instantaneous density and velocity field in the Lighthill stress tensor T_{ij} (defined in Section 4) in order to account for flow-acoustic interactions in compressible flow. On the other hand, an asymptotic analysis by Crow (1970) and its discussion in the review by Crighton (1975) shows that T_{ij} in the Lighthill analogy should not be used in the exact form but rather $\rho_0 \mathbf{u}^I \mathbf{u}^I$ where ρ_0 is the mean density (perhaps spatially varying) and $\mathbf{u}^I$ is the rotational component of the velocity field. The physical rationale behind this approximation is that the Lighthill analogy by its formulation separates the acoustic source and its effect – the far-field sound – into two distinct spatial regions, and the inclusion of unsteady fluctuations in the dilatational velocity field and density into T_{ij} violates this separation of source and effect. Since $\mathbf{u}^I$ decays like $1/r^3$ while the sound velocity and pressure decay like $1/r$, the spatial separation of source and effect is maintained. Mathematically, the inclusion of the full velocity field leads to an asymptotic inconsistency. The two criteria – i) capture of compressibility effects on the turbulence and ii) preserving the distinct nature of sound-generating turbulence and far-field acoustics – can be met, if a full compressible Navier-Stokes simulation is performed but only the rotational part of the fluctuating velocity and the mean density are used for evaluating T_{ij}.

Since the computational resources required for the direct computation of sound in turbulent flows seem inordinately large, we choose to use the hybrid DNS approach to compute the sound as mentioned earlier. The Lighthill analogy is used with $T_{ij} = \bar{\rho} u_i^I u_j^I$ as the source.

3. Implementation of the Lighthill Analogy

After reducing the aeroacoustics problem to an inhomogeneous wave equation, Lighthill (1952) obtained the following expression for the acoustic pressure radiated from a bounded volume of turbulence embedded in a quiescent medium

$$p_A(\mathbf{x},t) = \frac{1}{4\pi}\int \frac{T_{ij,ij}(\mathbf{y},t-r/c_0)}{r}\,d\mathbf{y}, \tag{3}$$

where c_0 is the ambient speed of sound, and $r = |\mathbf{x}-\mathbf{y}|$. We approximate the source term by $T_{ij,ij} \simeq \partial^2(\bar{\rho}u_iu_j)/\partial y_i\partial y_j$, since low Mach numbers which correspond to quasi-incompressible turbulence are considered in the present work. For the acoustic far-field, Lighthill (1952) obtained the following simplification

$$p_A(\mathbf{x},t) = \frac{1}{4\pi c_0^2}\frac{x_ix_j}{r^3}\int \ddot{T}_{ij}(\mathbf{y},t-r/c_0)\,d\mathbf{y} \tag{4}$$

which is an alternate form of the analogy. In Eq. (4), $\ddot{T}_{ij}(\mathbf{y},t-r/c_0)$ denotes $\partial^2 T_{ij}/\partial t^2$ evaluated at retarded time $t-r/c_0$. Since we are interested in the far-field pressure, either Eq. (3) or (4) can be used. The memory requirements for the spatial form, Eq. (3), is smaller by a factor of 1/6 than the temporal form, Eq. (4), because in the former case a scalar $T_{ij,ij}$ has to be stored rather than the symmetric tensor $\ddot{T}_{ij}$. Therefore, our preliminary choice was to use $T_{ij,ij}$.

Due to the retarded time effect in the acoustic analogy, the velocity field at time τ is associated with the observer point at time $t = \tau + r/c_0$ where r is the distance between source and observer. Two methods have been used to account for the retarded time effect – i) time accumulation and ii) spatial interpolation. In the time accumulation method, the observer time t was approximated by $t \simeq [(\tau + r/c_0)/\Delta t]\Delta t$ where [.] denotes the integer value function, and Δt denotes the time step used for time advancement of the flow. Such a procedure may be viewed as a zeroth-order interpolation in time of the flow field and causes an error of $O(\Delta t)$ in the observer time t. The advantage of the method is that it does not add to the memory requirements of the computation. In the spatial interpolation method, one finds the surfaces in the computational volume that lie at distances $nc_0\Delta t$ from the observer (n is an integer), computes the source at these points by spatial interpolation of the fluid velocity on the computational grid, and then obtains the contribution from these source points at appropriate time delays to compute the sound at the observer.

In order to determine the temporal resolution required if (3) is used along with the time accumulation method, we performed two test problems where the acoustic source was represented by a monopole and a quadrupole respectively, each located in the center of the computational domain.

For a monopole source located at $\mathbf{x}_0$ of the form

$$q(\mathbf{x}, t) = a\cos(\omega t)\delta(\mathbf{x} - \mathbf{x}_0)$$

the acoustic pressure is

$$p_A(\mathbf{x}, t) = \frac{a}{4\pi r}(cos[\omega(t - r/c_0)]). \tag{5}$$

Fig. 1 compares the exact solution, Eq. (5), with the solution computed using the strategy of the preceding paragraph. The computed solution which was obtained with 6 points per time period of the oscillation has the correct amplitude but the phase has an error of $O(\omega\Delta t)$.

The next test problem was a localized, lateral quadrupole composed of four monopoles $\pm a\cos\omega t$ placed at the vertices of a small square of length $2d$ centered at $\mathbf{x}_0$ and in the $x - y$ plane. The acoustic pressure is

$$p_A = \frac{2a\omega^2 d^2}{4\pi c_0^2 r}\sin 2\phi \sin\theta(cos[\omega(t - r/c_0)]) \tag{6}$$

where (r, ϕ, θ) are the spherical coordinates of the observer point with respect to $\mathbf{x}_0$. Fig. 2 compares the analytical result Eq. (6) and the computed solution for different time steps when non-dimensional values of $\omega = 2$, $d = 0.78$, $c_0 = 25$. The quadrupole is located at the Cartesian coordinates (π, π, π), while the sound is measured at $(100, 100, 100)$. With a $\Delta t = 0.1$ (30 points per time period), the computed acoustic pressure is zero; about 3000 points per time period are required to obtain close agreement with the analytical result! The $O(\omega\Delta t)$ phase error in the time accumulation method has disastrous consequences in the case of a quadrupole, because the sound amplitude is very sensitive to phase cancellation unlike the case of monopole. In particular, with 30 time steps, the signals from the four components of the quadrupole are recorded by the far-field observer at the same discrete point in time which leads to exact cancellation and therefore zero sound. Since broadband turbulence can be considered as a collection of quadrupole sound sources, the implication is that the time step restriction is very severe. A simple estimate for the required time step is now derived. Recall that the time-integration scheme has a phase error of $\omega(\Delta t)$ for a single monopole. The fractional error $(\delta p_A)/p_A$ in the acoustic pressure calculated by Eq. (6) is then of the order $(\omega \Delta t)/(\omega d/c_0)^2$. Since the frequency of sound emitted by unsteady fluid motion is $\omega = O(u/d)$, we obtain $(\delta p_A)/p_A = \omega \Delta t / M_t^2$. Thus the number of time points required per oscillation of the source is $O(1/M_t^2)$ which is a very severe requirement. Note that if a time-interpolation method of order α is used, the required number of time intervals is $O(M_t^{-2/(\alpha+1)})$

We now consider the spatial interpolation technique to account for retarded time effects. When such a procedure is applied to fluid motion of the quadrupole type, it is easy to show that the number of spatial grid points encompassing the smallest acoustically important scale of the turbulent flow must be of the order $M_t^{-2/(\beta+1)}$ where β is the order of accuracy of the interpolation scheme. This implies that a high-order (3 or higher) spatial interpolation scheme must be used for low Mach number turbulence.

Since the time derivative form Eq. (4) does not have the severe time step restriction in low Mach number turbulence of the space derivative form, we then decided to use the Lighthill analogy in the form of Eq. (4).

4. Simulation Method

The turbulent flow inside a cubical domain is computed by solving the Navier-Stokes equations numerically. Homogeneous turbulence in the cube permits periodic boundary conditions in all three directions. It is of course necessary that the length of the computational domain be much larger than the integral length scale of the turbulence for obtaining realistic flow fields. Spectral accuracy is obtained by using a Fourier collocation method for the spatial discretization of the governing equations. In order to avoid expensive evaluations of convolutions in Fourier space, the nonlinear terms are directly evaluated in physical space as products of derivatives. The ensuing aliasing error is negligible if the significant spatial scales of the computed variable are resolved on the grid. A third order, low storage Runge-Kutta scheme is used for advancing the solution in time.

Initial conditions have to be prescribed for u_i', ρ, p and T. The initial velocity field is split into two components, that is, $u_i' = u_i^{I'} + u_i^{C'}$, where each component has a zero average. The solenoidal velocity field $u_i^{I'}$ which satisfies $\nabla.\mathbf{u}^{I'} = 0$ is chosen to be a random Gaussian field with the power spectrum

$$E(k) = k^4 \exp(-2k^2/k_m^2) \tag{7}$$

where k_m denotes the wave number corresponding to the peak of the power spectrum. The compressible velocity $u_i^{C'}$ which satisfies $\nabla \times \mathbf{u}^{C'} = 0$ is also chosen to be a random Gaussian field satisfying the same power spectrum, Eq. (7). The power spectra of the two velocity components are scaled so as to obtain a prescribed $u_{\text{rms}} = \sqrt{u_i'u_i'}$, and a prescribed $\chi = u_{\text{rms}}^C/u_{\text{rms}}$ which is the compressible fraction of kinetic energy. The pressure $p^{I'}$ associated with the incompressible velocity is evaluated from the Poisson equation

$$\nabla^2 p^{I'} = -\bar{\rho} {u^{I'}}_{i,j} {u^{I'}}_{j,i}. \tag{8}$$

It remains to specify the initial values of the thermodynamic variables. The mean density $\bar{\rho}$ is chosen equal to unity, and $\overline{p}$ is chosen so as to obtain a prescribed Mach number $u_{\text{rms}}/\sqrt{\gamma\overline{p}/\bar{\rho}}$ characterizing the turbulence. The fluctuating density ρ' and compressible pressure $p^{C'}$ are chosen as random fields with the power spectrum Eq. (7) and prescribed ρ_{rms} and p_{rms}^C. The pressure then becomes $p = \overline{p} + p^{I'} + p^{C'}$, the density is $\rho = \bar{\rho} + \rho'$, and the temperature T is obtained from the equation of state $p = \rho RT$.

5. Results

DNS has been used to obtain the acoustic output from isotropic turbulence. Due to the reasons discussed in Section 4, the temporal form of the Lighthill analogy, Eq. (4), has been used with $T_{ij} = \bar{\rho} u_i u_j$ as the source. The initial turbulent Mach number is small. After checking the consistency of $\mathbf{u}^I$ obtained from the compressible Navier-Stokes solution with the incompressible solution for small Mach numbers, we used the incompressible Navier-Stokes equations for the rest of the simulations for expediency.

We present the results on the rms acoustic pressure p_A and the acoustic power P_A for the two cases defined in Table 1. Case 2 has the same physical parameters but a higher temporal resolution than Case 1. Here ν is the kinematic viscosity, M_t is the turbulent Mach number, and R_λ is the Taylor microscale Reynolds number.

Case	ν_0	$M_{t,0}$	$R_{\lambda,0}$	$\Delta t\epsilon_0/K_0$
1	1/225	0.01	38	0.0075
2	1/225	0.01	38	0.00375

Table 1: Parameters for the hybrid DNS computation of sound

Note that $R_\lambda = q\lambda/\nu$ where $q = \sqrt{\overline{u_i'u_i'}}$ and $\lambda = q/\sqrt{\overline{\omega_i'\omega_i'}}$, while $M_t = q/\bar{c}$ where $\bar{c}$ is the mean speed of sound.

In the simulations of unforced isotropic turbulence, statistical moments such as K and ϵ decay with time. In addition, R_λ also decreases with time; it can decay by a factor of 4 within an eddy turnover time. It was found that the rapid decrease of R_λ leads to Reynolds numbers so low that the temporal oscillations of the acoustic source are damped out. At sufficiently low R_λ, the exponential decay of Fourier modes due to viscosity dominates over the non-linear transfer and consequently the temporal chaos of the velocity field is annihilated. A possible solution is to force the turbulence so as to eliminate decay of turbulence statistics. However, there is the danger that such a procedure will unduly influence the acoustic radiation since the forcing is applied to the large scales which are believed to emit most of the sound. We choose the alternative of keeping R_λ constant by decreasing the viscosity ν as a function of time so as to compensate for the decay of kinetic energy and dissipation rate.

In the case of unforced isotropic turbulence, the acoustic source term, $\int \ddot{T}_{ij}(\mathbf{y}, t - r/c_0)\, d\mathbf{y}$, decays in time. Consequently, the statistics of the acoustic pressure are determined by averaging over an appropriate time interval T after subtracting out the non-oscillatory contribution due to the temporal decay of the volume integral of the Lighthill source. The contribution due to the statistical unsteadiness of the turbulence does not have temporal oscillations and is not acoustics; it must be subtracted out of the acoustic pressure. The rms acoustic pressure is denoted by $p_{\rm rms}$, while the acoustic power emitted from the volume V of fluid is

$$P_V = \frac{p_{\rm rms}^2}{\bar{\rho} c_0} 4\pi r^2$$

and the acoustic power per unit mass of turbulent fluid is

$$P_A = \frac{P_V}{\bar{\rho}V}$$

Fig. 3a shows the evolution of the acoustic pressure at a far-field point as a function of eddy turnover time in Case 1. The acoustic pressure decreases with time since the turbulent source itself decays with time. We have tried to factor out the decay in turbulence statistics by dividing the random pressure signal $p_A(t)$ by an appropriate deterministic function $g(t)$. The choice of $g(t)$ was guided by the analysis of Proudman (1952) who considered the generation of noise by isotropic turbulence and used statistical models of various two-point moments within the framework of the Lighthill analogy to obtain the following expression for the acoustic power P_A

$$P_A = 38\epsilon\frac{u_{\rm rms}^5}{c_0^5} .= 2.44\epsilon M_t^5 \tag{9}$$

where $M_t = \sqrt{2K}/\bar{c}$. Eq. (9) implies that $g(t) = \sqrt{\epsilon M_t^5}$ should factor out the temporal decay. The evolution of $p_A(t)/\sqrt{\epsilon(t)M_t^5(t)}$ in Fig. 3b shows that most of the decay of the pressure amplitude is factored out. However, the simulations do not strictly confirm the scaling in Eq. (9) because the peak amplitude of the scaled pressure signal, instead of being approximately constant, shows a systematic decay with time. Fig. 4 shows the evolution of acoustic pressure for Case 2 which uses half the time step of Case 1. The result from the higher resolution simulation of Case 2 shows some differences in details with respect to Case 1, although the broad features of the pressure time series are similar.

The frequency content and acoustic power associated with the pressure signal were computed. Since the pressure is statistically unsteady, the time series analysis was performed on the scaled pressure in Fig. 4b whose statistical unsteadiness is much smaller. The spectrum E_P of the normalized power $P_A/\epsilon M_t^5$ is plotted as a function of normalized frequency in Fig. 5. The peak acoustic power is at a non-dimensional frequency of 3.5 with well-defined higher harmonics. Thus, most of the radiated acoustic energy is associated with the energy containing scales of motion as anticipated. The jaggedness of the spectrum could be due to the short record of the pressure time series.

The scaled acoustic power $P_A/\epsilon M_t^5$ associated with the pressure signal in the time interval $0.5 < \epsilon_0 t/K_0 < 4.5$ is 0.10 in Case 1 and 0.12 in Case 2. These values are significantly smaller than those typical of jet acoustics. In subsonic jets, the acoustic efficiency (sound power/jet power) according to Lighthill (1954) is approximately $10^{-4}M^5$, where M is the jet Mach number. The turbulence intensity u_{rms}/U_{jet} can be roughly estimated to be 0.17, and ϵ is roughly 0.1 of the jet power, which implies that $P_A/\epsilon M_t^5 \simeq 7$. Thus the acoustic efficiency of isotropic, homogeneous turbulence is significantly smaller than the efficiency typical of jet acoustics. We argue below that one of the reasons for this dissimilarity is that homogeneous turbulence is a less efficient radiator of sound than inhomogeneous turbulence.

The acoustic pressure fluctuation is

$$p'_A(\mathbf{x},t) = \frac{1}{4\pi c_0^2}\frac{x_i x_j}{r^3}\int [\ddot{T}_{ij}] - \overline{[\ddot{T}_{ij}]}\, d\mathbf{y}. \tag{10}$$

Here, [.] denotes evaluation at retarded time, while the overbar denotes time average. Consider the integral in Eq. (10) when $M_t \to 0$.

$$\begin{aligned} & \int [\ddot{T}_{ij}] - \overline{[\ddot{T}_{ij}]}\, d\mathbf{y} \\ = & \int \ddot{T}_{ij} - \overline{\ddot{T}_{ij}}\, d\mathbf{y} \\ = & \; V < \ddot{T}_{ij} > -V\overline{< \ddot{T}_{ij} >}. \end{aligned} \tag{11}$$

The second line in Eq. (11) follows because retarded time can be neglected in the limit of $M_t \to 0$, and the third line follows from the definition of the volume average denoted by $< . >$. In the case of homogeneous turbulence, the volume average, $< \ddot{T}_{ij} >$, is equivalent to an ensemble average, and since further time averaging does not change a quantity which is already an average $\overline{< \ddot{T}_{ij} >} =< \ddot{T}_{ij} >$. Consequently the integral in Eq. (10) is zero in the limit of $M_t \to 0$. However, in the case of inhomogeneous turbulence $< \ddot{T}_{ij} >$ is not a temporally smooth quantity, and the integral is not necessarily zero. Thus, when $M_t \to 0$, the far-field acoustic pressure radiated by the homogeneous turbulence is $O(M_t^\alpha)$ smaller than that in the inhomogeneous case ($\alpha > 0$). For small but finite Mach number, when retarded time effects cannot be neglected, it is clear that the volume integration smoothes out the temporal oscillation in $[\ddot{T}_{ij}]$ to a greater extent in homogeneous turbulence than in inhomogeneous turbulence, which thus causes smaller fluctuations in the radiated sound from homogeneous turbulence relative to the inhomogeneous case.

The acoustic power obtained in our simulations is significantly smaller than predicted by Proudman's analysis. This may be so because the approximate function used by Proudman for the second-order correlation $\overline{\ddot{T}_{xx}(0)\ddot{T}_{xx}(r)}$ does not agree with the DNS data. We intend to investigate the DNS database in order to model the amplitude and shape of the correlation function and provide the required input to Proudman's statistical analysis.

6. Conclusions

We have considered two approaches to the computation of the sound radiated from turbulent flows – the direct approach and the hybrid DNS approach. Both approaches share the feature of performing a model-free simulation of the turbulence but differ in the treatment of the far-field acoustics. It appears that the hybrid approach which combines an acoustic analogy with the simulation of the turbulent flow is the computationally viable approach. We have applied the hybrid DNS approach to a simple benchmark problem – sound produced by isotropic turbulence. The Lighthill acoustic analogy was used both in the spatial derivative form and the temporal derivative form. Although the former requires less computer memory, we find that, when applied to the problem of sound emitted by a single imposed quadrupole, the time step required is $O(M_t^2)$ smaller than the latter. Therefore, the temporal derivative form of the Lighthill analogy is adopted.

We have been able to demonstrate that the determination of instantaneous acoustic pressure radiated from isotropic turbulence is computationally feasible with the hybrid DNS approach. Preliminary results indicate that the amplitude, frequency and power of the acoustic signal are representative of the energy-containing range of turbulence. It also appears that the acoustic power radiated from homogeneous, isotropic turbulence, which is a model for well-mixed turbulence, is significantly less than the power associated with sound from turbulent jets.

Acknowledgements

The authors wish to thank Geoffrey Lilley and Jay Hardin for helpful comments on a preliminary draft of the manuscript.

References

Crighton, D. G., 1975. "Basic principles of aerodynamic noise generation," *Prog. Aerospace Sci.* **16**, pp. 31-96.

Crow, S. C., 1970. "Aerodynamic sound generation as a singular perturbation problem," *Stud. Appl. Math.* **XLIX**, pp. 60-83.

Doak, P. E., 1968. "An introduction to sound radiation and its sources," *Noise and Acoustic Fatigue in Aeronautics*, E. J. Richards and D. J. Mead, eds., pp. 1-42, John Wiley and Sons.

Ffowcs Williams, J. E., 1977. "Aeroacoustics," Ann. Rev. Fluid Mech. **9**, pp. 447-468.

Gatski, T. B. "Sound production due to large-scale coherent structures," *AIAA Paper 79-4081.*

Goldstein, M. E., 1974. "Aeroacoustics," *NASA SP-346.*

Howe, M. S., 1975. "Contributions to the theory of aerodynamic sound with applications to excess jet noise and the theory of the flute," *J. Fluid Mech.* **71**, pp. 625-673.

Leonard, A., 1985, "Computing three-dimensional flows with vortex elements", in Annual Reviews of Fluid Mechanics, vol. 17, pp.523-59.

Lighthill, M. J., 1952. "On sound generated aerodynamically I. General theory," *Proc. Roy. Soc. A* **211**, pp. 564-587.

Lighthill, M. J., 1954. "On sound generated aerodynamically II. Turbulence as a source of sound," *Proc. Roy. Soc. A* **222**, pp. 1-32.

Lilley, G. M., 1974. "On the noise from jets," *AGARD-CP-131*, pp. 13.1-13.12.

Phillips, O. M., 1960. "On the generation of sound by supersonic shear layers," *J. Fluid Mech.* **9**, pp. 1-28.

Powell, A., 1964. "Theory of vortex sound," *J. Acoust. Soc. America* **36**, pp. 177-195.

Proudman, I., 1952. "The generation of noise by isotropic turbulence," *Proc. Roy. Soc. A* **214**, pp. 119-132.

Ribner, H., 1964. "The generation of sound by turbulent jets," *Advances in Applied Machanics* **8**, H. L. Dryden and Th. von Karman, eds., pp. 103-182.

Sarkar, S., Erlebacher, G., Hussaini, M. Y., and Kreiss, H. O., 1991a. "The analysis and modelling of dilatational terms in compressible turbulence," *J. Fluid Mech.* **227**, p. 473.

Sarkar, S., Erlebacher, G., and Hussaini, M. Y., 1991b. "Direct simulation of compressible turbulence in a shear flow," *Theor. Comput. Fluid Dyn.* **2**, p. 291.

Sarkar, S., 1992. "The pressure-dilatation correlation in compressible turbulence," *Phys. Fluids A*, in press.

Seiner, J., 1991. "Fluid dynamics and noise emission associated with supersonic jets," *Studies in Turbulence*, T. B. Gatski, S. Sarkar and C. G. Speziale, eds., pp. 297-323.

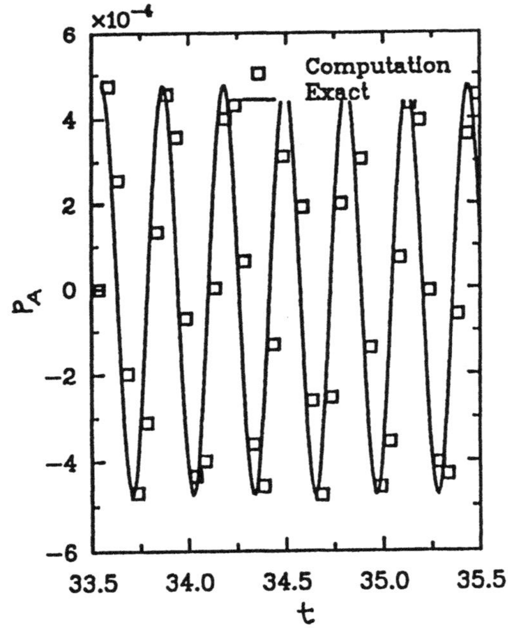

Fig. 1. Acoustic pressure due to a localized monopole.

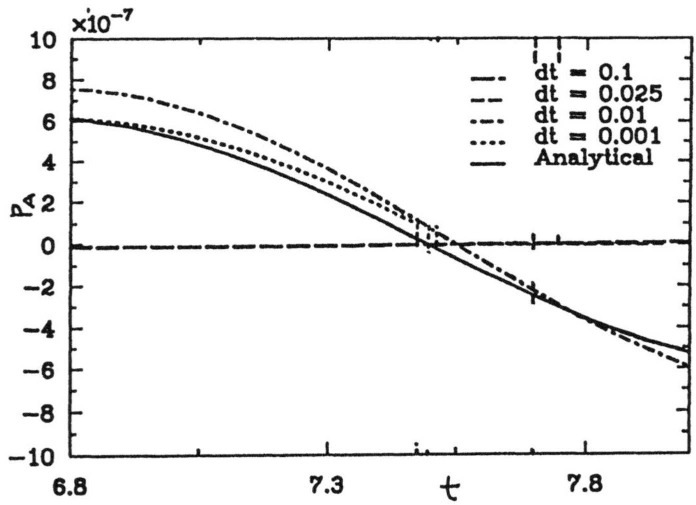

Fig. 2. Acoustic pressure due to a localized quadrupole.

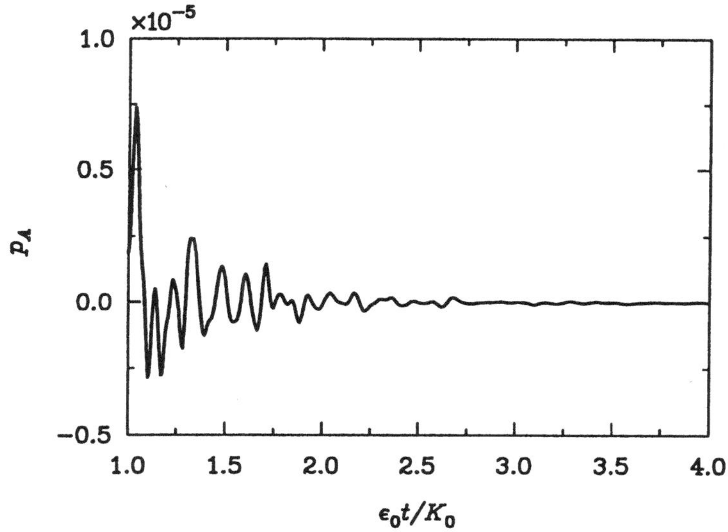

Fig. 3a. Acoustic pressure at a far–field point in Case 1.

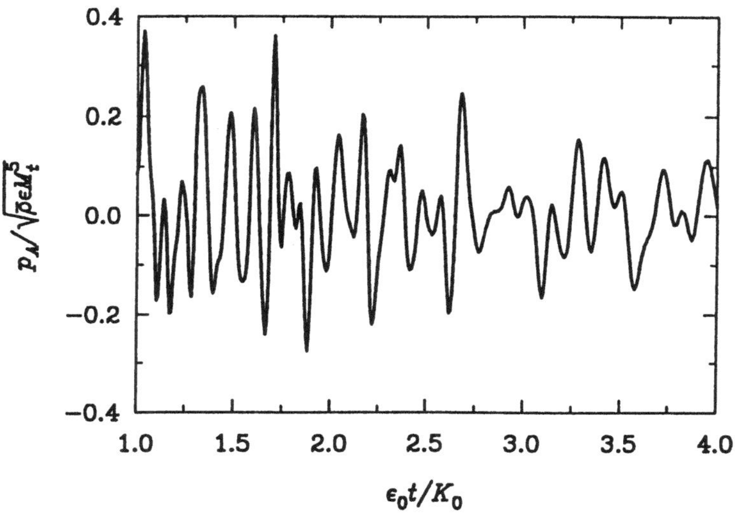

Fig. 3b. Acoustic pressure at the far–field point scaled to factor out the decay.

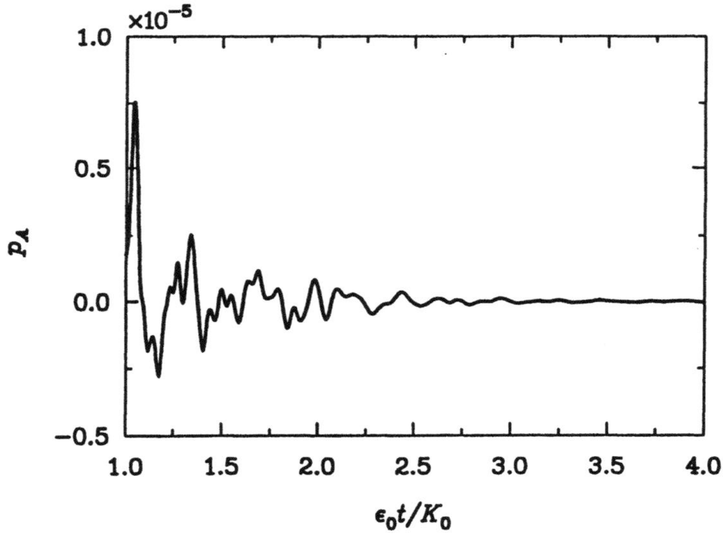

Fig. 4a. Acoustic pressure at a far–field point in Case 2.

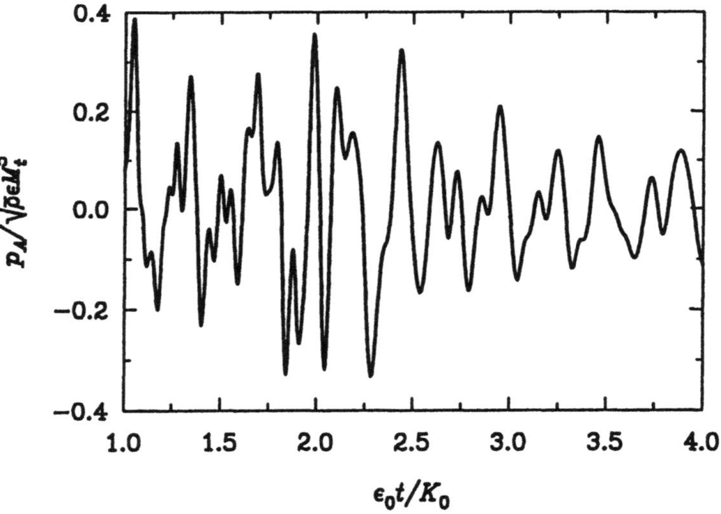

Fig. 4b. Acoustic pressure at the far–field point scaled to remove the decay of turbulence.

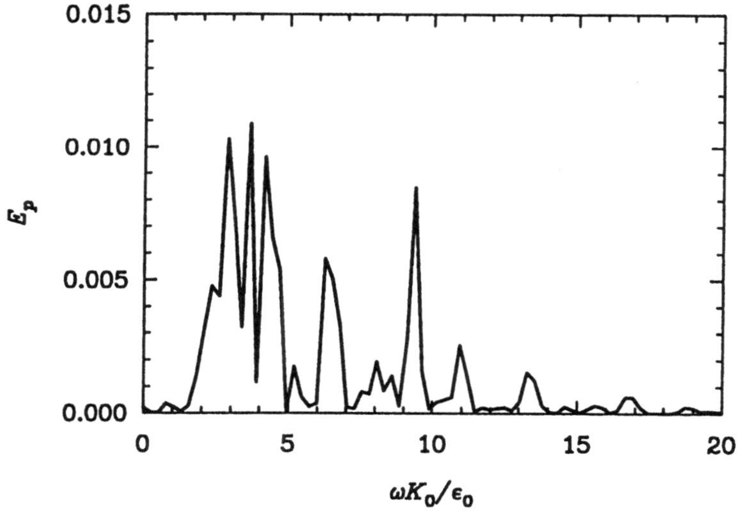

Fig. 5. Frequency spectrum of the scaled acoustic power.

ON THE DEVELOPMENT OF A TIME DOMAIN METHOD FOR CAA "THE SUCCESSES AND FAILURES"

Willie R. Watson

Mail Stop 460
NASA Langley Research Center
Hampton, Virginia 23681-0001

ABSTRACT

A summary of the successes and failures encountered in developing a time domain steady state numerical method for aeroacoustics in two space dimensions is presented. The equations and boundary conditions governing nonlinear acoustic disturbances propagating through a viscous compressible internal steady flow field are presented. These equations and boundary conditions are evolved to a steady state on a supercomputer using an unsplit MacCormack method. The capabilities of the method are demonstrated by presenting solutions to several linear and nonlinear problems for which the expected trends were known. Results calculated from the method are shown to be superior to those obtained using direct numerical simulation. Also included is the development of a new strategy for evolving acoustic systems to a steady state. This strategy is shown to be very effective in eliminating transient acoustic waves. Failure to obtain the expected steady state solution was observed to occur only in cases where the boundary conditions were inadequate.

1. Introduction

Understanding the effects of a steady flow field on the propagation of steady state sound waves has long been an important goal of the aeroacoustics community. It is becoming generally accepted that a fully adequate treatment of this phenomenon will be achieved only through application of computational methods. Four recent papers (Watson, 1991; Watson and Myers, 1990; Watson and Myers, 1991; and Watson and Myers, 1992) have discussed several aspects of research aimed at applying modern computational techniques to compute acoustic fields propagating through quite general two-dimensional subsonic steady flows. The major emphasis was on the determination of periodic acoustic fields propagating through exact and numerically computed steady flows generated by periodic sound sources. The purpose of this paper is to highlight the major results of those papers and present several other aspects of this work which have not been discussed previously.

2. Governing Equations

Figure 1 illustrates the geometry and coordinate system used in this analysis. The fluid flows subsonically from left to right and is considered viscous and compressible. The boundaries located at $y = y_1(x)$ and $y = y_2(x)$ are rigid and nonconducting and only two-dimensional flow is considered. An acoustic source pressure, $p_s(y, t)$, is prescribed at the outflow boundary and the objective is to determine the periodic acoustic disturbances elsewhere within the flow due to the source pressure.

The fundamental equations governing the fluid flow are the Navier Stokes equations. In direct numerical simulation of the acoustic process these equations are solved once for the steady flow field and again for the sum of the acoustic and steady flow field together. Acoustic disturbances in the flow are obtained by subtracting these two flow fields. Dissipation and dispersion errors may dominate the small amplitude sound waves obtained in this manner (Watson and Myers, 1992). A more accurate acoustic field can be obtained by solving the equations obtained by splitting the small amplitude sound waves from those of the steady flow (Watson and Myers, 1992). The equations governing the inviscid acoustic waves can be written as:

$$\frac{\partial\{\tilde{\Phi}\}}{\partial t} + [A]\frac{\partial\{\tilde{\Phi}\}}{\partial x} + [B]\frac{\partial\{\tilde{\Phi}\}}{\partial y} + \{\tilde{F}\} = \{0\} \tag{1}$$

where $\{\tilde{\Phi}\}$ is the vector of unknown acoustic variables. All terms and assumptions used in deriving (1) can be found in several previous works (Watson, 1991; Watson and Myers, 1990; Watson and Myers, 1991; and Watson and Myers, 1992) and are not reproduced here.

3. Boundary Conditions

Inflow/outflow boundary conditions for this problem utilize the characteristic form of equation (1). This can be written as

$$\frac{\partial\{\tilde{\Phi}\}}{\partial t} = \{F_S^+\} + \{F_S^-\} - \{F\}, \quad \{F\} = [B]\{\frac{\partial\tilde{\Phi}}{\partial y}\} + \{\tilde{F}\} \quad (2)$$

All vectors in (7) are explained in detail elsewhere (Watson, 1991; Watson and Myers, 1990; Watson and Myers, 1991; and Watson and Myers, 1992) and are not repeated here. At the inflow boundary, $x = 0$, three quantities must be specified. These can be taken to be

$$b(y,t) = -[\frac{\partial\tilde{p}}{\partial x} - c^2\frac{\partial\tilde{\rho}}{\partial x}], \quad \varsigma(y,t) = \frac{\partial\tilde{v}}{\partial x} - \frac{\partial\tilde{u}}{\partial y}, \quad z(y,t) = \frac{\tilde{p}}{\tilde{u}} \quad (3)$$

where $\tilde{p}$ is the acoustic pressure, c the local speed of sound, $\tilde{\rho}$ the acoustic density, $\tilde{u}$ the axial acoustic velocity component, and $\tilde{v}$ the components of transverse acoustic velocity. The quantities $b(y,t)$, $\varsigma(y,t)$ and $z(y,t)$ may be interpreted as a condition on the fluid entropy perturbation, the fluid vorticity perturbation, and acoustic impedance respectively (Watson and Myers, 1991). At the outflow boundary, on the other hand, only one condition can be specified. This is taken to be the known source pressure

$$\tilde{p} = p_s(y,t) \quad (4)$$

Finally, the lateral boundaries, $y = y_1(x)$ and $y = y_2(x)$, are rigid; the proper physical boundary condition on these is the flow tangency condition

$$\tilde{\mathbf{v}} \bullet \mathbf{n} = 0 \quad (5)$$

in which $\tilde{\mathbf{v}}$ and $\mathbf{n}$ are the acoustic velocity vector and the unit normal vector to the wall, respectively.

4. The Numerical Method

The computations were performed with a second order explicit unsplit MacCormack predictor-corrector method. Equations (1) are first transformed to an appropriate computational domain (ξ, η). The grid is uniform in the computational domain to maintain the required order of accuracy and to implement boundary conditions. However, the sound waves in the physical (x, y) plane are obtained on a grid which may be compressed in both x and y near regions where high concentrations of flow gradients exist. Further details concerning the numerical method are described in detail elsewhere (Watson, 1991; Watson and Myers, 1990; Watson and Myers, 1991; and Watson and Myers, 1992) and not repeated here.

5. Steady State Solution Strategy

Dissipative effects of real physical viscosity cause acoustic systems with periodic sources to evolve to a periodic state. However, in this as well as most other acoustic analysis, it has been desirable to neglect the viscous forces. Although this greatly simplifies the governing equations and boundary conditions, their absence may cause the transient wave field to be trapped within the computational domain so that a steady-state will never be achieved.

In the current effort, when transient acoustic waves become trapped within the computational domain the two-step iterative strategy developed in Watson and Myers (1990) is employed to eliminate them. The objective of this strategy is to obtain a solution to the system of equations (1) which, by definition, is periodic with the same period as the source. The specific procedure utilized is a straightforward application of standard Fast Fourier Transform (FFT) techniques. Starting from a particular initial state, the total acoustic field $\{\tilde{\Phi}(x, y, t)\}$ is computed for 512 time steps utilizing MacCormack's method. This time series is then transformed into discrete frequency components by use of the FFT calculation. This yields the complex frequency components $\{\hat{\Phi}(x, y, \omega)\}$ of the total field. The periodic field is then obtained by filtering the components at the transient frequencies from $\{\hat{\Phi}\}$ and inverting the transform; i.e., by performing the inverse FFT on the spectrum including only the components at the harmonics of the driving source frequency $n\tilde{\omega}$, $n = 0, 1, 2.....$ (this is referred to as an iteration). This new field provides the initial condition for subsequent iterations on the method (Watson, 1991).

6. Results And Discussion

A computer code implementing the numerical method has been developed and programmed to run on a CRAY-2S computer. Results were obtained with the underlying objective of establishing confidence in the boundary conditions, the two-step steady state strategy, and the accuracy of the approach. In each case considered, the steady flow is inviscid and is either specified exactly or is computed numerically using MacCormack's algorithm. For purposes of this paper, the sound source is considered to be a pure tone oscillating sinusoidally at a frequency of $\frac{1100}{\pi}$ hertz. As mention earlier, only the most important results of the previous work are highlighted along with others which have not been presented before. A detailed discussion of parameter values used here has been presented elsewhere (Watson, 1991; Watson and Myers, 1990; Watson and Myers, 1991; and Watson and Myers, 1992) and is not reproduced here.

6.1. Test of boundary conditions

In order to test the integrity of the boundary conditions a series of simulations were performed in a straight channel with uniform flow for which the source pressure was of the form, $p_s(y,t) = \epsilon p_0 \cos(\frac{\pi y}{H})$ $\sin(\tilde{\omega} t)$. Numerical results are compared to exact and approximate modal solutions which are obtainable for outgoing acoustic waves in a channel.

The inflow boundary conditions were chosen to simulate a reflection-free steady state inflow. In order to simulate such an inflow the following expressions derived from linearized steady state theory were used

$$z = -\frac{(\tilde{\omega} + K_1 u_0)\rho_0}{K_1}, \quad \varsigma = 0, \quad b(y) = 0 \qquad (6)$$

where K_1 is the axial propagation constant obtained from linear theory. The rationale behind employing conditions such as these is that they simulate the behavior of a nonvortical, isentropic, locally reacting surface, a common acoustic boundary model. Because of the resistive term in the impedance condition (6) these boundary conditions are expected to gives rise to dissipation of transient energy while allowing the steady state to evolve without reflecting. Prior to the development of the current steady state strategy, this boundary model was applied to nonplanar acoustic problems and was virtually the only one other than nonreflection which led successfully to steady state acoustic predictions.

Figure 2 summarizes the type of comparisons obtained with the use of these boundary conditions. Plots of the acoustic pressure distribution at $x = \frac{L}{4}$ for a uniform .5 Mach number flow are shown.

The distribution across the channel has been calculated with $\epsilon = 10^{-3}$ in order that a comparison with linear theory could be made and the two-step strategy for evolving the solution to a steady state has not been applied. Two important points should be observed. First, although the initial disturbance has reached the inflow boundary, the periodic solution has not been reached after 400 time steps. Thus the boundary conditions (6) can not be thought of as nonreflecting to transient acoustic waves. Secondly, the numerical solution and the exact periodic solution are nearly identical at the 1000th time step. Since no evidence of the initial transient was observed after 1000 time steps, it can be concluded that it was dissipated by the boundary conditions as expected. Similar trends were observed for other sound sources and when a sheared mean flow was employed.

6.2. Application of the steady state strategy

The integrity of the two-step method has been thoroughly tested by demonstrating that it reproduces the periodic solution to several acoustic problems for which the chosen boundary condition is not dissipative to transient acoustic waves. Both linear and nonlinear solutions have been extracted in a noflow channel using the method. The three critical inflow parameters were chosen as follows:

$$b(y) = 0, \quad \varsigma = 0, \quad z = 0 \tag{7}$$

Note that the impedance condition selected in (7) does not have a resistive component and will not dissipate the transient wave field.

Results in Figure 3a for which $x = L/4$ and $y = H/4$ are representative of the time histories obtained by a straightforward integration of (1) by MacCormack's method without application of the two-step strategy. The source pressure was selected as, $p_s(y,t) = \epsilon p_0 \cos \frac{\pi y}{H} \sin \tilde{\omega} t$, where ϵ was taken as 10^{-3} in order that a comparison could be made with linear theory. Good agreement between the full exact and MacCormack time histories has been obtained. Unfortunately, this acoustic solution is contaminated with transient acoustic waves and should not be use to compute the periodic noise. The solid line in the figure is the exact periodic solution that should be used instead. Note that the MacCormack time history is significantly different from that of this exact periodic solution. Because the chosen boundary conditions do not dissipate the transient wave field the two-step method will be required to extract this periodic field.

Figures 3b show the results of applying the two-step method for two iterations. Numerical results cannot be distinguished from the exact periodic state. A number of additional calculations were presented in Watson (1991). In these, it was determined that two iterations could be considered as a rule of thumb for linear periodic disturbances with amplitudes of the same order as the transient acoustic wave. Further, it was determined that approximately four iterations were needed to obtain the same accuracy when the periodic disturbance amplitude was several orders of magnitude smaller than the transient wave.

A nonlinear periodic solution has been obtained for a planar wave source of the form, $p_s(y,t) = \epsilon p_0 \sin \tilde{\omega} t$. In order to simulate nonlinear acoustic propagation, ϵ was chosen as .3. Figure 4 show results after seven iterations. The solid curve in the figure depicts the exact periodic solution from linear theory and is included as a reference against which to compare the nonlinear solution. Both the shape and peak amplitude of the time history for the disturbance changed rather significantly from one to six iterations evidence that a periodic state had not been achieved prior to seven iterations. Results were computed for up to forty iterations. Both the peak amplitude and shape of the time history did not change from those shown in Figure 4. Thus the solution has reached a periodic state after seven iterations. It is emphasized that these nonlinear results cannot be obtained analytically.

6.3. Effect of numerically computed flows

One of the critical considerations for more general flows is whether the degree of accuracy with which the steady flow can be determined numerically is sufficient for direct numerical simulation. Small errors in a calculated steady flow can be expected to have an impact on the prediction of small amplitude acoustic disturbances. This is especially true in multidimensional steady flows since these can be computed with only a limited degree of accuracy even with modern computational techniques. It was found during the studies leading to Watson and Myers (1990) and Watson and Myers (1991) that direct numerical simulation and the current method are equally successful when the steady flow is specified exactly. With a computed flow of limited accuracy, however, there is reason to suspect that this conclusion is no longer valid. This aspect of computation is addressed here in the context of quasi-one dimensional theory. The equations used to obtain the quasi-one dimensional results are the quasi-one dimensional counterpart of equations (1). These equations are written explicitly elsewere (Watson, 1991) and are not repeated here.

Quasi-one-dimensional acoustic results have been calculated for two different Crocco-Tsien nozzles. One nozzle was designed to have .6 throat Mach ($M_t = .6$) number while the other was designed to have nearly sonic flow at the throat ($M_t = .95$). Entrance and exit geometry of both nozzles was chosen to have .3 Mach number at the entrance and exit. Calculations were carried out using both direct numerical simulation and the current method. Unwanted transients were eliminated by using a nonreflecting boundary condition (Watson, 1991) at the inflow boundary. The source pressure was a pure tone of the form, $p_s(t) = \epsilon p_0 \sin(\omega t)$. During the disturbance calculations only the machine zero steady flow was employed.

Figures 5a and 5b show the periodic acoustic solution obtained from both the current method and direct numerical simulation respectively. The throat Mach number is .6 and the two values of ϵ shown in the graphs were used (i.e. $\epsilon = 10^{-3}$ and $\epsilon = 10^{-8}$) for each prediction. These values of ϵ were selected because they correspond to the upper and lower decibel level respectively of human hearing. For the highest value of ϵ both methods give comparable results. As expected, both periodic disturbances are amplified in the throat region and are symmetrical about a point near the throat for the higher value of ϵ. When the disturbance amplitude is reduced further ($\epsilon = 10^{-8}$) the two methods give different periodic solutions particularly upstream of the throat. The solution using direct simulation (Figure 5b) shows unrealistically large sound pressures at the junction between the uniform and variable area portion of the nozzle ($x = \frac{L}{4}$ and $x = \frac{3L}{4}$). Further, for this smaller source amplitude, the periodic solution obtained using direct simulation is not symmetrical about a point near the throat as it should be. This is a consequence of the small numerical error in the steady flow calculation. In contrast, the results using the current method appear nearly symmetrical about a point near the throat as expected and do not predict the unrealistically large sound pressure levels at the junction between the uniform section and variable area portion of the nozzle.

In order to further demonstrate the superiority of the current method over direct simulation the nozzle with nearly sonic flow (i.e. $M_t = .95$) was used to compute the periodic sound field. Figure 6 shows plots similar to those of Figure 5a using the current method. Even at this high Mach number the method appears to be successful. On the other hand the results using direct simulation were considerably worse than those predicted in Figure 5b and are not shown.

6.4. 2-D Variable area results with compressible flow

Results were computed with 2-D flow in a quartic nozzle. Figure 7 shows the nozzle geometry along with contour plots of the steady flow Mach number calculated after 40,000 time steps. The steady flow Mach number varies from a minimum value of .25 near the entrance to a maximum value of .41 near the throat. As expected, the largest value of Mach number occurs at the wall in the vicinity of the throat. The flow is nearly symmetrical about both the centerline and the throat. Thus, the steady flow exhibits the expected trends.

Steady state acoustic disturbances in the nozzle were computed using only the current method. The outflow source pressure was a planar acoustic wave of the form $p_s(t) = \epsilon p_0 \sin(\omega t)$. The inflow boundary condition was that given in equation (6) where K_1 was determine from an equivalent uniform duct. Figure 8 shows computed normalized acoustic pressure time histories along the lower wall at three different axial locations with $\epsilon = 10^{-3}$. The acoustic pressure has reached an apparent periodic state at all three axial locations. Note that the amplitude of the disturbance at the wall is highest in regions where the Mach number is highest as expected.

7. Conclusions

Based upon the results of this work the following specific conclusions are drawn:

(1) The computational approach as applied here was successful.

(2) The current choice of boundary conditions give stable and consistent results.

(3) The two-step method represents a powerful, efficient and stable method for evolving acoustic systems to a periodic state.

(4) The current method is superior to direct numerical simulation for acoustic problems.

Two problems encountered during numerical calculations not presented here have not been discussed. First, in 2-D cases which utilized significant reductions in nozzle throat area, the use of the boundary conditions for an equivalent uniform duct (6) did not achieve a steady state. Secondly, the periodic sound field in most commercial aircraft engines propagate over boundary surfaces which are sound absorbing and not rigid as considered here. Thus there is a need to develop boundary conditions which are applicable to these situations.

The above conclusions are based upon sample calculations presented both in this paper and on others performed by the author. Although these calculations were of limited scope, they have pointed up some issues which warrant further discussion and future work. However, one deserves reemphasis here. The results presented show that direct numerical simulation performed poorly in comparison to the current method. Of course, there exists a class of problems for which the effects of viscosity on the acoustic disturbance cannot be neglected and to which a splitting of the flow field is not possible. Direct numerical simulation may be the only alternative for acoustic problems of this type. This work therefore supports a recommendation that efforts should be devoted to the development of more accurate methods capable of solving acoustic problems using direct numerical simulation.

References

Watson, Willie R., 1991. *A Time Domain Numerical Theory for Studying Steady-State Acoustic Disturbances In Flow*, Doctoral Dissertation, The George Washington University, Washington D.C.

Watson, Willie R. and Myers, M. K., 1990. "A two-step iterative method for evolving nonlinear acoustic systems to a steady-state," to appear in *AIAA Journal*, May 1992. Also see AIAA Paper No. 90-3946.

Watson, Willie R. and Myers, M. K., 1991. "Inflow-outflow boundary conditions for two-dimensional acoustic waves in channels with flows," *AIAA Journal* **29**(9), pp. 1383-1389.

Watson, Willie R. and Myers, M. K., 1992. "Numerical computational of steady-state acoustic disturbances in flow," AIAA Paper No. 92-02-074.

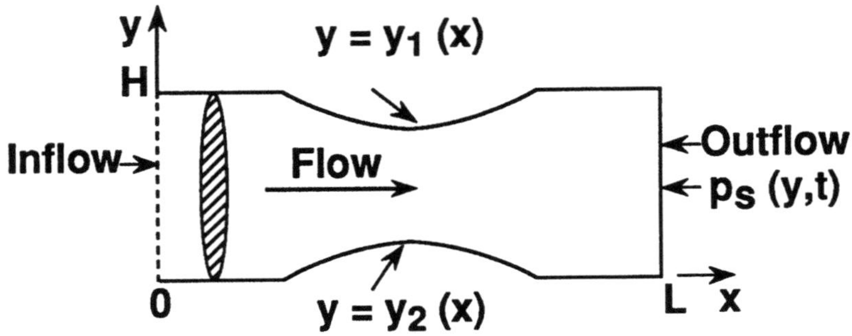

Figure 1. Two dimensional geometry and coordinate system for studying the interaction of sound and flow.

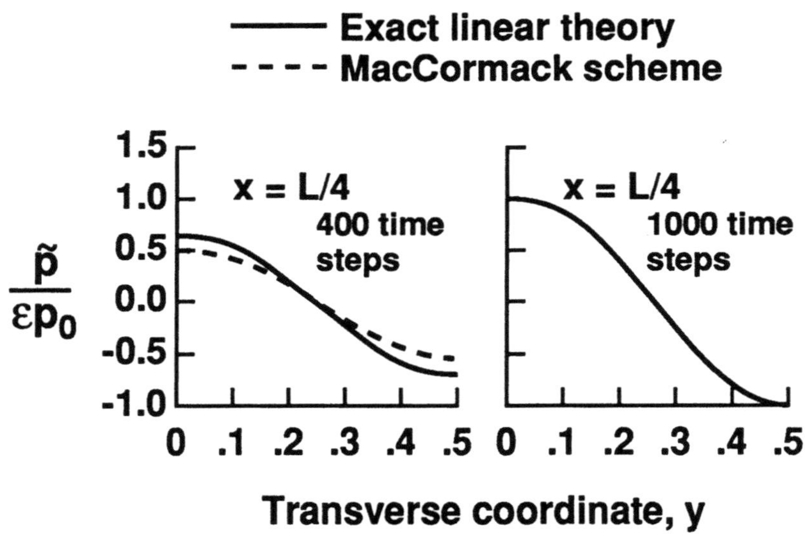

Figure 2. Comparison of the MacCormack and exact solution with $\epsilon = 10^{-3}$.

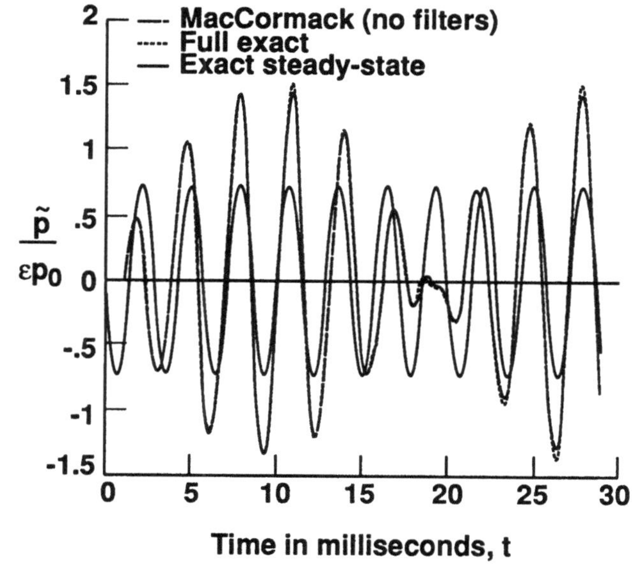

Figure 3a. Comparison of the MacCormack and exact solutions with $\epsilon = 10^{-3}$.

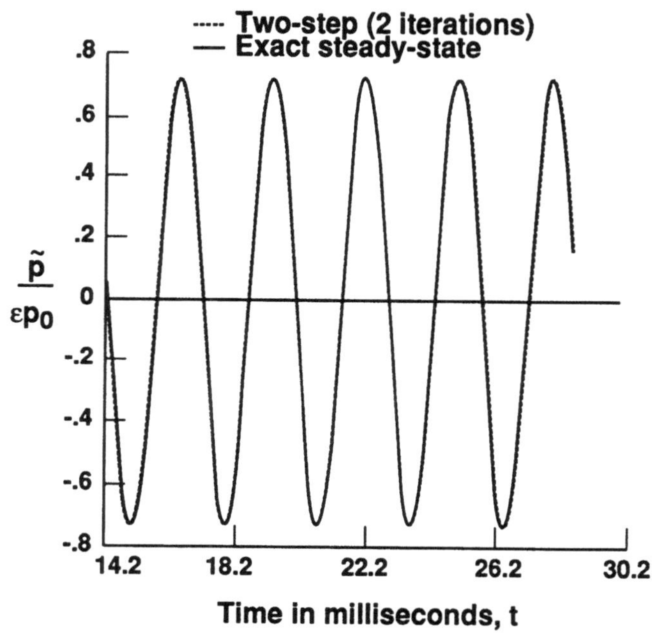

Figure 3b. Comparison of the Two-step and exact periodic solution with $\epsilon = 10^{-3}$.

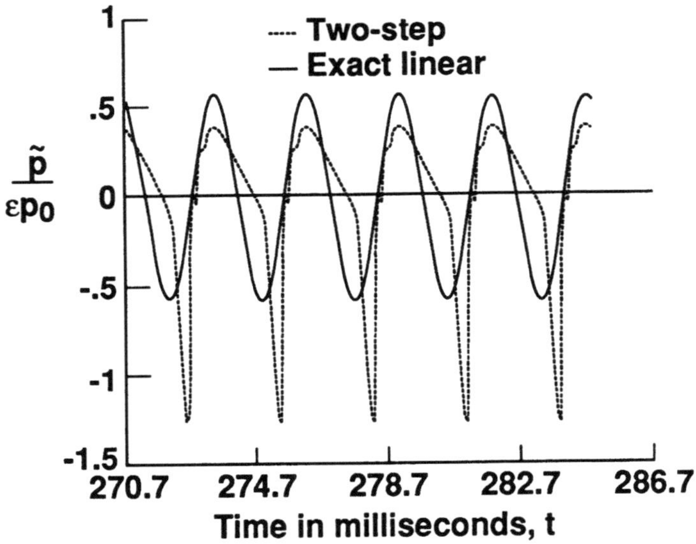

Figure 4. Nonlinear periodic solution for a planewave source with $\epsilon = .3$ and seven iterations.

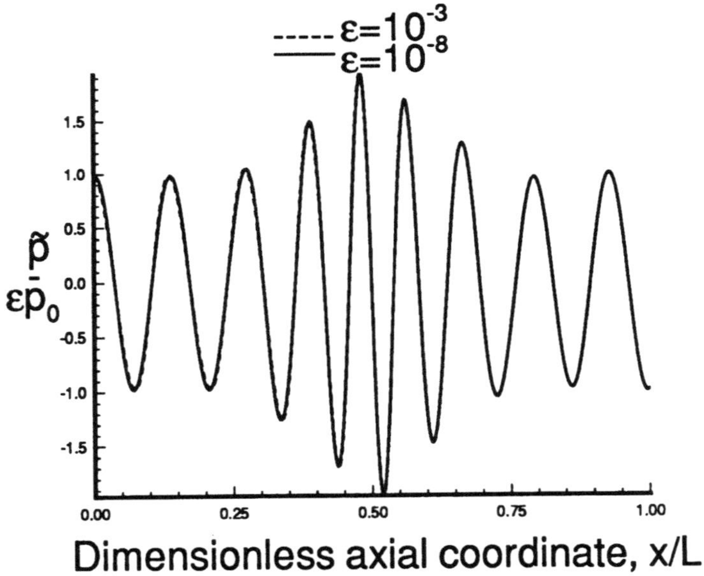

Figure 5a.- Periodic solution using the current method ($M_t = .6$, 8,000 time steps).

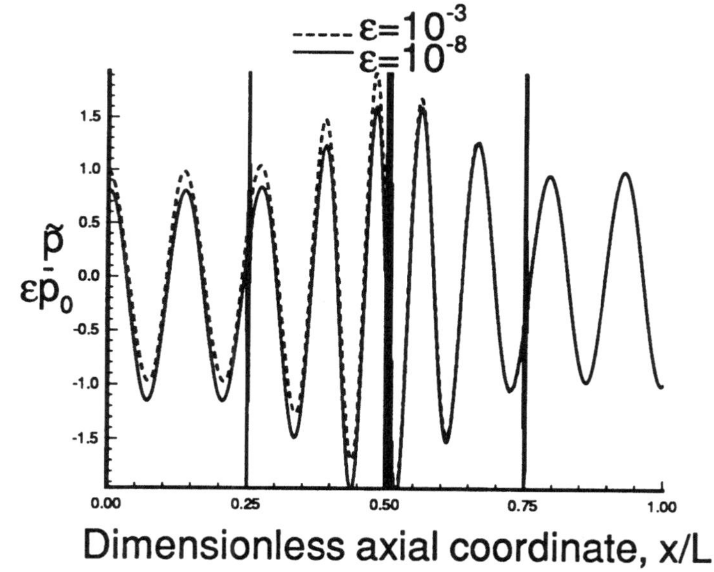

Figure 5b.- Periodic solution using direct simulation (M_t = .6, 8,000 time steps).

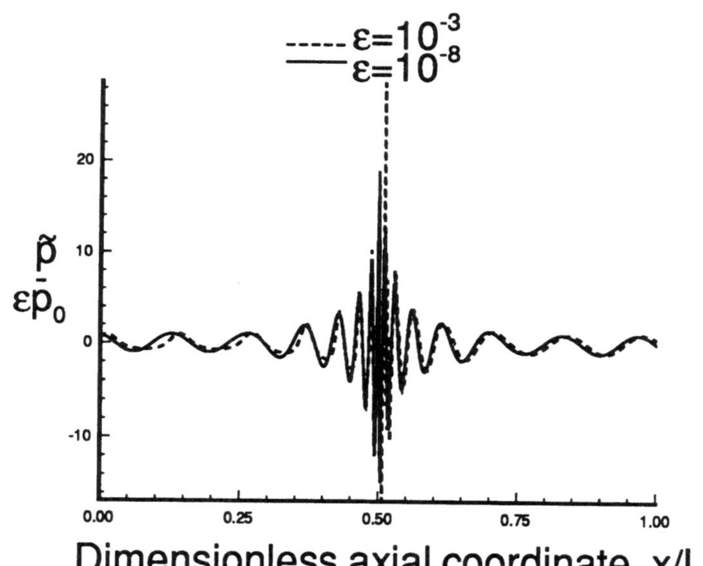

Figure 6.- Periodic solution using the current method (M_t = .95, 8,000 time steps).

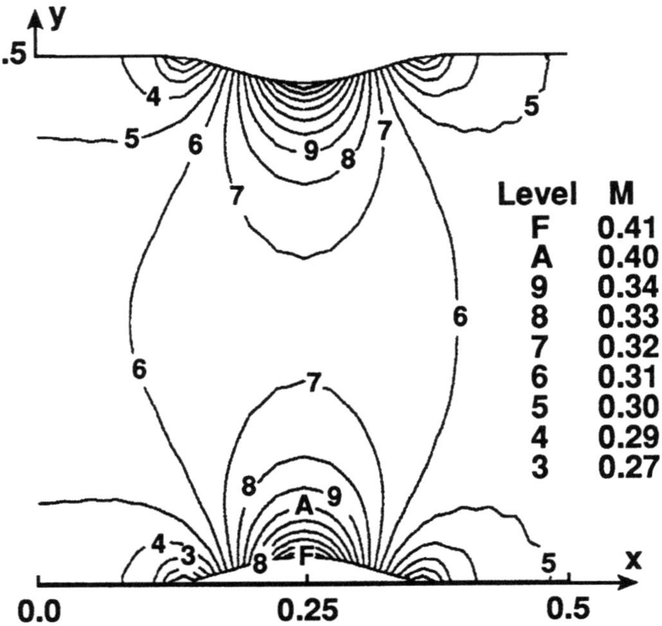

Figure 7.- Mean flow Mach number contours for the quartic nozzle at 40,000 time steps.

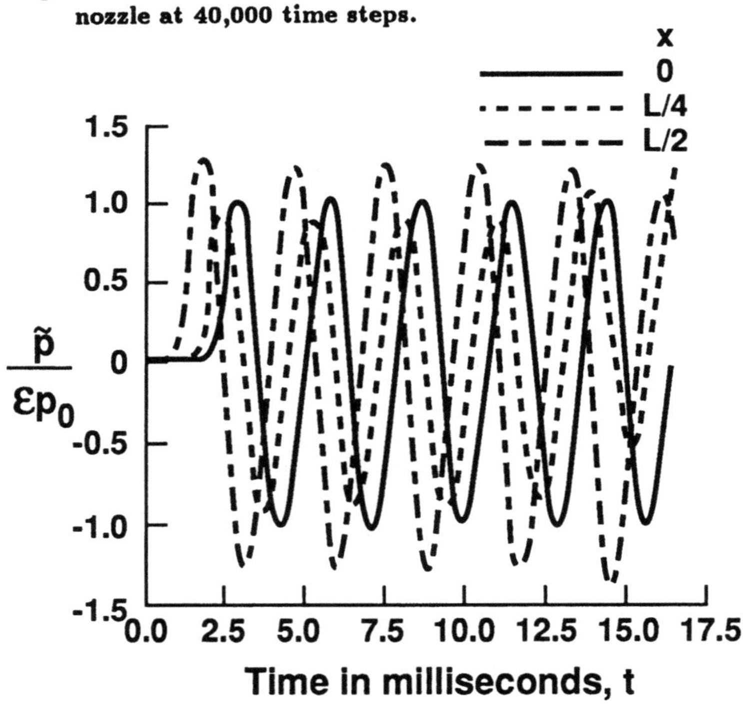

Figure 8.- Lower wall disturbance history for the quartic nozzle with $\epsilon = 10^{-3}$.

UNSTEADY COMPRESSIBLE FLOW COMPUTATIONAL METHODS

CFD METHODS DEVELOPMENT CONSIDERATIONS FOR UNSTEADY AERODYNAMIC ANALYSIS

John T. Batina

NASA Langley Research Center
Hampton, Virginia 23681-0001

ABSTRACT

The development of computational fluid dynamics (CFD) methods for unsteady aerodynamic analysis is described. Special emphasis is placed on considerations that are required for application of the methods to unsteady aerodynamic flow problems. Two broad categories of topics are discussed including grid considerations and algorithm development considerations, and example calculations are presented to illustrate the major points. Although the primary application of these CFD methods is to relatively low-frequency oscillatory phenomena such as flutter, the ideas that are presented may be of value to developers of computational aeroacoustics methods for predicting high-frequency acoustics.

1. Introduction

Considerable progress in developing computational fluid dynamics (CFD) methods for aerodynamic analysis has been made over the past two decades (Jameson, 1987; Edwards and Thomas, 1987; and Edwards and Malone, 1991). Although the vast majority of this work has been on the development of methods for steady-state aerodynamic applications, significant progress also has been made in developing CFD methods for unsteady aerodynamic and aeroelastic applications (Edwards and Thomas, 1987; and Edwards and Malone, 1991). This latter work has been focused primarily on potential flow methods (Edwards, 1986), either at the transonic small-disturbance (TSD) or full-potential equation levels, although research is concentrated currently on developing advanced codes for numerical solution of the Euler or Navier-Stokes equations (Edwards and Malone, 1991).

The development of methods for unsteady applications generally has lagged the development of steady methods, primarily because of additional complicating considerations that arise for unsteady applications. Therefore, the purpose of the paper is to describe the development of CFD methods for unsteady aerodynamic analysis with special emphasis on the considerations that are required because of the unsteady application of the methods. These considerations may be divided into two broad categories including grid considerations and algorithm development considerations. In the category of grid considerations, the paper discusses (1) the type of grid, (2) generation details, (3) boundary treatment, and (4) mesh movement. In the category of algorithm development, the paper describes (1) spatial discretizations, (2) temporal discretizations, and (3) adaption techniques. Also, although the primary application of the unsteady aerodynamic methods described herein is to relatively low-frequency oscillatory phenomena such as flutter, the ideas that are presented may be of value to developers of computational aeroacoustics (CAA) methods for predicting high-frequency acoustics.

2. Grid Considerations

2.1. Type of grid

The first topic in the category of grid considerations is whether a structured or unstructured grid is used (Batina, 1989a). Generally, either type of mesh topology is applicable to steady or unsteady problems, and the use of each has advantages and disadvantages. The majority of work that has been done in CFD over the years has been on developing methods for use on computational grids that have an underlying geometrical structure and thus are referred to as "structured" grids. For example, Fig. 1 shows a structured grid for the NACA 0012 airfoil. The grid is of C-type topology, has 159 points in the wraparound direction, and 49 points in the outward direction. Unsteady applications of methods developed for structured grids generally have been limited to relatively simple geometries such as airfoils, wings, and wing-body configurations (Edwards and Malone, 1991). Extensions to more complex configurations often require more sophisticated meshing methodologies such as blocked, patched, chimera, or hybrid type grids. These extensions, in turn, significantly complicate the solution algorithms. Other difficulties arise in moving the grid for unsteady or aeroelastic motion where the grid must conform to the instantaneous shape of the geometry being considered.

An alternative approach is the use of unstructured grids (Batina et al., 1991). In two dimensions, these grids are constructed from triangles, and in three dimensions, they consist of an assemblage of tetrahedral cells. The triangles or tetrahedra are oriented in an arbitrary way to conform to the geometry, thus making it possible to treat very complicated shapes. Unsteady aerodynamic and aeroelastic applications of these methods to complete aircraft configurations already have been made (Rausch, Batina, and Yang, 1992). An unstructured grid for the NACA 0012 airfoil is shown in Fig. 2 (Batina, 1990). The total grid has 3300 nodes and 6466 triangles. From the figure it is obvious that the cells are arranged in an arbitrary manner to create a mesh about the airfoil. Figure 3 shows an example of a surface mesh representing the Boeing 747 transport configuration from an unstructured tetrahedral mesh (Batina, 1991). The total mesh has 101,475 cells for the half-span airplane. There are also over 8000 triangles which lie on the boundaries of the mesh, which include the half-span airplane, the symmetry plane, and the far-field boundaries. The figure demonstrates that additional aircraft components such as pylons and engine nacelles can be modeled easily with an unstructured grid. To further illustrate the point, Fig. 4 shows the details of the grid and steady pressure coefficient contours on the outboard pylon and engine nacelle of the 747 airplane (Batina, 1991). The pressure coefficients were computed using an Euler code for a freestream Mach number M_∞ of 0.84 and an angle of attack α_0 of 2.73°.

2.2. Generation details

Concerning grid generation details, for steady-state external-flow applications, grids generally are constructed to be fine near the body and coarse near the outer boundaries with some type of stretching in between. This gridding philosophy is reasonably accurate (because of mesh fineness near the body) and efficient (because of few cells in the far field where flow gradients are small) for steady problems but may be inadequate for unsteady problems. For unsteady problems, waves that are propagating through the mesh may be reflected internally if the grid stretches too rapidly (Seidel, Bennett, and Whitlow, 1982; Bland, 1989; and Bland, 1991). To illustrate this problem, Fig. 5 shows an example of internal grid reflections. The calculations were performed for a flat plate airfoil at $M_\infty = 0.85$ using the TSD equation on two Cartesian meshes (Seidel, Bennett, and Whitlow, 1982). The results shown in Fig. 5(a) were obtained using a grid where the points in the vertical direction were determined by an exponential stretching, whereas the results of Fig. 5(b) were obtained using a vertical grid with a more gradual quadratic stretching. As shown in Fig. 5(a), the real and imaginary parts of the unsteady lift-curve slope as a function of reduced frequency k have oscillations which are due to internal grid reflections. As shown in Fig. 5(b), however, the oscillations have virtually disappeared due to the more gradual stretching of the vertical grid. From these results, it is evident that careful attention should be paid to the stretching of the grid to alleviate or eliminate internal grid reflections for unsteady aerodynamic applications.

2.3. Boundary treatment

In addition to possible internal grid reflections, waves may also reflect off of the outer boundaries of the mesh, propagate back into the interior of the computational domain, and contaminate the near-field solution (Seidel, Bennett, and Whitlow, 1982). To demonstrate this problem, Fig. 6 shows an example of the effects of boundary reflections. Again the calculations were performed for a flat plate airfoil at $M_\infty = 0.85$ using the TSD equation. As shown in Fig. 6(a), the real and imaginary parts of c_{ℓ_α} have oscillations for low values of k due to boundary reflections. These oscillations, as well as those that occur from internal grid reflections, are of concern because they occur in the range of reduced frequency where the calculations need to be most accurate since this is the frequency range where flutter typically occurs. As shown in Fig. 6(b), however, when so-called nonreflecting far-field boundary conditions (Whitlow, 1984) are used, the oscillations no longer occur. This is because the non-reflecting conditions tend to absorb waves that are incident on the boundaries. From these results, similar to the treatment of internal grid reflections, it is evident that careful attention also must be given to an accurate treatment of the far-field boundary conditions to alleviate or eliminate boundary reflections.

2.4. Mesh movement

A final grid consideration in the development of CFD methods for unsteady aerodynamic and aeroelastic analysis is how to move or deform the mesh so that it continuously conforms to the instantaneous shape or position of the vehicle. The mesh movement procedure must be general enough to treat realistic aeroelastic motions of complex aircraft configurations. This can be accomplished by modeling the mesh with a network of springs as depicted in Fig. 7 (Batina, 1989b). Each edge of each cell is modeled using a spring, where the spring stiffness is inversely proportional to the length of the edge. Points on the outer boundary of the grid are held fixed and the locations of the points on the inner boundary (aircraft) of the grid are specified either through the aeroelastic equations of motion or the prescribed unsteady motion. The displacements of the interior nodes then are determined by solving the static equilibrium equations which result from a summation of forces in each coordinate direction. In practice, these equations are solved using a predictor-corrector procedure. The displacements of the nodes are first predicted using a simple linear extrapolation in time of displacements from previous grids given by (Batina, 1989b)

$$\tilde{\delta}_{x_i} = 2\delta^n_{x_i} - \delta^{n-1}_{x_i} \quad \tilde{\delta}_{y_i} = 2\delta^n_{y_i} - \delta^{n-1}_{y_i} \quad \tilde{\delta}_{z_i} = 2\delta^n_{z_i} - \delta^{n-1}_{z_i}. \tag{1}$$

The displacements of the nodes are then corrected by solving the static equilibrium equations defined by (Batina, 1989b)

$$\delta^{n+1}_{x_i} = \frac{\sum k_m \tilde{\delta}_{x_m}}{\sum k_m} \quad \delta^{n+1}_{y_i} = \frac{\sum k_m \tilde{\delta}_{y_m}}{\sum k_m} \quad \delta^{n+1}_{z_i} = \frac{\sum k_m \tilde{\delta}_{z_m}}{\sum k_m} \tag{2}$$

using several Jacobi iterations. To demonstrate mesh movement, instantaneous meshes are presented for a half-span airplane pitched nose up 15° in Fig. 8(a) and pitched nose down 15° in Fig. 8(b). With the spring network the mesh moves smoothly to conform to the instantaneous position of the pitching airplane. The mesh movement procedure using the network of springs is a general method that can treat realistic motions as well as aeroelastic deformations of complex aircraft configurations.

3. Algorithm Development Considerations

3.1. Spatial discretizations

For either steady or unsteady flow applications, the residual (right-hand-side of the governing fluid flow equations) needs to be discretized in space. Generally speaking, there are two types of spatial discretizations including central differencing and upwind differencing

(Batina, 1991). Either type of differencing has advantages and disadvantages depending upon a number of things such as the problem being solved and the density of the mesh. Specifically, central differencing uses straightforward central differences to approximate all of the spatial derivatives. This type of approach, though, requires explicitly added artificial dissipation terms to include dissipation in the solution. The unsteady Euler equations, for example, are a set of nondissipative hyperbolic partial differential equations which require some form of dissipation to allow convergence to steady state. Furthermore, the explicitly added dissipation terms involve free parameters to control the level of dissipation in the problem. In contrast, upwind differencing accounts for the local wave-propagation characteristics of the flow and thus is naturally dissipative. Consequently, the upwind-approach does not require the adjustment of free parameters to control the dissipation.

Advantages of the central-difference type spatial discretizations are that they are easier to code and take less memory than upwind discretizations (Batina, 1991). A disadvantage is that they tend to smear shock waves and contact discontinuities and consequently require finer meshes to achieve similar accuracy. Advantages of the upwind-difference type spatial discretizations are that they tend to minimize the artificial dissipation in the problem that is being solved, since they are naturally dissipative, and consequently discontinuities such as shocks and contacts are captured sharply. This attribute of the upwind discretizations in fact may be important for CAA calculations to help ensure that the dissipation is small and thus does not destroy the relatively weak acoustic field. A disadvantage of the upwind methods, however, is that they are generally more difficult to code and require more memory than central-difference methods.

To demonstrate the sharp shock-capturing features of the upwind approach for the spatial discretization of the Euler equations, steady and unsteady results are presented for the NACA 0012 airfoil (Batina et al., 1991). Both sets of results were obtained using the flux-difference splitting of Roe. The steady calculation was performed for $M_\infty = 0.8$ and $\alpha_0 = 1.25°$ with the resulting pressure distribution shown in Fig. 9. For this case, there are shock waves on the upper and lower surfaces of the airfoil. The shocks are sharply captured with only one grid point within the shock structure on either surface. Additionally, these sharp shock capturing features of the upwind method carry over to unsteady cases as well. For example, for the NACA 0012 airfoil pitching harmonically at $M_\infty = 0.755$, $\alpha_0 = 0.016$, with an oscillation amplitude of $\alpha_1 = 2.51°$ at $k = 0.0814$, instantaneous pressure distributions at eight points in time during a cycle of motion are shown in Fig. 10. This is a very interesting case since the shock waves on the upper and lower surfaces of the airfoil periodically appear and disappear during the cycle of motion. It is clear from the results of Fig. 10, that similar to the steady-state example, the calculated shock waves are sharply captured with only one point within the shock structure.

3.2. Temporal discretizations

For unsteady applications, the temporal accuracy and efficiency of the numerical scheme that is used to integrate the governing flow equations are of significant importance. Generally, there are two types of time-integration methods referred to as explicit and implicit (Batina, 1991). Either type of integration method has advantages and disadvantages depending upon a variety of factors including the type of unsteady problem, the density of the mesh, the characteristic frequency of the problem, etc. Specifically, the most commonly used explicit temporal discretization (for the integration of the Euler or Navier-Stokes equations) is a multi-stage Runge-Kutta time integration. Codes that are based on a Runge-Kutta integration also typically use local time-stepping, implicit residual smoothing, and multi-grid techniques to accelerate convergence to steady state. Local time-stepping uses the maximum allowable step size at each grid point as determined by a local stability analysis. Implicit residual smoothing permits the use of local time steps that are larger than those required by the CFL (Courant–Friedricks–Lewy) condition. This is accomplished by averaging the residual with values from surrounding grid points. Multi-grid uses corrections that are determined on a sequence of grids of different density to also accelerate convergence to steady state. As for implicit temporal discretizations, factored methods are typically used on structured grids, whereas relaxation procedures, usually of the Gauss-Seidel form, are used on unstructured grids. The implicit discretizations may be time-accurate for unsteady problems and they allow large CFL numbers for rapid convergence to steady state. Generally speaking, codes based on an implicit integration do not require the above-mentioned techniques for convergence to steady-state, although they usually use local time-stepping and sometimes use multi-grid.

A typical four-stage Runge-Kutta time-integration scheme that is used to solve the Euler equations is given by (Batina, 1991)

$$\begin{aligned} Q^{(0)} &= Q^n \\ Q^{(1)} &= Q^{(0)} - \frac{1}{4}\frac{\Delta t}{V} R(Q^{(0)}) \\ Q^{(2)} &= Q^{(0)} - \frac{1}{3}\frac{\Delta t}{V} R(Q^{(1)}) \\ Q^{(3)} &= Q^{(0)} - \frac{1}{2}\frac{\Delta t}{V} R(Q^{(2)}) \\ Q^{(4)} &= Q^{(0)} - \frac{\Delta t}{V} R(Q^{(3)}) \\ Q^{n+1} &= Q^{(4)}. \end{aligned} \tag{3}$$

(A five-stage scheme is typically used to solve the Navier-Stokes equations.) With this type of scheme the residual at any stage of the scheme is evaluated using the flow variables computed in the previous stage. Although the integration constants 1/4, 1/3, 1/2, and 1 are somewhat arbitrarily defined, the last constant must be unity, and the next-to-last constant is 1/2 for second-order temporal accuracy. When time accuracy is not important, such as when marching the equations to steady state, the constants may be defined otherwise, provided that the resulting scheme has good stability and damping properties. Advantages of the explicit temporal discretization such as that represented by Eq. (3) are that it is simple to code and takes less memory than an implicit time integration. A disadvantage, however, is that the explicit method generally is inefficient for unsteady problems because of the very small time steps that are required for numerical stability.

A typical implicit temporal discretization for structured grids is a three-factor, spatially-split method given by

$$\left[I + \frac{\Delta t}{\mathrm{vol}}\delta_x A\right]\left[I + \frac{\Delta t}{\mathrm{vol}}\delta_y B\right]\left[I + \frac{\Delta t}{\mathrm{vol}}\delta_z C\right]\Delta Q = \frac{-\Delta t}{\mathrm{vol}} R. \tag{4}$$

Equation (4) is written in the so-called delta form where ΔQ represents the change in flow variables from one time step to the next.

The equation as written is first-order-accurate in time (a second-order-accurate version simply adds a time derivative term to the right-hand side) and is efficient because each of the implicit factors on the left-hand side of the equation involves a spatial derivative in only one coordinate direction. The equation is solved by performing three sweeps through the mesh, each of which uses one of the implicit factors.

A typical implicit temporal discretization for unstructured grids is a Gauss-Seidel relaxation procedure given by (Batina, 1991)

$$\left[\frac{\mathrm{vol}}{\Delta \mathrm{t}} I + \sum_{m=1}^{4} A^{+}(Q_j)\Delta S\right] \Delta Q_j + \sum_{m=1}^{4} A^{-}(Q_m)\Delta S \Delta Q_m = -R \quad (5)$$

which is also first-order accurate in time. The relaxation procedure is implemented by first ordering the cells that make up the unstructured mesh, usually from upstream to downstream. Equation (5) is then solved by performing two sweeps through the mesh, one in the forward direction and one in the backward direction. As the sweeps are performed, the equation is solved directly for ΔQ_j while the values for ΔQ_m are updated in a Gauss-Seidel fashion.

Either implicit method (Eq. (4) or Eq. (5)) requires the calculation of the flux jacobian matrices represented by A, B, and C in Eq. (4) and by A^+ and A^- in Eq. (5). These matrices are defined as the derivative of the respective flux that appears in the residual by the conserved variables. Advantages of the implicit temporal discretizations are that they are numerically stable for large CFL numbers and consequently enable rapid convergence to steady state (Batina, 1991). Furthermore for unsteady applications, they allow the selection of the step size based on the physical problem that is being solved rather than on numerical stability considerations. A disadvantage though, is that they require more memory than an explicit method, primarily due to having to store the flux jacobians. Also, the linearization and either factorization (Eq. (4)) or relaxation (Eq. (5)) errors associated with the implicit methods may be too large for a given step size and thus contaminate the solution. To illustrate this problem, Fig. 11 shows the effects of step size on instantaneous pressure distributions using the Gauss-Seidel relaxation procedure of Eq. (5) (Batina, 1991). The calculations are for the same pitching NACA 0012 airfoil case presented in Fig. 10. Here though, three sets of results were obtained corresponding to using 250, 1000, and

2500 steps per cycle of motion. The figure shows the instantaneous pressure distribution at one instant in time (instantaneous angle of attack $\alpha(\tau)$ equal to 2.34° which is 69°($k\tau$) into a cycle of motion). These results indicate that large errors in the strengths and locations of the shock waves on the upper and lower surfaces of the airfoil can occur when too large of a step size is used (corresponding to 250 steps per cycle). However, when an appropriately small step size is used (corresponding to 2500 steps per cycle), the correct solution is obtained with a shock of moderate strength on the upper surface and subcritical flow (no shock wave) about the lower surface.

3.3. Adaption techniques

Other considerations in the category of algorithm development involve spatial (Rausch, Batina, and Yang, 1991) and temporal (Kleb, Batina, and Williams, 1990) adaption techniques. These techniques are usually implemented on unstructured grids to produce solutions of high spatial and temporal accuracy at minimal computational cost. The procedures are applicable to either steady or unsteady aerodynamic problems, although for the unsteady case, special attention is required to ensure the time-accuracy of the solution. Also, although temporal adaption is normally applied only to unsteady cases, that need not necessarily be the case.

As for spatial adaption (Rausch, Batina, and Yang, 1991) the technique involves an enrichment procedure to add points in high gradient regions of the flow. The technique also involves a coarsening procedure to remove points where they are not needed. This is accomplished by "marking" cells for either enrichment or coarsening using some type of error indicator to judge the local accuracy of the solution. The objective is to produce solutions of high spatial accuracy at minimal computational cost by simply minimizing the total number of cells in the grid. Figures 12(a) and 12(b) show the various combinations of cells that are possible for an unstructured grid of triangles when using enrichment and coarsening procedures, respectively. It is noted that in three dimensions, when using unstructured grids of tetrahedral cells, many more combinations of enrichment and coarsening are possible, all of which must be accounted for in the coding of the spatial adaption techniques.

As an example of mesh enrichment for a steady flow (Batina, 1989a) Fig. 13 shows conical vortex-dominated (Euler) flow solutions for a 75° swept flat plate delta wing at a supersonic freestream Mach number of 1.4. The wing is at 20° angle of attack and has 10° of yaw. The solution was obtained by adapting the original coarse mesh of Fig. 13(a) three times to the instantaneous flow. The final result shown in Fig. 13(b) is a highly accurate solution of the conical Euler equations, produced by using an order of magnitude fewer grid points than if a globally fine mesh were used. To demonstrate the spatial adaption procedures for an unsteady case, results were again obtained for the pitching NACA 0012 airfoil case presented previously. Figure 14(a) shows the instantaneous adapted meshes obtained using three levels of enrichment on a coarse background mesh and Fig. 14(b) shows the corresponding instantaneous density contour lines. The instantaneous meshes and density contours are plotted at the same eight points in time as before. The meshes (Fig. 14(a)) clearly indicate the enrichment in regions near the shock waves and near the stagnation points. They also show coarsened regions where previously enriched regions have relatively small flow gradients. The density contours during the cycle (Fig. 14(b)) demonstrate the ability of the spatial adaption procedures to produce sharp transient shock waves. The results were obtained with a computational savings of a factor of seventeen in comparison to using a globally fine mesh for the same case.

Analogous to spatial adaption, temporal adaption (Kleb, Batina, and Williams, 1990) may be used to improve the computational efficiency of explicit time-integration methods for unsteady aerodynamic applications. Temporal adaption can be thought of as time-accurate local time-stepping where each grid cell is integrated according to the local flow physics and numerical stability. The efficiency of the method comes from using small time steps where they are needed and large time steps where they are not. The "trick" is to accomplish this in a time-accurate manner. Simply stated, the method integrates small cells with small time steps and large cells with large time steps, as depicted in Fig. 15. Time accuracy is maintained by bringing all of the cells to the same time level as dictated by the step size of the largest cell.

To demonstrate the temporal adaption procedure, results were once more obtained for the same pitching NACA 0012 airfoil problem as before (Kleb, Batina, and Williams, 1990). Figure 16 shows calculated results obtained using temporal adaption and global time-stepping as well as comparisons with the experimental pressure data of Landon (1982). The two sets of calculated pressures agree very well with each other. This excellent agreement verifies the time accuracy of the solution computed using temporal adaption, which was obtained at one-fourth of the CPU time that the global time-stepping solution required. Also, both sets of calculated results agree reasonably well with the experimental data.

4. Concluding Remarks

The development of CFD methods for unsteady aerodynamic analysis was described. Special emphasis was placed on considerations that are required for application of the methods to unsteady aerodynamic flow problems. Two broad categories of topics were discussed including grid considerations and algorithm development considerations, and example calculations were presented to illustrate the major points. Although the primary application of these methods is to relatively low-frequency oscillatory phenomena such as flutter, the ideas that were presented may be of value to developers of CAA methods for predicting high-frequency acoustics.

References

Batina, J. T., 1989a. "Unsteady aerodynamics methods for transonic aeroelastic analysis," Paper No. 89–002, presented at the European Forum on Aeroelasticity and Structural Dynamics 1989, Aachen, Germany, April 17–19, 1989.

Batina, J. T., 1989b. "Unsteady Euler algorithm with unstructured dynamic mesh for complex-aircraft aeroelastic analysis," AIAA Paper No. 89–1189.

Batina, J. T., 1990. "Implicit flux-split Euler schemes for unsteady aerodynamic analysis involving unstructured dynamic meshes," AIAA Paper No. 90–0936.

Batina, J. T., 1991. "Unstructured-grid methods development – lessons learned," presented at the 4th International Symposium on Computational Fluid Dynamics, Davis, California, September 9–12, 1991.

Batina, J. T., Lee, E. M., Kleb, W. L., and Rausch, R. D., 1991. "Unstructured-grid methods development for unsteady aerodynamic and aeroelastic analyses," Paper No. 2, presented at the AGARD Structures and Materials Panel Specialist's Meeting on Transonic Unsteady Aerodynamics and Aeroelasticity, San Diego, California, October 9–11, 1991.

Bland, S. R., 1989. "Personal computer study of finite-difference methods for the transonic small disturbance equations," NASA TM 102582.

Bland, S. R., 1991. "Suggestions for CAP-TSD mesh and time step input parameters," NASA TM 104083.

Edwards, J. W., 1986. "Applications of potential theory computations to transonic aeroelasticity," ICAS-86-2.9.1.

Edwards, J. W. and Malone, J. B., 1991. "Current status of computational methods for transonic unsteady aerodynamic and aeroelastic applications," Paper No. 1, presented at the AGARD Structures and Materials Panel Specialist's Meeting on Transonic Unsteady Aerodynamics and Aeroelasticity, San Diego, California, October 9–11, 1991.

Edwards, J. W. and Thomas, J. L., 1987. "Computational methods for unsteady transonic flows," AIAA Paper No. 87-0107.

Jameson, A., 1987. "Successes and challenges in computational aerodynamics," AIAA Paper No. 87-1184.

Kleb, W. L., Batina, J. T., and Williams, M. H., 1990. "Temporal adaptive Euler/Navier-Stokes algorithm for unsteady aerodynamic analysis of airfoils using unstructured dynamic meshes," AIAA Paper No. 90-1650.

Landon, R. H., 1982. "NACA 0012. Oscillating and transient pitching," Data Set 3 in AGARD-R-702, Compendium of Unsteady Aerodynamic Measurements.

Rausch, R. D., Batina, J. T., and Yang, H. T. Y., 1991. "Spatial adaption procedures on unstructured meshes for accurate unsteady aerodynamic flow computation," AIAA Paper No. 91-1106.

Rausch, R. D., Batina, J. T., and Yang, H. T. Y., 1992. "Three-dimensional time-marching aeroelastic analyses using an unstructured-grid Euler method," AIAA Paper No. 92–2506.

Seidel, D. A., Bennett, R. M., and Whitlow, W., Jr., 1982. "An exploratory study of finite difference grids for transonic unsteady aerodynamics," NASA TM 84583.

Whitlow, W., Jr., 1984. "Characteristic boundary conditions for three-dimensional transonic unsteady aerodynamics," NASA TM 86292.

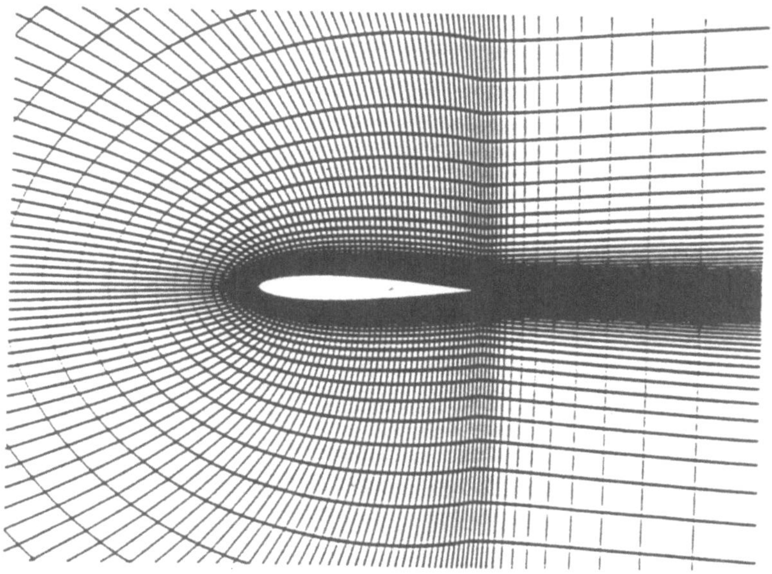

Fig. 1. Structured grid about the NACA 0012 airfoil.

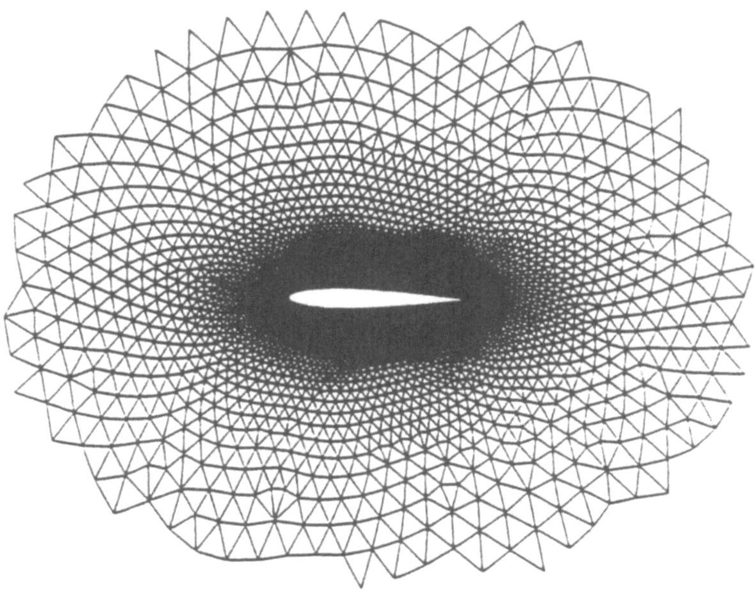

Fig. 2. Unstructured grid about the NACA 0012 airfoil.

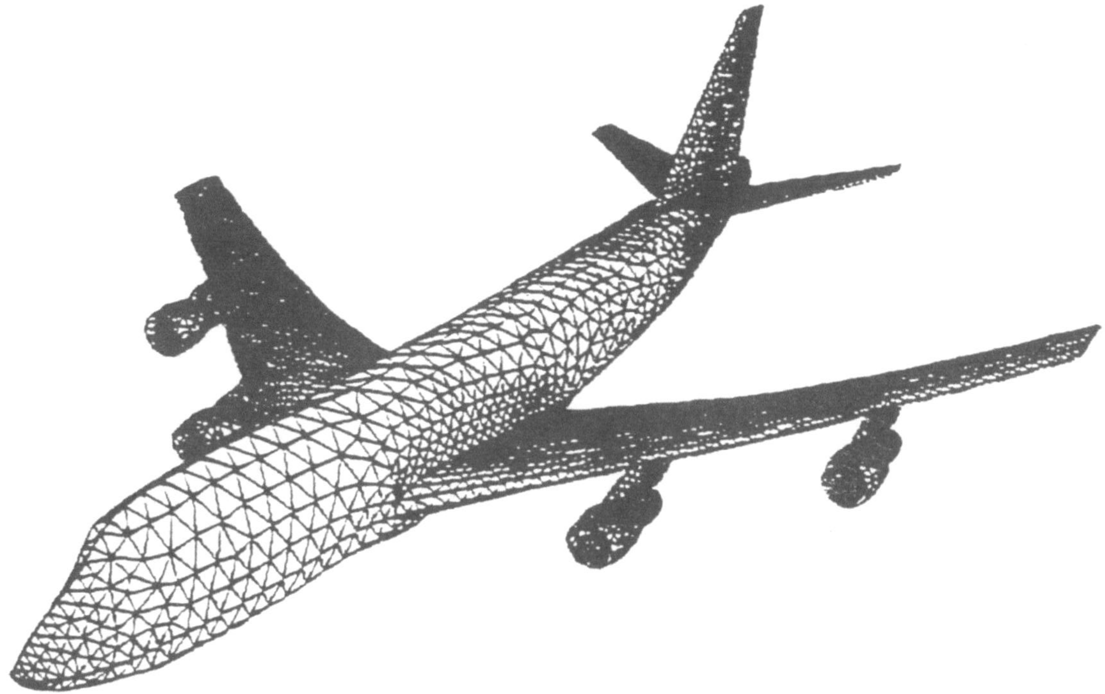

Fig. 3 Surface mesh from an unstructured tetrahedral mesh about the Boeing 747 transport configuration.

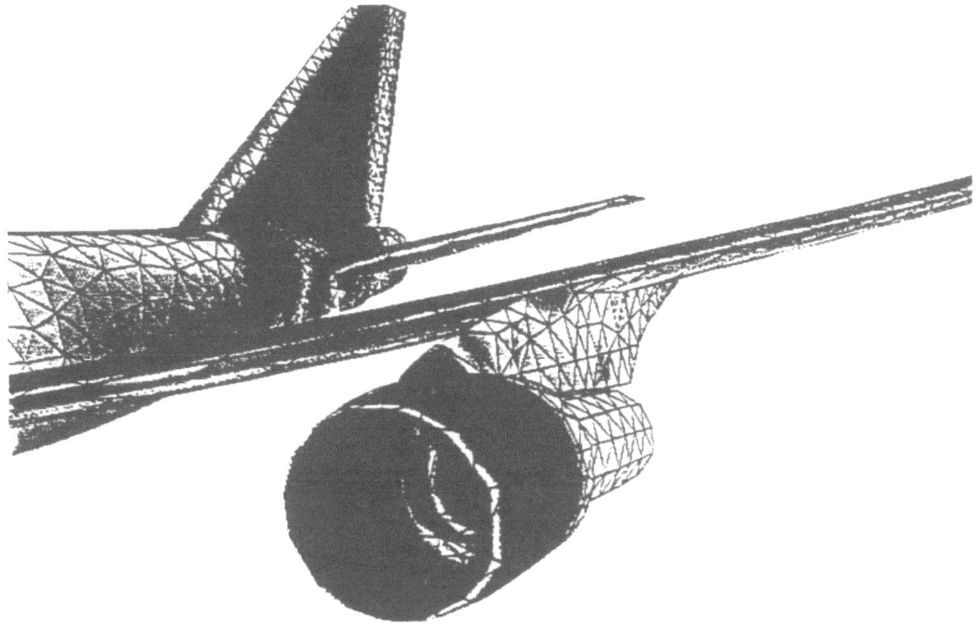

Fig. 4 Details of surface grid and steady pressure contours on the outboard pylon and engine nacelle of the Boeing 747 transport configuration computed using the Euler equations at $M_\infty = 0.84$ and $\alpha_0 = 2.73°$.

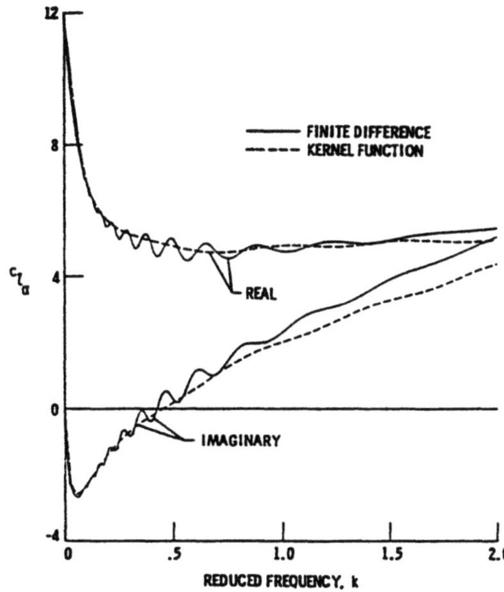

(a) exponential stretching.

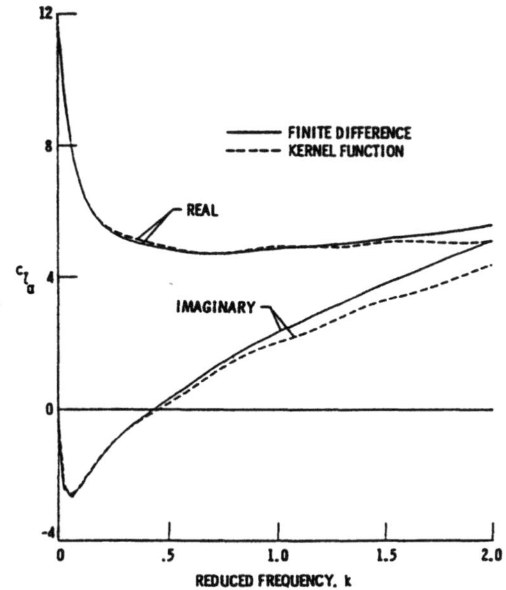

(b) quadratic stretching.

Fig. 5 Effects of grid stretching on the unsteady lift-curve slope of a flat plate airfoil at $M_\infty = 0.85$ and $\alpha_0 = 0°$ computed using the TSD equation on Cartesian meshes.

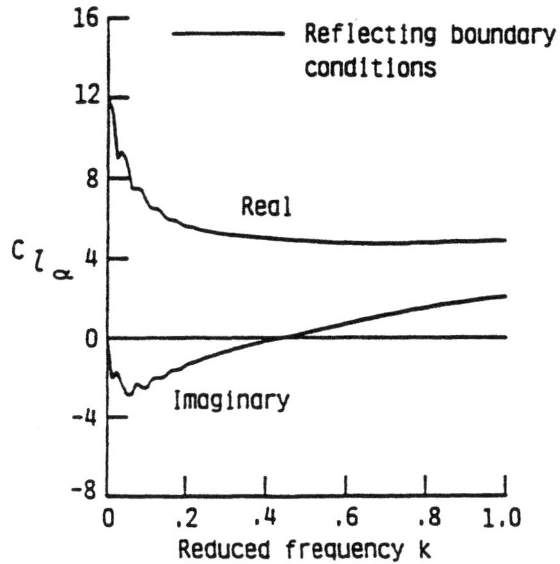

(a) reflecting boundary conditions.

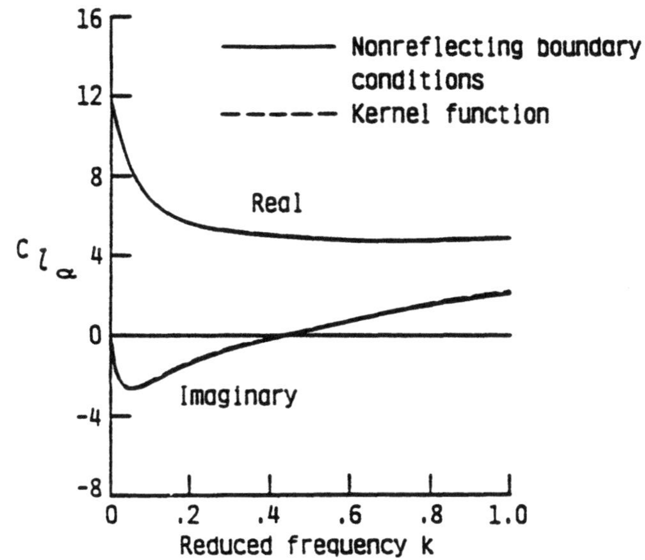

(b) nonreflecting boundary conditions.

Fig. 6 Effects of far-field boundary conditions on the unsteady lift-curve slope of a flat plate airfoil at $M_\infty = 0.85$ and $\alpha_0 = 0°$ computed using the TSD equation.

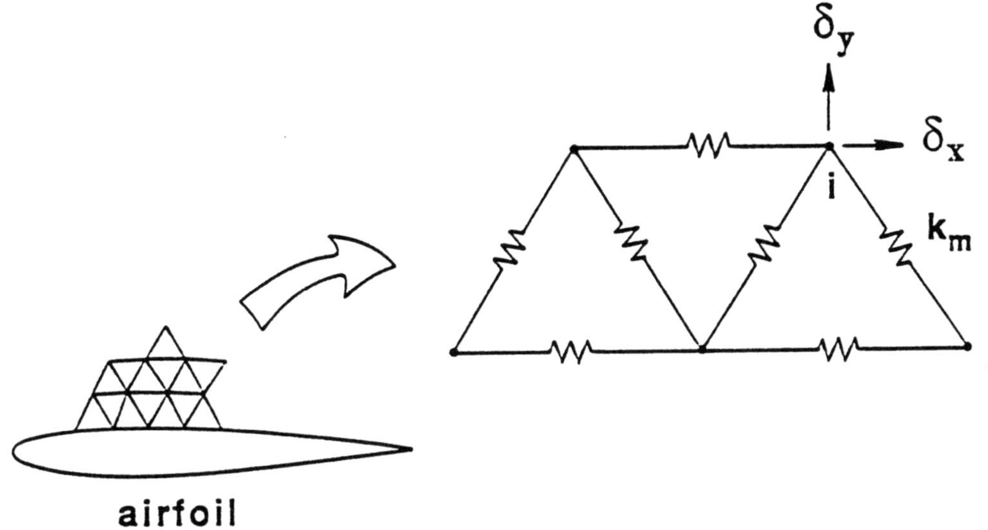

Fig. 7 Mesh modeling by a network of springs.

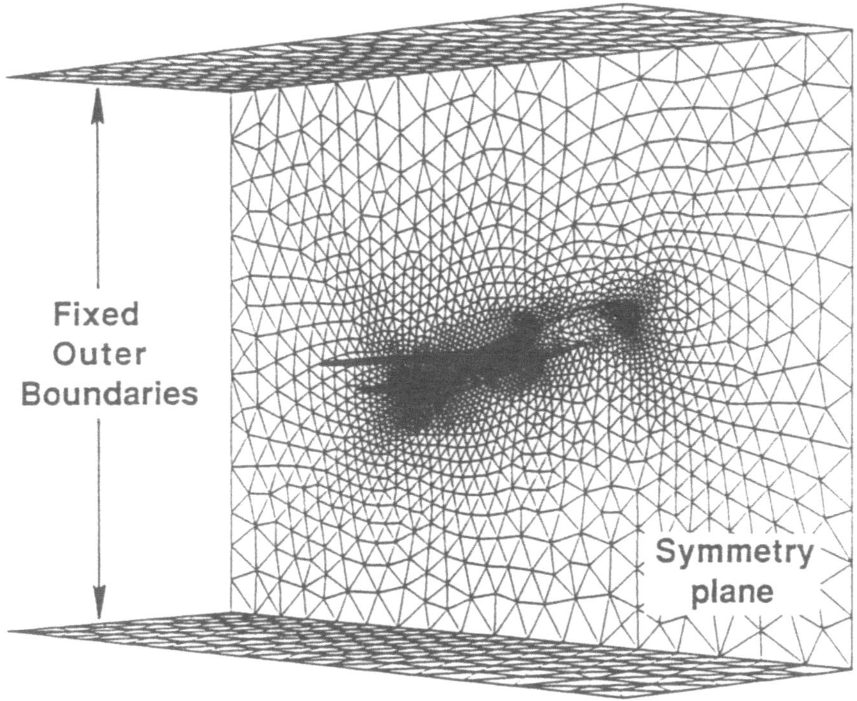

(a) nose up 15°.

Fig. 8 Instantaneous meshes for a pitching airplane.

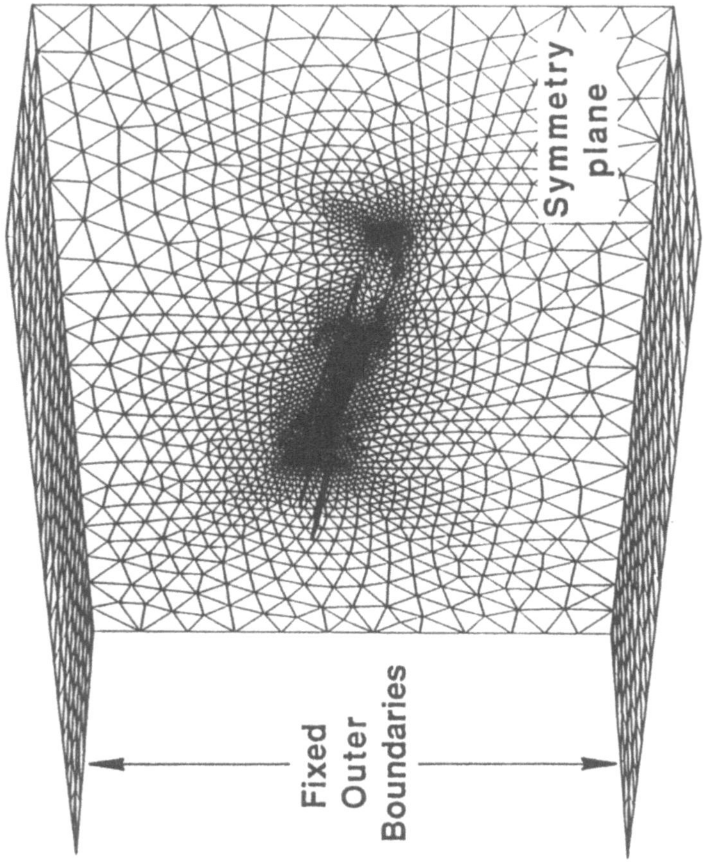

(b) nose down 15°.

Fig. 8 Concluded.

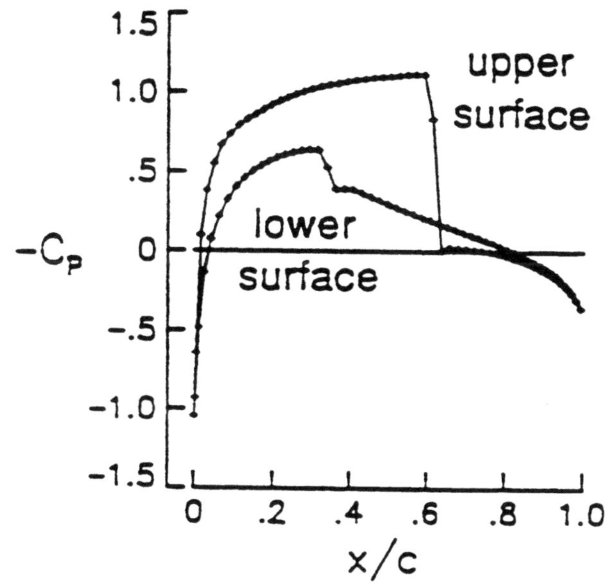

Fig. 9 Steady pressure distribution for the NACA 0012 airfoil at $M_\infty = 0.8$ and $\alpha_0 = 1.25°$ computed using an upwind-type Euler solution algorithm with flux-difference splitting.

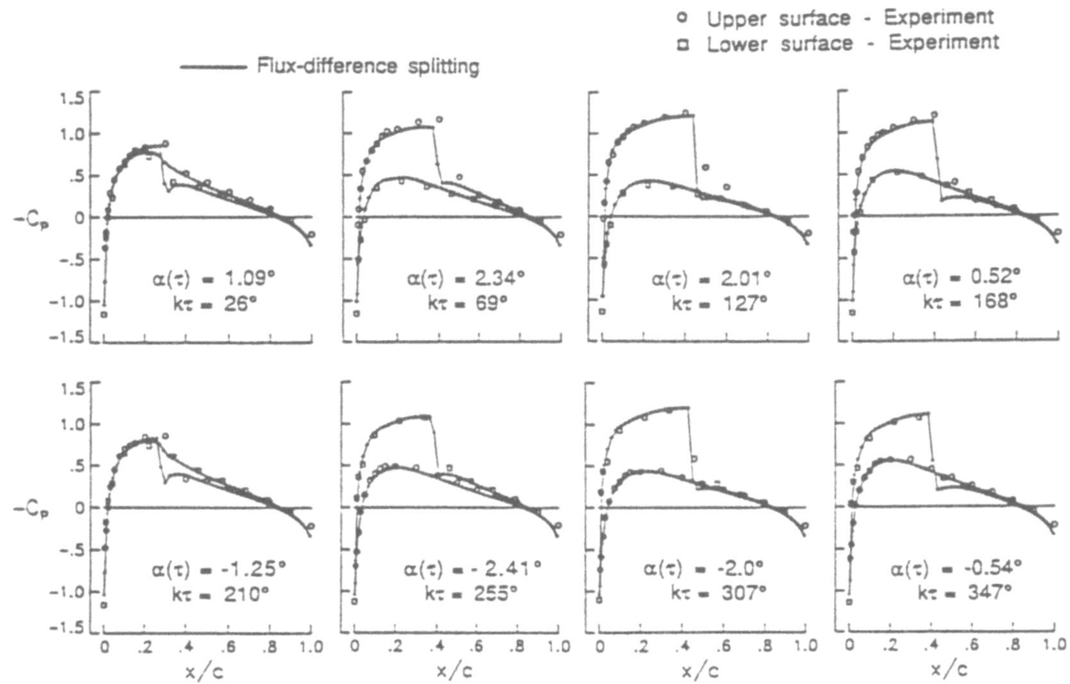

Fig. 10 Instantaneous pressure distributions for the NACA 0012 airfoil pitching harmonically at $M_\infty = 0.755$, $\alpha_0 = 0.016°$, $\alpha_1 = 2.51°$, and $k = 0.0814$ computed using an upwind-type Euler solution algorithm with flux-difference splitting.

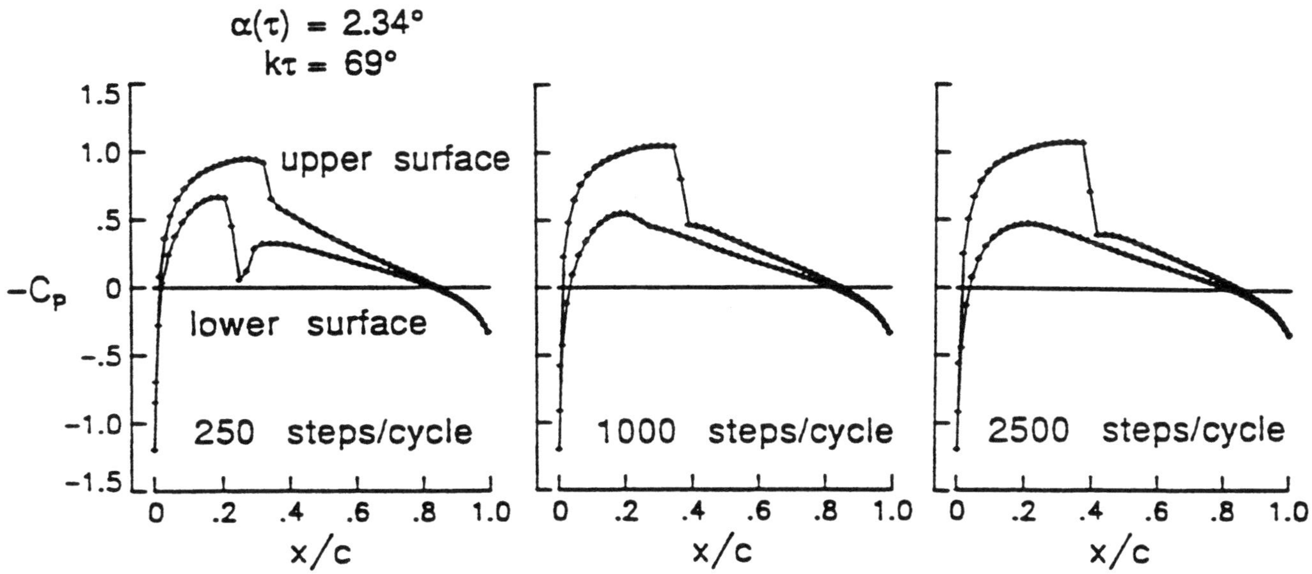

Fig. 11 Effects of step size on the instantaneous pressure distribution at $k\tau = 69°$ (corresponding to $\alpha(\tau) = 2.34°$) for the NACA 0012 airfoil pitching harmonically at $M_\infty = 0.755$, $\alpha_0 = 0.016°$, $\alpha_1 = 2.51°$, and $k = 0.0814$ computed using an implicit Euler solution algorithm.

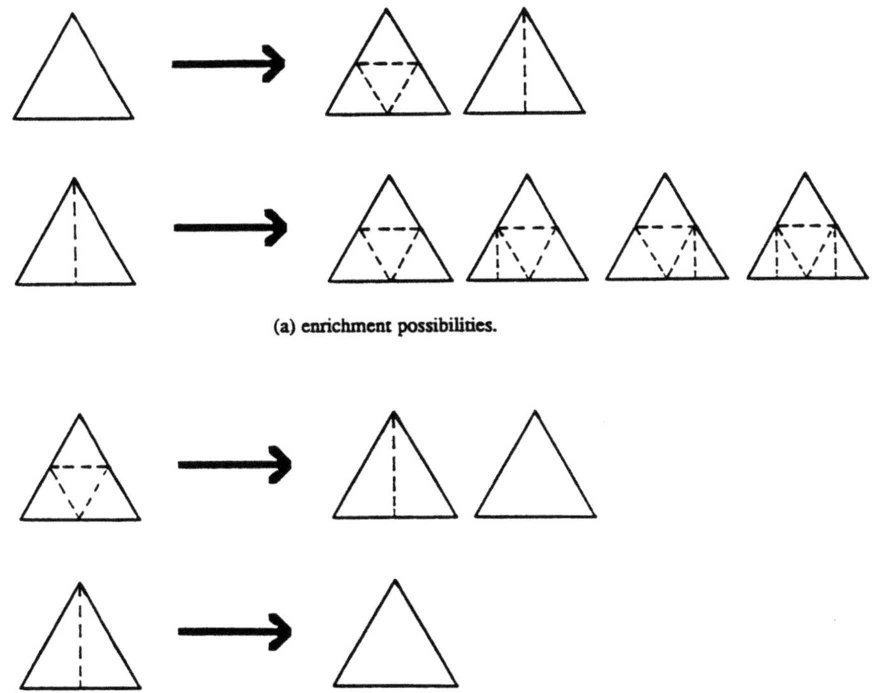

(a) enrichment possibilities.

(b) coarsening possibilities.

Fig. 12 Details of the spatial adaption procedures.

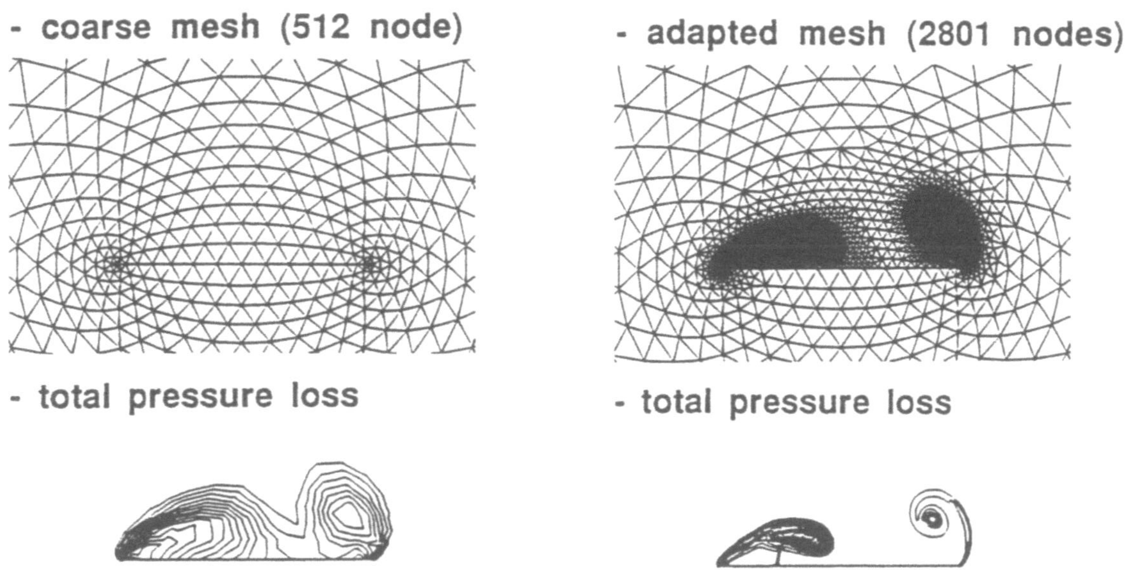

(a) starting solution.

(b) final solution.

Fig. 13 Steady example of mesh enrichment for a 75° swept flat plate delta wing at $M_\infty = 1.4$, $\alpha = 20°$, and $\beta = 10°$ computed using the conical Euler equations.

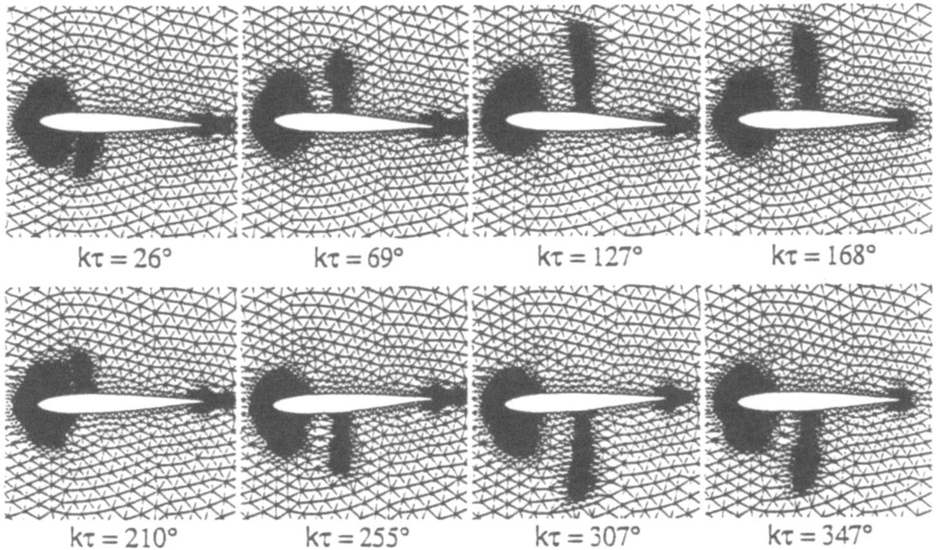

(a) instantaneous meshes.

Fig. 14 Spatial adaption calculation for the NACA 0012 airfoil pitching harmonically at $M_\infty = 0.755$, $\alpha_0 = 0.016°$, $\alpha_1 = 2.51°$, and $k = 0.0814$ computed using an upwind-type Euler solution algorithm with flux-vector splitting.

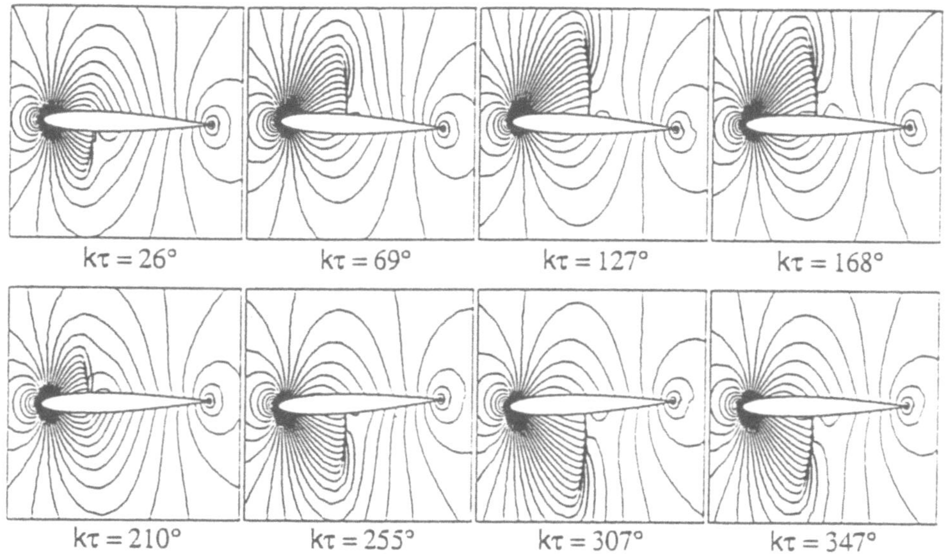

(b) density contour lines.

Fig. 14 Concluded.

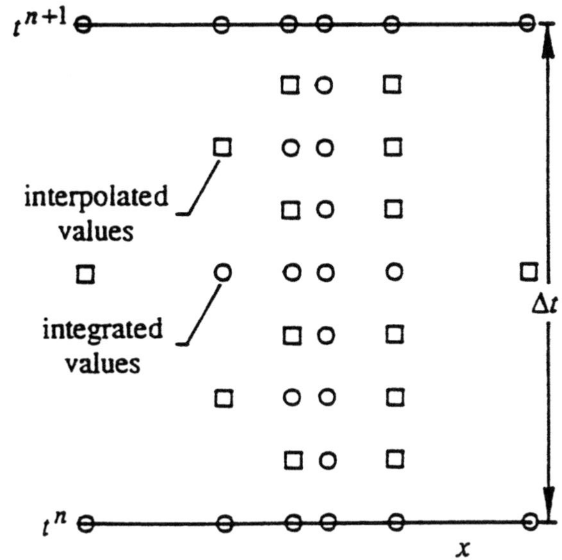

Fig. 15 Temporal stencil illustrating the details of the temporal adaption technique.

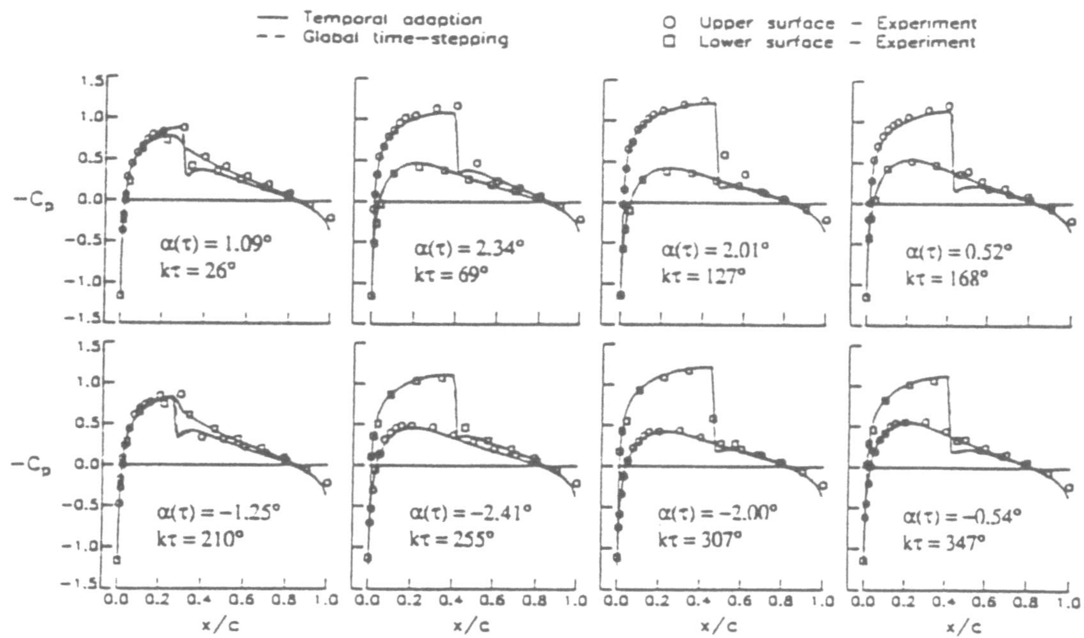

Fig. 16 Comparison of instantaneous pressure distributions for the NACA 0012 airfoil pitching harmonically at $M_\infty = 0.755$, $\alpha_0 = 0.016°$, $\alpha_1 = 2.51°$, and $k = 0.0814$ computed using an upwind-type Euler solution algorithm with flux-vector splitting.

NUMERICAL STUDY OF FLOW/STRUCTURE INTERACTION OVER A FLEXIBLE PLATE[1]

A. Frendi

Analytical Services and Materials, Inc.
Hampton, Virginia 23665

L. Maestrello

NASA Langley Research Center
Hampton, Virginia 23681-0001

A. Bayliss

Northwestern University
Evanston, Illinois 60208

ABSTRACT

We study the coupling between an acoustically loaded flexible plate and the surrounding fluid. We consider two cases: (i) a viscous boundary layer flow is established over the flexible plate, and (ii) the fluid is assumed to be inviscid and excited only by the motion of the flexible plate. In both cases, the evolution of the fluid flow determines the pressure in the fluid which in turn is employed as a source term for the motion of the flexible plate. In the case that the flexible plate oscillates in a viscous boundary layer, we find an increase in disturbance amplitude and vorticity production as well as the formation of transient inflection points in the streamwise velocity field. In the case of an inviscid fluid excited by the motion of the plate, we find that the coupling acts as a stabilizing mechanism so that for a given acoustic loading the nonlinear response of the plate is reduced over the case where the fluid motion is decoupled from the plate response. When the acoustic loading is large, high frequencies tend to predominate as the pressure disturbance is propagated into the far field, leading to a progressive steepening in the pressure disturbance with distance from the flexible plate.

---

[1]Support for A. Frendi was provided by NASA Langley Research Center under Contract NAS1-19317. A. Bayliss was partially supported by NASA Contract Nos. NAS1-18107 and NAS1-18605 while in residence at the Institute for Computer Applications in Science and Engineering (ICASE). Additional support was provided by NSF Grant MSS-91-02981.

1. Introduction

In this paper, we consider the problem of an acoustically loaded flexible plate vibrating in a fluid. The fluid motion is coupled to the vibration of the plate. We consider two cases. In the first case, we assume that the plate is vibrating in a viscous, supersonic, laminar boundary layer. In the second case, we assume that the fluid is inviscid and is initially at rest under ambient conditions. The objective is to determine the effect of fluid/structure interaction under different flow conditions, in particular the interaction between structural vibrations and fluid viscosity and the nonlinear propagation of acoustic waves generated by plate vibration.

The problem of boundary layer stability over rigid plates has been studied for both subsonic and supersonic flows (e.g. Mack, 1990). However, little work has been done on the stability of a supersonic boundary layer over a flexible plate. It is known that the pressure field developed in the boundary layer can lead to significant structural vibration which in turn can be a source of noise (Maestrello, 1970). The effect of these additional sound sources on the underlying flow field may be an important factor in the design of high speed transports.

Recently, Frendi (Frendi *et al.*, 1992a) found that the stability of a supersonic boundary layer is not affected by the presence of a flexible plate when the plate is excited only by two dimensional instability waves. This result was due to the large difference between the vibration frequencies of the plate and the instability frequencies. However, when the plate was forced to vibrate by plane waves at normal incidence emanated by a source on the opposite side of the boundary layer, the stability of the boundary layer was altered due to the large amplitude of the plate vibrations. In particular, there was an increase in disturbance level and the appearance of a transient inflection point in the streamwise velocity profile. The flow parameters chosen for that calculation were such that the boundary layer was stable over the entire length of the flexible plate. In this paper, the flow parameters have been changed in such a way that the streamwise disturbance undergoes significant growth before it decays.

The second problem addressed here is that of coupling between the vibration of a flexible plate and the acoustic radiation. This work was motivated by a recent experimental and numerical work, (Maestrello *et al.*, 1992), in which a flexible plate was anchored between two long rigid plates. In the experiments, the frequency of the acoustic source was near a resonant frequency of the plate. The amplitude of the acoustic source was progressively increased. Stress measurements on the flexible plate and some limited flow field measurements revealed that, as the level of the acoustic loading was increased, the response of the plate underwent the following sequence of transitions: (i) nearly linear response (ii) development of harmonics - nonlinear response (iii) development of subharmonics indicating a possible bifurcation and (iv) development of a fully chaotic response.

A numerical investigation of this problem was also carried out. The numerical computations were restricted to solving the equations for the plate itself. No coupling between the fluid and plate was accounted for, rather the pressure above the plate, a source term for the plate equations, was specified as the ambient pressure. The behavior indicated by the computations was qualitatively similar to that observed in the experiments.

Many other studies in structural acoustics have employed models which do not account for the coupling with the underlying fluid. These models, while significantly less costly than models which do account for the coupling, cannot generally produce an accurate pressure both in the near field and the far field. In addition, models which neglect the coupling may not accurately predict the plate response, as accounting for the inertia of the underlying fluid changes its total damping and therefore its response (Robinson et al., 1992). Simplified models which do account for this coupling have been studied analytically. In a paper by Foda (Foda, 1990), the behavior of an infinitely long, periodically clamped vibrating plate was analyzed. The analysis predicted a steepening of the pressure field as the acoustic disturbance propagated out to the far field, and in addition indicated that shocks could form under certain conditions.

We now summarize our results. In the first problem, we find that a relatively low level source can generate large disturbances both within and beyond the boundary layer. For a given level of forcing, the response of the flow field is highly frequency dependent. As the frequency is lowered, lower modes of vibration are excited on the flexible plate. These lower modes are more efficient in transmitting acoustic energy into the flow field and consequently in increasing the level of the fluctuating disturbances within the flow field. For sufficiently low frequencies, we find low level acoustic disturbances resulting in a significant increase of the disturbance level and relatively large transient changes in the structure of the boundary layer. In contrast, for sufficiently high excitation frequencies, there is very little coupling between the boundary layer and the flexible plate.

In the second problem, we consider the coupled behavior for different levels of acoustic forcing. We compare our results with computations which do not account for coupling between the fluid and structure. We consider excitations at a resonant frequency of the flexible plate. We find that the fluid coupling acts as a stabilizing mechanism for the plate, in that the response of the plate is reduced. Thus the result of neglecting the coupling with the fluid is to overpredict the level of nonlinearity on the plate. We also find that there is a progressive dominance of high frequencies in the fluctuating pressure field as the far field is approached. For the parameters that we consider, the largest radiation is due to the clamped supports of the flexible plate and appears to be beamed at low angles measured from the direction of the plate (Frendi et al., 1992b).

2. Numerical Method

In this section, we describe the numerical method. The discussion will be brief, more details can be found in the references. The computations are in two-dimensions so that the solution is assumed independent of the spanwise coordinate. The basic numerical scheme for the flow equations (either the Euler or Navier-Stokes equations) is a 2-4 version of the MacCormack scheme (Gottlieb and Turkel, 1976). This scheme is fourth order accurate on the convective terms of the equations, and second order accurate in time and for the viscous terms (for the Navier-Stokes equations). The scheme has been extensively studied and found to be significantly less dispersive than the second order MacCormack scheme for several different problems in fluid dynamics, including problems involving instability waves and other nonsteady phenomenon (Bayliss et al., 1985). The equations are solved by operator splitting, alternating the x and y directions. Further details were given by Bayliss (Bayliss et al., 1985).

The motion of the flexible plate is described by a one dimensional nonlinear equation for the displacement $w(x,t)$. This equation is solved using an implicit finite difference scheme (Hoff and Pahl, 1988). The pressure difference across the flexible plate acts as a source term in this equation. On one side, the pressure is specified by the acoustic source, while on the other side of the plate the pressure is determined dynamically from the equations governing the fluid motion (either the Euler or Navier-Stokes equations). In turn, the normal velocity is computed from the plate equation and is a boundary condition for the equations governing the fluid motion. In the case of boundary layer flow, the displacement of the plate is much smaller than the boundary layer thickness, so that for the fluid equations, the grid is assumed to be independent of time and the boundary is taken as $y = 0$.

The only physical boundary in the problem is the infinitely long plate at $y = 0$ which has both rigid and flexible regions. The remaining boundaries are artificial. Boundary conditions which minimize reflections must be imposed at these boundaries. We have tested two different boundary conditions. In one case, we impose the normal characteristic entering or exiting the computational domain. This characteristic is obtained from linearizing the equations about the state either at the preceding time step or free stream conditions (Bayliss et al., 1985). In the second case, we have implemented a boundary condition which assumes that the outgoing pressure has the functional form of an outgoing, cylindrically spreading wave in the far field (Bayliss and Turkel, 1980). The Navier-Stokes solutions were obtained using the normal characteristics, while for the Euler equations both boundary conditions have been implemented and tested. For the results presented here, there is very little difference in the computed solutions away from the boundaries for either choice of boundary conditions. In addition, our results have been validated by computing in a larger computational domain and comparing the results of the two computations.

3. Results

For both the viscous and inviscid calculations, the flexible part of the plate is 10 inches long and 0.02 inches thick. The properties of the flexible plate are taken as those of aluminum and are stiffness $D = 15.2$ lbf x in, density $\rho = 10^{-5}$ lbf/in^2, Poisson ratio $\nu = 0.3$ and a variable damping. We first describe the calculations using the Navier-Stokes equations. In this case, we studied a boundary layer with a free stream Mach number (M_∞) of 4.5, Reynolds number per foot (Re/ft) of 2.4 x 10^6, and stagnation temperature (T_t) of 560°R. In our computations, the flow field is excited both by an acoustic source below the flexible part of the plate and by an instability wave generated at the inflow. For the parameters chosen here, however, the contribution of the instability wave to the fluctuating velocity field is small compared to that of the acoustic excitation. The nondimensional frequency of the inviscid, second mode disturbance is $F = 2\pi f\nu/U_\infty^2 = 1.2\text{x}10^{-4}$ corresponding to a dimensional frequency of $f = 10^5$ Hz. In the relation giving F, ν is the kinematic viscosity and U_∞ is the free stream velocity. The sound source frequencies used are subharmonics of f ($f/10$, $f/100$).

The configuration of the two computational domains corresponding to the two cases studied is shown on Fig. 1a-b. Figure 1a shows the computational domain used in the study of coupling between plate vibration and acoustic radiation. The domain is 20 ft long in the x-direction and 10 ft in the y-direction. There were 481 and 241 points in both directions, respectively. The computational domain used in the study of the interaction between a supersonic laminar boundary layer and a flexible plate is shown in Fig. 1b. The dimensions of this domain are one foot in the x-direction and 0.1 ft in the y-direction. There were 751 and 151 points in both directions, respectively. Mappings were employed in both directions in order to enhance the resolution of the boundary layer and regions of large streamwise gradients.

Figure 2 shows the variation of the root-mean-square growth of the fluctuating streamwise mass flux disturbance with downstream distance. Two cases are shown on the figure, the dashed line with squares corresponds to the fluctuations obtained over a rigid plate while the solid line with triangles corresponds to the case of a vibrating flexible plate. The figure shows that the growth over the flexible plate is higher than that over the rigid plate and that the location of the maximum shifts downstream of the leading edge of the flexible plate. This result suggests that the boundary layer is more receptive to plate vibrations near the two clamped edges of the plate. Figure 3 shows the variation of the streamwise velocity profile with time near the region of maximum growth. As the plate motion goes through a cycle, a transient inflection point appears in the velocity profile suggesting a destabilization due to the vibration of the flexible plate.

The results related to the coupling between a flexible plate and an inviscid fluid are obtained for air at ambient conditions and a flexible plate having the same properties as above. The frequency of the acoustic source corresponds to the fifth natural frequency of the plate which is $f = 580$ Hz. Figure 4 shows the power spectral density of the plate displacement at the center for an acoustically coupled and uncoupled plate. For the acoustically uncoupled case (dashed line), the spectrum shows the presence of the fundamental frequency (f) and one subharmonic ($f/4$) along with their harmonics. In the presence of coupling (solid line), the spectrum shows a significant reduction in the level of the subharmonic and its harmonic. The level of the fundamental and its harmonics is also reduced. This result indicates that, for the present input conditions, the response of the flexible plate to a harmonic excitation is quasi-linear in the presence of coupling while it is nonlinear when coupling is neglected. This indicates that coupling with the fluid is a stabilizing mechanism.

Figure 5 shows the pressure contours in the fluid below the plate. As the wave fronts move away from the plate, they propagate spherically into the far field. The contours indicate that the radiation is more intense near the corners of the flexible plate than at the center as a result of a strong scattering due to the clamped boundaries. The power spectra of the input pressure, the radiated surface and far field pressures are shown on Fig. 6a-c, respectively. The power spectrum of the input pressure shows a single peak at a frequency of 580 Hz and with a sound pressure level of 160 dB (Fig. 6a). The spectrum of the radiated pressure at the surface shows a strong peak (140 dB) at the excitation frequency and two lower peaks corresponding to the odd harmonics. These harmonics are the effect of nonlinearities due to the high level of forcing. Similarly, the power spectrum of the far field pressure shows the fundamental frequency and its harmonics. However, as a result of nonlinear wave propagation, the difference in level between the fundamental and its first harmonic is reduced. Thus as the pressure is radiated into the far field the spectral content is weighted more toward the high frequencies. This results in a steepening of the pressure profiles in both space and time.

4. Conclusions

The results obtained from the calculations of boundary layer/flexible plate interaction indicate that when the plate is excited by Tollmien-Schlichting type instability waves, the coupling between plate vibration and the instabilities is weak. However, when a sound source from the opposite side of the boundary layer is used to force the plate, the disturbance level in the boundary layer is increased and transient inflection points appear in the streamwise velocity profile.

Results from the second case studied show that, when acoustic coupling is accounted for, the response of the flexible plate is reduced. Specifically, for a given forcing level, the development of subharmonics is reduced. As the pressure disturbance propagates into the far field, the spectral content shifts toward higher frequencies.

References

Bayliss, A., Parikh, P., Maestrello, L., and Turkel, E., 1985. "A fourth-order scheme for the unsteady, compressible Navier-Stokes equations," AIAA Paper 85-1694.

Bayliss, A. and Turkel, E., 1980. "Radiation boundary conditions for wave-like equations," *Comm. Pure Appl. Math.* **33**, p. 703.

Foda, M. A., 1990. "Analysis of non-linear propagation of waves induced by a vibrating flat plate," *Acustica* **22**, p. 118.

Frendi, A., Maestrello, L., and Bayliss, A., 1992a. "On the coupling between a supersonic laminar boundary layer and a flexible surface," Submitted to *AIAA Journal.*

Frendi, A., Maestrello, L., and Bayliss, A., 1992b. "Coupling between plate response and acoustic radiation," Abstract submitted to the 31st AIAA Aerospace Sciences Meeting and Exhibit, 1993.

Gottlieb, D. and Turkel, E., 1976. "Dissipative two-four methods for time-dependent problems," *Math. Comp.* **30**, p. 703.

Hoff, C. and Pahl, P. J., 1988. "Development of an implicit method with numerical dissipation from a generalized single-step algorithm for structural dynamics," *Comput. Methods Appl. Mech. Engrg.* **67**, p. 367.

Mack, L. M., 1990. "On the inviscid acoustic mode instability of supersonic shear flows, Part I: Two-dimensional waves," *Theoret. Comput. Fluid Dynamics* **2**, pp. 97-123.

Maestrello, L., 1969. "Radiation from a panel response to a supersonic turbulent boundary layer," *J. Sound Vibration* **10**(2), pp. 261-295.

Maestrello, L., Frendi, A., and Brown, D. E., 1992. "Nonlinear vibration and radiation from a panel with transition to chaos induced by acoustic waves," Accepted for publication in *AIAA Journal.*

Robinson, J., Rizzi, S., Clevenson, S., and Daniels, E., 1992. "Large deflection random response of flat and blade stiffened Carbon-Carbon panels," Work presented at the 33rd Annual Meeting on Structures, Structural Dynamics and Materials Conference organized by AIAA-ASME-ASCE-AHS-ASC. Paper AIAA-92-2390.

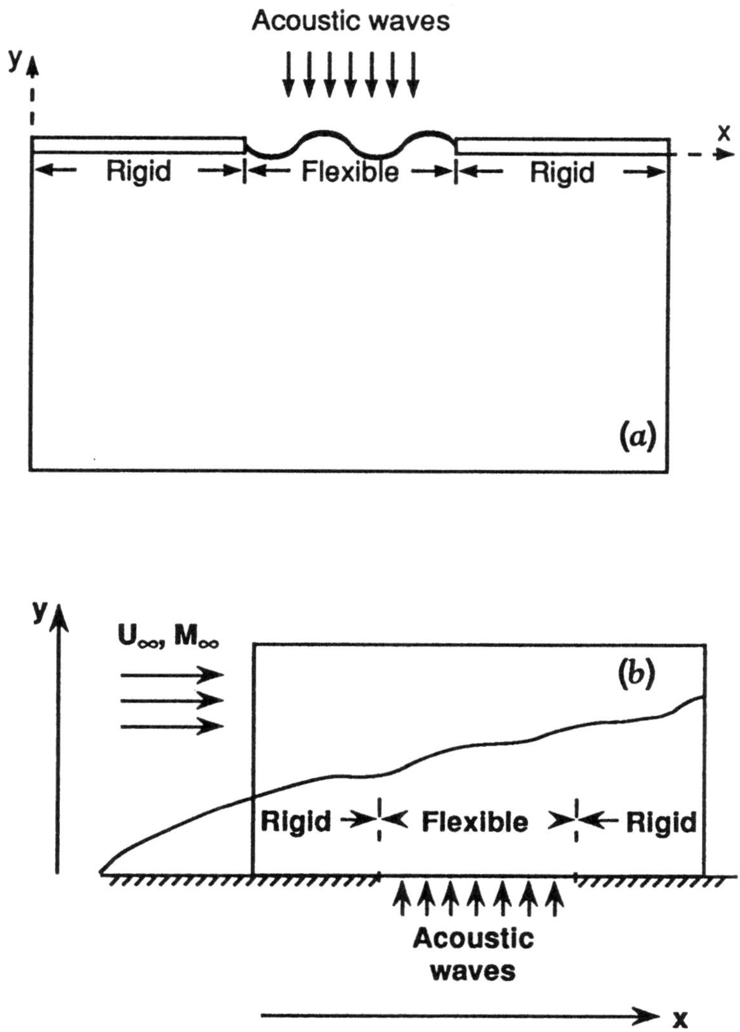

Figure 1. Computational domains (a) inviscid fluid (b) viscous boundary layer.

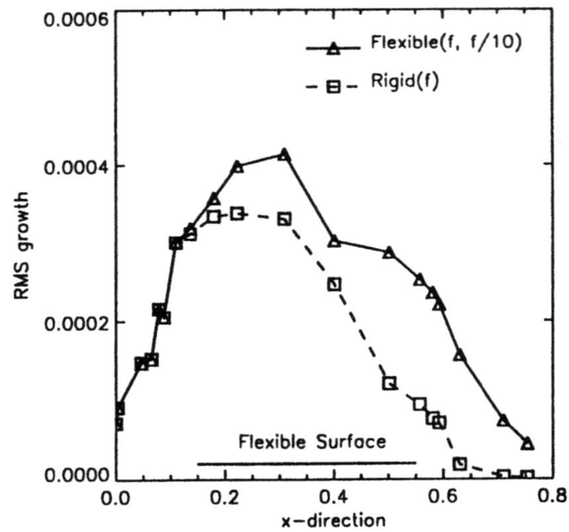

Figure 2. Root–mean–square growth of the streamwise mass flux disturbance.

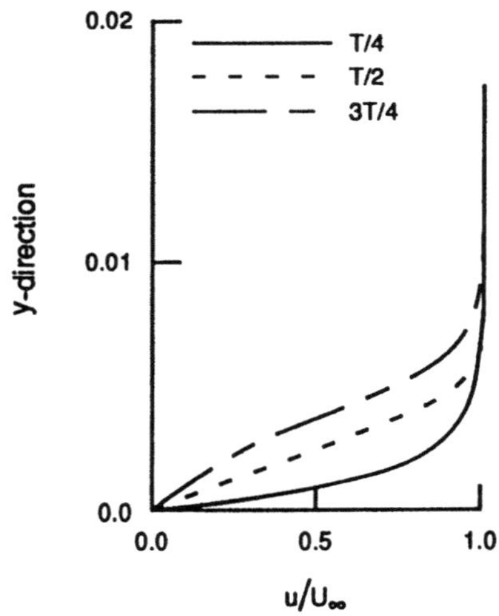

Figure 3. Variation of the streamwise velocity profile with time.

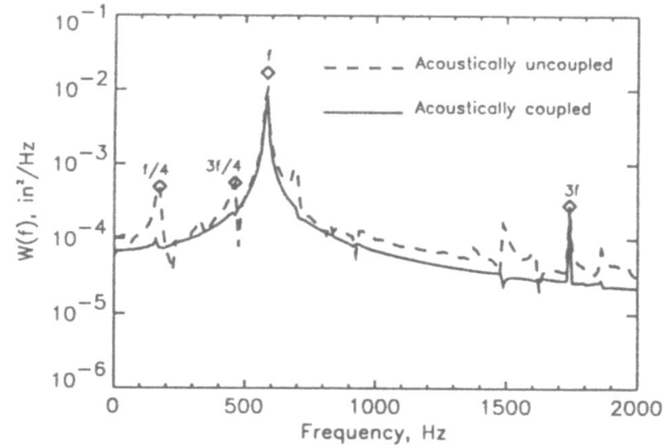

Figure 4. Power spectra of the displacement response for an acoustically coupled and uncoupled panel.

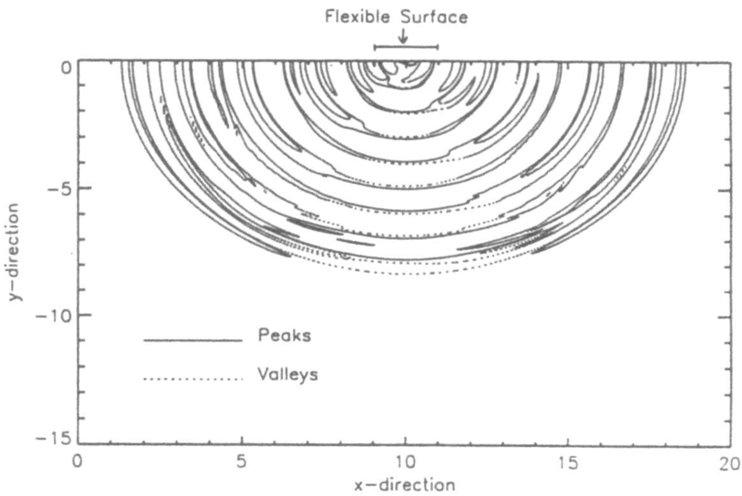

Figure 5. Radiated pressure contours.

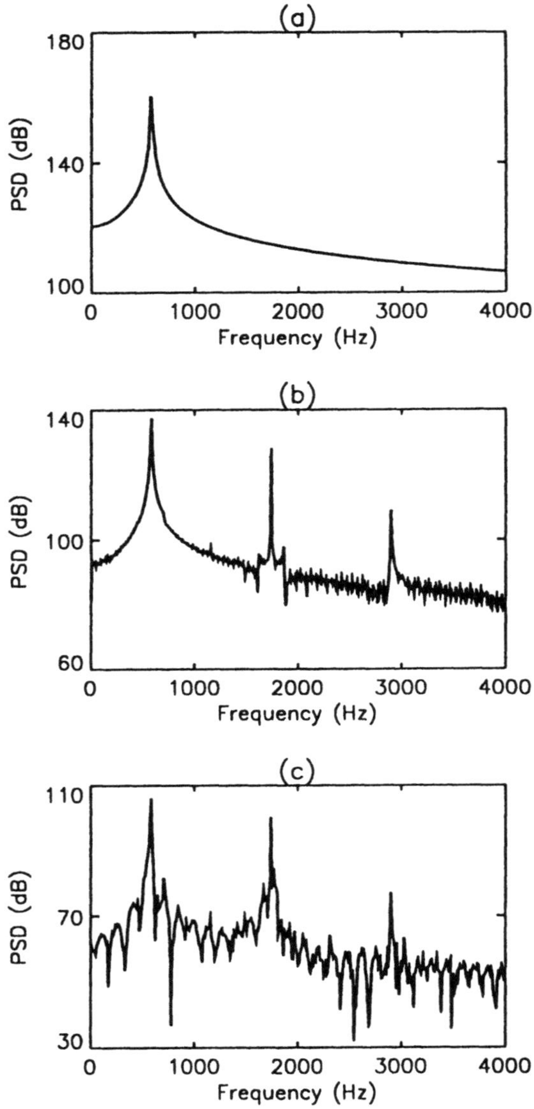

Figure 6. Pressure power spectra: (a) input pressure, (b) radiated surface pressure, and (c) radiated farfield pressure.

NAVIER-STOKES SIMULATIONS OF BLADE-VORTEX INTERACTION USING HIGH-ORDER-ACCURATE UPWIND SCHEMES

Man Mohan Rai

NASA Ames Research Center
Moffett Field, California 94035

ABSTRACT

Conventional second-order-accurate finite-difference schemes are much too dissipative for computations involving vortices that travel large distances (relative to some measure of the size of the vortex). This study uses a fifth-order-accurate upwind-biased scheme that preserves vortex structure much longer than second-order-accurate, central or upwind difference schemes. Vortex computations demonstrating this aspect of the fifth-order scheme are presented. The method is then applied to a blade-vortex interaction problem. The computed results are in good agreement with earlier numerical computations that required a priori information regarding vortex structure and position throughout the interaction period in order to sustain the vortex. High-order-accurate methods such as the one used in this study may prove essential to computing both nearfield aerodynamics and farfield acoustics associated with blade-vortex interactions as well as other acoustical phenomena.

1. Introduction

Many commonly encountered flowfields involve interactions between vortices and airfoils. A typical example is the helicopter rotor where the advancing blades encounter the tip vortices of preceding blades. A similar situation exists in turbomachinery where tip leakage and passage vortices generated by upstream rows of airfoils interact with the downstream rows of airfoils. Blade-vortex interactions result in unsteady loading of airfoils, and, noise. A good understanding of blade-vortex interaction is therefore necessary in order to compute the unsteady loads on airfoils and noise levels. Such a predictive capability is crucial to improving blade designs and minimizing noise.

Previous work in computing blade-vortex interactions has ranged from solving the simple small disturbance equation to the complex, compressible formulation of the thin-layer, Navier-Stokes equations. Srinivasan et al. (1985) present unsteady, thin-layer, Navier-Stokes results for a two-dimensional analog of the experimental configuration of Caradonna, et al. (1984). A Lamb-type vortex is released several chords upstream of the airfoil and its progression through the grid and interaction with the airfoil flowfield is computed using the Beam-Warming scheme (Beam and Warming, 1977). One of the problems with such computations, where the vortex is calculated as a part of the solution and not imposed on the solution, is that the numerical dissipation in second-order-accurate finite-difference schemes distorts the vortex rapidly. Since vortex initialization must necessarily be done several chords upstream of the airfoil, the structure of the vortex is distorted considerably by the time it has traveled close to the airfoil. In general, it can be said that the coarser the finite-difference grid used to perform the computation, the greater the distortion of the vortex. To perform an accurate blade-vortex simulation with existing schemes, a very fine grid extending all the way to the vortex initialization region would have to be used. Such a computation would be rather expensive. Srinivasan et al. (1985) use the 'perturbation' approach where an estimate of the position and structure of the vortex is used to generate source terms in the governing equations that counteract the effect of the artificial dissipation of the numerical scheme. The computed results are in good agreement with the experimental data of Caradonna et al. (1984).

Sankar and Tang (1985) also present results obtained with the unsteady Navier-Stokes equations and the Beam-Warming scheme. They develop an alternate approach that removes the dissipation due to the vortex structure and preserves the vortex for longer distances. However, as in the work of Srinivasan et al. (1985), this approach also requires knowledge of the structure and position of the vortex at each instant in time.

Although the two computational efforts described above do not impose a known vortex flowfield on the flow associated with the airfoil, they do require some knowledge of the vortex in order to negate the effects of excessive numerical dissipation. For this reason, these approaches are restricted to 'weak interaction' cases, for example, when the vortex is at a moderate distance from the airfoil throughout its travel. In 'strong interaction' cases where the vortex comes close to the airfoil, the trajectory and structure of the vortex change considerably and hence cannot be specified beforehand. Obviously a general computational procedure is required to solve these more difficult problems.

A straightforward approach to the blade-vortex interaction problem is to use a numerical scheme which is more accurate than second-order schemes. Such a high-order-accurate scheme can be used routinely for both strong and weak interactions and, hence, may be the most appropriate algorithm for a general purpose computer program.

The use of higher-order-accurate methods alleviates the numerical dissipation problem. However, there will always be a lower limit on the number of grid points required per core radius of the given vortex. The smaller the core radius of the vortex the greater will be the number of grid points required for the computation. In situations where the trajectory of the vortex is known approximately (to within a chord or two), special high density grids can be patched into coarser grids (see for example, Rai (1986)).

While both central and upwind difference schemes can be made high-order-accurate, there are two advantages to choosing upwind schemes for this purpose. First, central difference schemes require the use of arbitrary smoothing parameters to stabilize them. The choice of these parameters is scheme and problem dependent, and is a skill that has been developed over many years. A learning process would be necessary to establish the right smoothing parameters for any new scheme. Upwind schemes on the other hand do not require the specification of such arbitrary parameters. Second, upwind schemes model the physics of the problem more accurately.

In this study a fifth-order-accurate upwind-biased scheme (Rai, 1987) that is set in an iterative implicit framework is used. The scheme is second-order-accurate in time and solves the fully implicit, nonlinear, finite-difference formulation of the Navier-Stokes equations at each time step. The scheme uses the Osher flux-differencing approach (Chakravarthy and Osher, 1983). In the present study the vortex preserving capability of the fifth-order scheme is tested by computing the motion of a vortex convecting in an uniform freestream and monitoring a measure of its decay rate. The improvement in solution accuracy with the use of the higher-order method is demonstrated for this case. The method is then used to compute blade-vortex interaction.

2. Vortex Preservation Tests

An understanding of the vortex preserving property of a scheme can be obtained by computing the flow associated with a vortex convecting in an uniform freestream. A dissipative scheme is incapable of maintaining the minimum value of the pressure (at the core of the vortex) at its original value; instead, the pressure at the center increases continually as the vortex convects with the flow. Therefore a good measure of the vortex preserving capability is the core pressure p_{core}.

In this test the flow associated with a vortex convecting in a freestream is computed using both a second-order-accurate central difference scheme (the Beam-Warming scheme) and the upwind-biased fifth-order scheme. A portion of the grid used for the computation is shown in Fig. 1. It is a simple rectangular grid that is 55 core radii in length with equal spacing in the x and y directions ($\Delta x = \Delta y$). The complete grid extends outward a hundred core radii from the center in both the positive and negative x and y directions. The exact solution corresponding to the vortex moving in a freestream is imposed on all the boundaries.

Figure 2 shows the variation of p_{core} with the length of vortex travel. Curve A was obtained with the standard Beam-Warming scheme, first-order accuracy in time and grid spacing values of $\Delta x = \Delta y = 1/3$. It is extremely dissipative and, hence, unsuitable for computations requiring vortex preservation. Curve B shows the amount of vortex decay obtained on the same grid with the Beam-Warming scheme and second-order accuracy in time. A considerable improvement is noticed. However, for the grid density chosen, the scheme is still inadequate for vortex computations. Curves A and B indicate the importance of second-order accuracy in time in performing unsteady vortex computations. Curve C shows results obtained with the fifth-order-accurate scheme and the same grid. There is a moderate amount of vortex decay and the scheme is barely adequate for blade-vortex interaction computations with this grid density. Curves D and E were obtained with the central and fifth-order schemes, respectively, and on a grid with $\Delta x = \Delta y = 1/4$. Curves C and D exhibit about the same amount of vortex decay. Curve E shows almost negligible decay and indicates the grid density required for an accurate computation with the fifth-order scheme.

3. Computation of Blade-Vortex Interaction

Having chosen an appropriate algorithm we now proceed to computing blade-vortex interactions. In keeping with the experiments of Caradonna et al. (1984), a NACA 0012 airfoil was used in the following computations. The computations were performed using the unsteady, thin-layer, Navier-Stokes equations. The Baldwin-Lomax model (Baldwin and Lomax, 1978) was used to compute the eddy viscosity and Sutherland's Law was used to determine the natural viscosity of the fluid. Other important aspects of the computation, such as the initialization procedure and the grids used, are discussed below.

3.1. Grid system for the blade-vortex computation

As the vortex preservation tests of the previous section indicate, a minimum grid density of about four grid cells per core radius of the vortex is required for accurate computations. For the relatively small vortices considered in this study (about 1 inch in core radius), a grid with a minimum grid density everywhere of four cells per core radius would result in an enormous number of grid points. However, such large grid densities are only required in the path of the vortex and in the boundary layer and wake associated with the airfoil (the grid densities in the boundary layer and wake are actually several orders of magnitude larger). The simplest approach to selectively refining the grid, and thus reducing the number of grid points required to perform the computation, is to use a multiple-zone method such as that of Rai (1986).

Figure 3 shows the three zones used to represent the region of interest. The region as a whole is a 'C' type of region surrounding the airfoil; the outer boundaries are approximately 10 chords away in every direction from the airfoil. Zones 1 and 2 are symmetric. Zone 3 is a tube like region that extends from the airfoil all the way to the outer boundary of the region. Vortex travel upstream of the airfoil takes place only in zone 3, and, hence only zone 3 contains the high grid density required to preserve the vortex. Figure 4 shows the grids in all three zones in the vicinity of the airfoil. The grids in all the zones were generated using a combination of algebraic and elliptic grid generation procedures (Steger and Sorenson, 1979). The grids are orthogonal to the surface of the airfoil (this feature cannot be discerned in Fig. 4 because of the closely spaced grid points in the boundary layer). Zones 1 and 2 are discretized with 121 × 151 point grids and zone 3 is discretized with a 91 × 311 point grid; the total number of grid points is approximately 65,000.

The use of multiple grids in simulating the flow over the airfoil configuration shown in Fig. 3 results in several computational boundaries. The boundary conditions used at each of these boundaries are described in Rai (1987).

3.2. Initialization procedures

The vortex structure obtained in the experimental investigation of Caradonna et al. (1984) is similar to that of a Lamb-type vortex. In order that the structure of the vortex used in the computation resemble closely the experimental one, the core radius, vortex strength and the peak circulatory velocity must all be nearly the same as in the experimental vortex. Srinivasan et al. (1985) use an analytically defined vortex structure wherein the necessary constants are determined from experimental data. The circulatory velocity component v_θ for this vortex is given by

$$\frac{v_\theta}{u_\infty} = \frac{\Gamma}{2\pi r}\left(\frac{r^2}{r^2 + a^2}\right) \tag{1}$$

where a is the core radius of the vortex (0.984 inches), Γ represents the strength of the vortex (Γ = -1.477), r is the distance from the center of the vortex and u_∞ is the freestream velocity. The negative value of Γ results in a vortex that rotates in the clockwise direction. The vortex structure used in this study was also obtained from Eq. 1 with the same values for Γ and the core radius.

The solution is initialized at all grid points to freestream values. The governing equations with the appropriate boundary conditions are integrated to convergence on the grid system presented earlier for the airfoil. The vortex is then initialized at a distance of approximately 7 chord lengths ahead of the airfoil. The convection of the vortex with the fluid and the subsequent interaction of the vortex with the airfoil is monitored.

3.3. Results

The three-zone, unsteady, thin-layer, Navier-Stokes code was first validated by computing the flow over a NACA 0012 airfoil at zero angle of attack. The freestream Mach number for this computation was 0.8 and the Reynolds number based on the chord was one million. Figure 5 shows the pressure contours obtained at convergence. The smooth transition of the contour lines from zone to zone at the interface boundaries is clearly seen. In order to capture the shock without obtaining spurious oscillations, the accuracy of the method was dropped to second-order in the vicinity of the shock and flux limiters (Chakravarthy and Osher, 1983) were employed in this region. The shock capturing capability of this procedure is evident from Fig. 5. Figure 6 shows a comparison of the surface pressure coefficients obtained with the present scheme and with the ARC2D code (private communication, Dr. T. H. Pulliam, NASA Ames). The two sets of results are in good agreement.

The blade-vortex interaction study consisted of simulating a two-dimensional analog of the experiment of Caradonna et al. (1984). A schematic of their experimental configuration is shown in Fig. 7. It consists of a two-bladed rotor in a uniform freestream. The angle ψ measures the azimuthal position of the blade. The rotor blade interacts with the tip vortex generated by a NACA 0015 airfoil that is positioned upstream of the rotor. At approximately $\psi = 180°$ the test section A-A of the rotor undergoes an interaction with the vortex commonly referred to as parallel blade-vortex interaction. The flow at this station can be roughly approximated as a two-dimensional blade-vortex interaction.

The airfoil flowfield was first computed without a vortex and with a freestream Mach number of 0.536 (corresponding to a rotor tip Mach number of 0.6). The vortex was then released at approximately 7 chord lengths from the leading edge of the airfoil and 2.4 inches below the axis of symmetry (the chord length of the airfoil being 6.0 inches). The Reynolds number used in the computation was 1.3 million based on the chord and the freestream temperature was $530° R$.

Figure 8a shows the surface pressure distribution at the rotor azimuthal position $\psi = 182.65°$. This azimuthal position corresponds to $x_v = 1.74$ inches, where x_v represents the distance from the leading edge of the airfoil to the center of the vortex (measured as positive in the positive x direction). The full and dashed lines are the results obtained in the present study and the symbols depict the results of Srinivasan et al. (1985). The area between the two curves represents the negative lift the airfoil experiences at this azimuthal position.

Figure 8b depicts the surface pressure distribution at $\psi = 185.95°$ ($x_v = 3.9$ inches). The lift on the airfoil has changed from negative to a small positive value at this azimuthal position (the numerically obtained transition point for the lift being 185°). Figure 8c shows the pressure distribution at $\psi = 192.85°$ ($x_v = 8.4$ inches). The agreement between the two sets of results in Figs. 8a-8c is good. Note that the perturbation approach of Srinivasan et al. (1985) is adequate for this case because of the relatively weak interaction between the airfoil and the vortex.

4. Summary

A major problem in performing vortex computations with second-order finite-difference schemes is the rapid decay of the vortex structure that occurs because of numerical dissipation. This study uses a fifth-order-accurate, upwind-biased scheme that preserves vortex structure much longer than second-order-accurate central or upwind difference schemes. Vortex computations demonstrating this aspect of the fifth-order scheme are presented.

A thin-layer, Navier-Stokes code for blade-vortex computations has been developed. The code employs the new fifth-order scheme to preserve vortices over large distances. The code also has a multi-zone capability to selectively refine the grid only in the path of the vortex so that grid point requirements are reduced. The code has been validated for a weak blade-vortex interaction case using results from an earlier computation. The new scheme made unnecessary the use of vortex preserving techniques such as perturbation methods (which require a good estimate of the vortex structure and location during the interaction process).

The strength of the high-order-accurate approach lies in the fact that it can be used routinely for 'strong interaction' cases without any modification. The computer program that uses the high-order-accurate scheme has successfully been applied to 'head-on' collisions between vortices and airfoils, and, blade-vortex interaction at transonic speeds, with the attendant vortex-shock interaction (Rai, 1987). High-order-accurate methods such as the one used in this study may prove essential to computing farfield acoustics associated with blade-vortex interactions as well as other acoustical phenomena.

References

Baldwin, B. S. and Lomax, H., 1978. "Thin layer approximation and algebraic model for separated turbulent flow," AIAA Paper 78-257, AIAA 16th Aerospace Sciences Meeting, Huntsville, Alabama, January 16-18, 1978.

Beam, R. M. and Warming, R. F., 1977. "An implicit factored scheme for the compressible Navier-Stokes equations," Proceedings of the AIAA 3rd Computational Fluid Dynamics Conference, Albuquerque, New Mexico, June 27-28, 1977.

Caradonna, F. X., Laub, G. H., and Tung, C., 1984. "An experimental investigation of the parallel blade-vortex interaction," Tenth European Rotorcraft Forum, The Hague, Netherlands.

Chakravarthy, S. R. and Osher, S., 1983. "High resolution applications of the Osher upwind scheme for the Euler equations," AIAA Paper 83-1943, AIAA 6th Computational Fluid Dynamics Conference, Danvers, Massachusetts, July 13-15, 1983.

Rai, M. M., 1986. "An implicit, conservative, zonal-boundary scheme for Euler equation calculations," *Computers and Fluids* **14**(3), pp. 295-319.

Rai, M. M., 1987. "Navier-Stokes simulations of blade-vortex interaction using high-order accurate upwind schemes," AIAA Paper No. 87-0543, AIAA 25th Aerospace Sciences Meeting, Reno, Nevada, January 12-15, 1987.

Sankar, N. L. and Tang, W., 1985. "Numerical solution of unsteady viscous flow past rotor sections," AIAA Paper 85-0129, AIAA 23rd Aerospace Sciences Meeting, Reno, Nevada, January 14-17, 1985.

Srinivasan, G. R., McCroskey, W. J., and Baeder, J. D., 1985. "Aerodynamics of two-dimensional blade-vortex interaction," AIAA Paper 85-1560, AIAA 18th Fluid Dynamics and Plasmadynamics and Lasers Conference, Cincinnati, Ohio, July 16-18, 1985.

Steger, J. L. and Sorenson, R. L., 1979. "Automatic mesh-point clustering near a boundary in grid generation with elliptic partial differential equations," *J. Comput. Phys.* **33**(3), pp. 405-410.

Figure 1. Grid for vortex preservation tests.

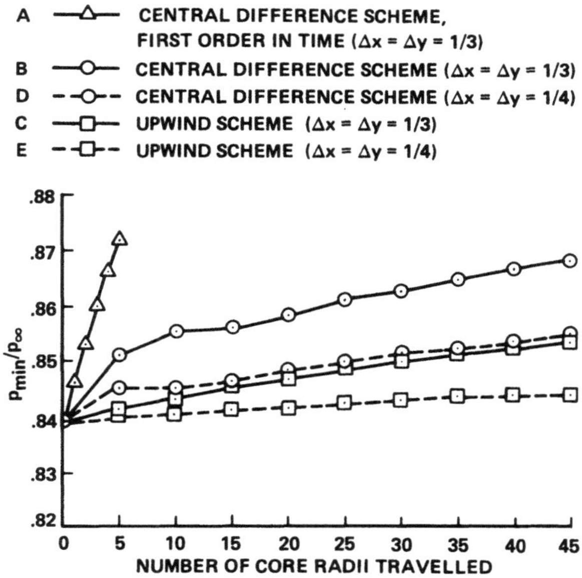

Figure 2. Vortex decay rates for various schemes.

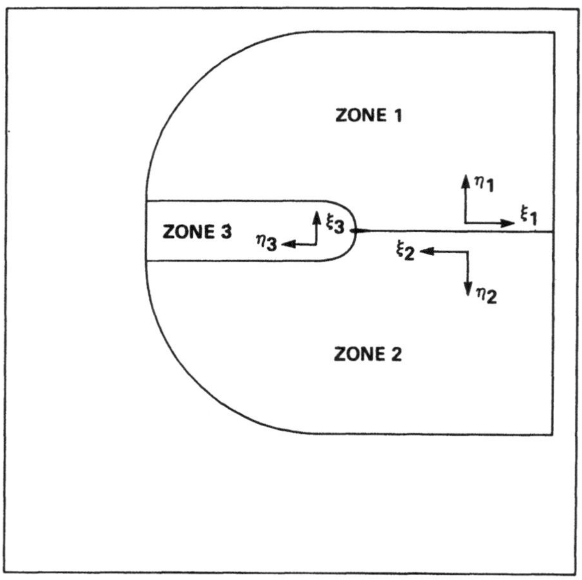

Figure 3. Zoning and outer boundaries used for the blade-vortex computation.

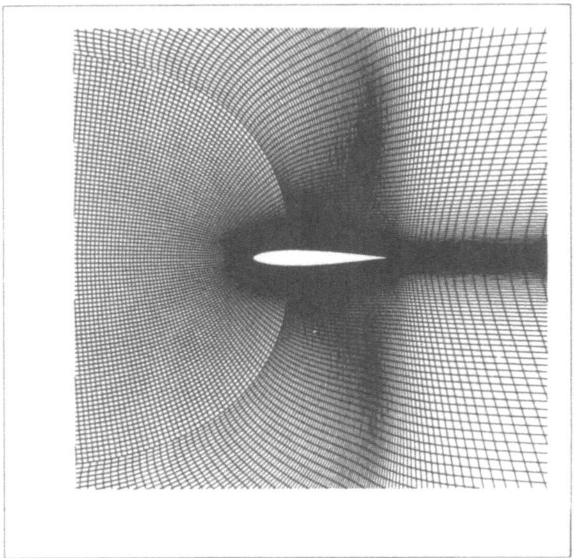

Figure 4. Grids for all three zones in the vicinity of the airfoil.

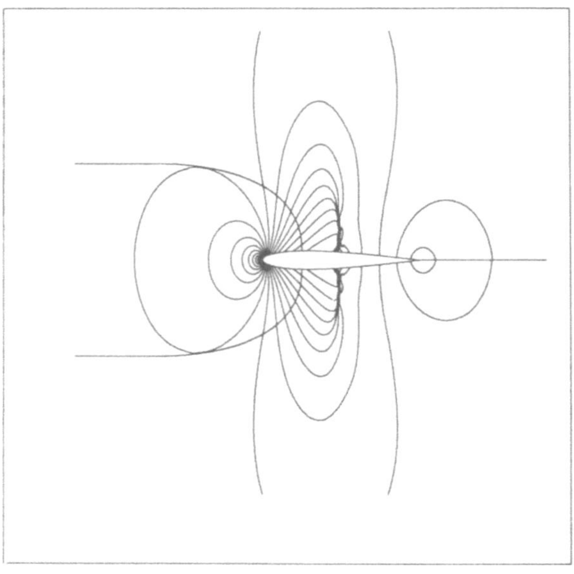

Figure 5. Pressure contours at convergence for the airfoil in isolation.

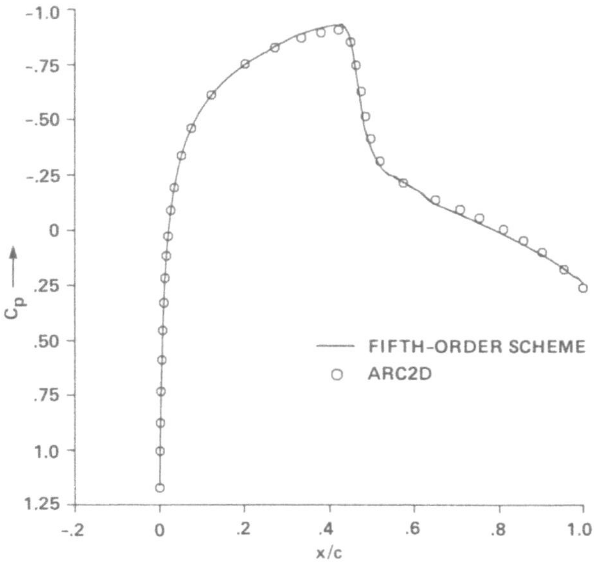

Figure 6. Surface pressure distribution for the airfoil in isolation.

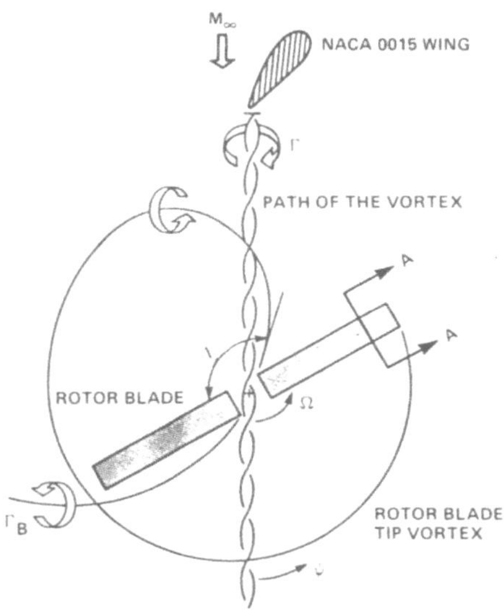

Figure 7. Schematic of the experimental configuration of Caradonna, et al. (1984).

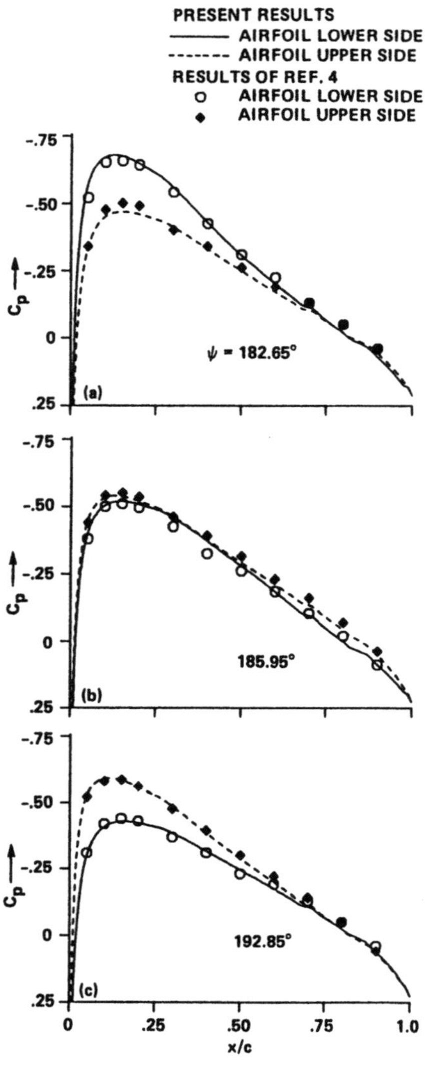

Figure 8. A comparison of surface pressure distributions obtained with the current scheme and the scheme of Srinivasan et al. (1985).

a) $\psi = 182.65^o$

b) $\psi = 185.95^o$

c) $\psi = 192.85^o$

CONTRIBUTED PAPERS

SCATTERING OF SOUND BY RIGID BODIES IN ARBITRARY FLOWS [1]

K. S. Huh

McDonnell Douglas Research Laboratories
St. Louis, Missouri 63166

Massachusetts Institute of Technology
Cambridge, Massachusetts 02139

S. E. Widnall

Massachusetts Institute of Technology
Cambridge, Massachusetts 02139

R. K. Agarwal

McDonnell Douglas Research Laboratories
St. Louis, Missouri 63166

ABSTRACT

A method for directly computing acoustic signatures without a wave equation analogy is presented. The governing acoustic equations are derived from the unsteady Euler equations by linearizing about a steady mean flow and by assuming a single frequency disturbance. A rigid, stationary body is assumed and a non–reflecting type of boundary condition is used at the far–field.

Scattering of plane sound waves is investigated in both lifting and non–lifting scenarios. Some comparisons are made with known analytic solutions where available.

1. Introduction

In principle an exact acoustic flowfield can be obtained given any boundary/initial conditions by applying the Lighthill acoustic analogy. An analogy type of approach affords a distinct advantage in that the solution process requires only an inversion of a wave equation operator with a source term. However, since all the interaction terms

[1]This research was supported by the McDonnell Douglas Independent Research and Development Program. Some computations were performed at the MIT Cray Supercomputer Facility.

in the Navier–Stokes equations are included as source terms, the Lighthill theory requires prior knowledge of the solution in order to specify the sources.

Recently there has been an effort by numerous researchers to directly compute the acoustic quantities by numerically integrating the Navier–Stokes equations (Colonius *et al.*, 1991) or the Euler equations (Khan *et al.*, 1986; Hariharan, 1984; Hall, 1987; Huh *et al.*, 1990), and thereby avoiding the source–term specification limitations in the analogy type of approach.

The current methodology computes the fluctuating acoustic quantities by directly integrating the linearized Euler equations using proven computational fluid dynamics (CFD) methods.

2. Governing Equations

The governing perturbation Euler equations are obtained by perturbing the aerodynamic quantities about the mean flow such that,

$$U(x,y,t) = u_0(x,y) + u'(x,y,t), \qquad V(x,y,t) = v_0(x,y) + v'(x,y,t)$$

$$\rho(x,y,t) = \rho_0(x,y) + \rho'(x,y,t), \qquad P(x,y,t) = p_0(x,y) + p'(x,y,t)$$

where U, V, ρ, and P denote the x and y velocity components, density, and pressure, respectively. The subscript $_0$ refers to the steady mean quantity, and the superscript $'$ refers to the time-varying fluctuation quantity. Then outside of a source singularity, the perturbation quantities obey the linearized Euler equations of the form,

$$\frac{\partial}{\partial t}\mathbf{Q} + \frac{\partial}{\partial x}\mathbf{F} + \frac{\partial}{\partial y}\mathbf{G} = 0. \tag{1}$$

Because of the homogeneity of the Euler equations, any unsteadiness must arise from the applied boundary conditions such as those due to upstream and/or downstream pressure perturbations. Suppose that these external excitations are harmonic in time with a frequency ω, then since the given equations are linear, the time dependent solution will also be harmonic with frequency ω. Thus the unsteady linearized Euler equations can be recast as a set of steady complex equations by the use of the single frequency assumption,

$$p' = \Re(\tilde{p}(x,y)e^{-i\omega t}), \qquad \rho' = \Re(\tilde{\rho}(x,y)e^{-i\omega t})$$

$$u' = \Re(\tilde{u}(x,y)e^{-i\omega t}), \qquad v' = \Re(\tilde{v}(x,y)e^{-i\omega t})$$

where the tilde $\tilde{}$ denotes a complex quantity and $\Re$ refers to the real part.

The governing equations then become,

$$-i\omega\mathbf{Q} + \frac{\partial}{\partial x}\mathbf{F} + \frac{\partial}{\partial y}\mathbf{G} = 0 \tag{2}$$

where

$$\mathbf{Q} = \begin{pmatrix} \tilde{\rho} \\ \widetilde{(\rho u)} \\ \widetilde{(\rho v)} \\ \widetilde{(\rho e)} \end{pmatrix} \tag{3}$$

$$\mathbf{F} = \begin{pmatrix} \widetilde{(\rho u)} \\ u_0[2\widetilde{(\rho u)} - u_0\tilde{\rho}] + \tilde{p} \\ v_0[\widetilde{(\rho u)} - u_0\tilde{\rho}] + \widetilde{(\rho v)}u_0 \\ u_0[\widetilde{(\rho h)} - h_0\tilde{\rho}] + \widetilde{(\rho u)}h_0 \end{pmatrix}, \mathbf{G} = \begin{pmatrix} \widetilde{(\rho v)} \\ u_0[\widetilde{(\rho v)} - v_0\tilde{\rho}] + \widetilde{(\rho u)}v_0 \\ v_0[2\widetilde{(\rho v)} - v_0\tilde{\rho}] + \tilde{p} \\ v_0[\widetilde{(\rho h)} - h_0\tilde{\rho}] + \widetilde{(\rho v)}h_0 \end{pmatrix} \tag{4}$$

A state equation is required for closure and is of the form,

$$\tilde{p} = (\gamma - 1)\left\{\widetilde{(\rho e)} - \frac{1}{2}\left[u_0(2\widetilde{(\rho u)} - u_0\tilde{\rho}) + v_0(2\widetilde{(\rho v)} - v_0\tilde{\rho})\right]\right\}. \tag{5}$$

The variables $\widetilde{(\rho u)}$, $\widetilde{(\rho v)}$, $\widetilde{(\rho e)}$, $\widetilde{(\rho h)}$ are defined as,

$$\widetilde{(\rho u)} = \rho_0\tilde{u} + \tilde{\rho}u_0 \quad \widetilde{(\rho v)} = \rho_0\tilde{v} + \tilde{\rho}v_0$$

$$\widetilde{(\rho e)} = \frac{\tilde{p}}{\gamma - 1} + \rho_0 u_0\tilde{u} + \rho_0 v_0\tilde{v} + \frac{1}{2}\tilde{\rho}(u_0^2 + v_0^2)$$

$$\widetilde{(\rho h)} = \widetilde{(\rho e)} + \tilde{p}.$$

It should be noted that no generality has been lost due to the single frequency assumption. Since the governing equations are linear, a general case of N number of frequencies can be solved as a superposition of N number of single frequencies by applying the Fourier theorem. A solution methodology to solve a single frequency problem also solves the problem of multiple frequencies.

The linearized complex Euler equations (2) can now be solved using an appropriate numerical method. In practice, however, it is more efficient to recast the equations in the "scattered" form. The scattered form requires less grid resolution in the far field where the scattered quantities are expected to be small, but where the incident quantities are not. Thus the total perturbation quantity is formed as a sum of the incident quantity and the scattered quantity,

$$\tilde{p} = \tilde{p}_i + \tilde{p}_s, \qquad \tilde{\rho} = \tilde{\rho}_i + \tilde{\rho}_s,$$
$$\tilde{u} = \tilde{u}_i + \tilde{u}_s, \qquad \tilde{v} = \tilde{v}_i + \tilde{v}_s,$$

where the subscripts i and s denote the incident and scattered parts, respectively. The governing equations (2) become,

$$\frac{\partial}{\partial t^*}\mathbf{Q}_s + \frac{\partial}{\partial x}\mathbf{F}_s + \frac{\partial}{\partial y}\mathbf{G}_s - i\omega\mathbf{Q}_s = \mathbf{S} \tag{6}$$

where

$$\mathbf{S} = i\omega\mathbf{Q}_i - \frac{\partial}{\partial x}\mathbf{F}_i - \frac{\partial}{\partial y}\mathbf{G}_i \tag{7}$$

and t^* is an artificial pseudo–time variable.

Note that since $\mathbf{S}$ contains all the incident-wave/mean-flow diffraction terms, $\mathbf{S} \to 0$ as $r \to \infty$ and $\mathbf{S} \equiv 0$ when there is no flow. The use of pseudo time t^* makes it possible to adopt all the convergence acceleration methods developed in CFD such as local time stepping and multigrid methods.

A more complete derivation of the governing equations is provided in Huh *et al.* (1990).

3. Energy Corollary

The set of governing equations is complemented by an energy corollary relation. The corollary is derived by algebraic manipulations of the governing equations. It is not an independent equation, and the equation need not be integrated. The utility of the relation is that it expresses a second–order balance of energy which involves only first–order perturbation quantities. The complete details of the derivation may be found in Myers (1986).

The second–order energy balance may be expressed as a time variation of the total energy–density E_a,

$$\frac{\partial E_a}{\partial t} + \nabla \cdot \mathbf{I}_a = D_a \tag{8}$$

where given a rotational but homentropic flow, E_a, $\mathbf{I}_a$, and D_a are defined as,

$$E_a = \frac{p'^2}{2\rho_0 c_0^2} + \frac{\rho_0(u'^2 + v'^2)}{2} + \rho'\mathbf{u}_0 \cdot \mathbf{u}'$$

$$\mathbf{I}_a = (\frac{p'}{\rho_0} + \mathbf{u}_0 \cdot \mathbf{u}')(\rho_0 \mathbf{u}' + \rho' \mathbf{u}_0)$$

$$D_a = \rho_0 \mathbf{u}_0 \cdot (\xi' \times \mathbf{u}') + \rho' \mathbf{u}' \cdot (\xi_0 \times \mathbf{u}_0)$$

and $\xi = \nabla \times \mathbf{u}$ the vorticity.

An equivalent form of energy conservation is obtained by taking the time average and then integrating the differential equation (8) over an area A bounded by a closed loop σ. The conservation equation becomes,

$$W \equiv \oint_\sigma \bar{\mathbf{I}}_a \cdot \mathbf{n} d\sigma = \iint_A \bar{D}_a dA \tag{9}$$

where the overbar $^-$ corresponds to a time-averaged variable, $\mathbf{n}$ is the unit normal vector to the surface $d\sigma$, and we define W as the power. The usefulness of power is that when σ_1 and σ_2 are two surfaces enclosing the same sources of sound, the same value of W is found for both surfaces.

4. Boundary Conditions

4.1. Wall condition

A rigid wall boundary condition is used for all calculations. For an inviscid, stationary wall the boundary condition is,

$$\tilde{\mathbf{u}} \cdot \hat{n} = 0 \tag{10}$$

where $\tilde{\mathbf{u}}$ is composed of both incident and scattered components, and $\hat{n}$ is the unit normal vector to the surface. In terms of the scattered quantities, the boundary condition becomes,

$$\tilde{\mathbf{u}}_s \cdot \hat{n} = -\tilde{\mathbf{u}}_i \cdot \hat{n} \tag{11}$$

where $\tilde{\mathbf{u}}_i$ is a given incident complex velocity.

4.2. Far–field conditions

A correct mathematical model would extend the domain to infinity, and would require only out–going wave modes (Radiation Condition) which vanish at infinity. This is impractical numerically, and a set of well–posed boundary conditions must be placed on a finite domain.

The current methodology uses the non–reflecting boundary conditions proposed by Giles (1988). The background theory is in Giles (1988), and the numerical details are in Huh (1992).

The boundary condition is, in theory, accurate to $O(l/\omega)^4$ where l is the Fourier decomposed wave number in the azimuthal direction; it has been found to be more effective than the commonly used one–dimensional radiation condition or the characteristic condition.

4.3. Kutta condition

When $M_0 > 0$, the method enforces a pressure continuity condition at a sharp trailing edge,

$$\tilde{p}_s^+ = \tilde{p}_s^- \tag{12}$$

where + and − refer to the upper and lower surfaces at the trailing edge.

5. Numerical Method

The set of conservative governing equations (6) is numerically solved using a globally conservative, finite volume, node–based scheme. A body–fitted coordinate system is used which offers the advantages of geometric generality and ease of boundary condition implementation.

The spatial discretization in the normal direction to the surface is a finite volume, node–based, central difference type and is second order accurate for a smooth grid, while the discretization in the azimuthal direction is based on the classical Padé scheme and is fourth order accurate for a uniformly spaced grid. The details of the central difference operator maybe found in, for example, Jameson *et al.* (1981), and the Padé scheme is explained in Lele (1990).

A four stage Runge–Kutta scheme is used to integrate the governing equations to convergence in the pseudo–time plane. The integration method is point implicit for the source term $-i\omega\mathbf{Q}$ which alleviates the stiffness for large values of ω; the stable time–step limit is independent of ω. A point implicit time integration method has been initially investigated by Bussing and Murman (1985), and its current implementation is detailed by Huh (1992).

6. Results

The current methodology has been applied to solve some model problems in aeroacoustics. The first two cases are intended primarily as test runs to ensure correct circulation specification and vortex shedding processes. The latter two cases illustrate the numerical method as applied to plane wave scattering problems in aeroacoustics.

For all cases c_0 and λ are the local speed of sound and the acoustic wavelength, respectively. The mean flow conditions were computed using a node–based Euler solver, and the computations were performed by pseudo–time marching in the complex domain until convergence was achieved. Convergence was defined as $L_2(\Delta|\tilde{\rho}|) \leq 5.0 \times 10^{-3}$. Local time–stepping was implemented to accelerate the convergence.

6.1. Linear flow at an angle of attack

Consider a steady flow of Mach number M_0 over an airfoil at an angle of attack α_0. If the principle of superposition were valid, this flow can be thought of as a combination of a mean flow of Mach number M_0 at zero angle of attack plus a perturbation flow at an angle of attack α'. Since in general $p \sim u^2$, the perturbation flow must be sufficiently weak or equivalently α_0 must be small for this summation of flows to be valid. Typically the acoustic perturbations are quite small ($|p'|/p_0 \sim 10^{-5}$) allowing the use of superpositioning for analysis.

The current case compares a full nonlinear flow of Mach number $M_0 = 0.3$ over a NACA0012 at $\alpha_0 = 0.1°$ and a flow which has a mean Mach number of $M_0 = 0.3$ at zero angle of attack plus a weak acoustic pressure wave of amplitude $|\tilde{p}_i| = 4.26 \times 10^{-3}$ at $\alpha' = 10°$. To make the comparison valid the acoustic wave has to be steady. Strictly speaking there are no steady acoustic waves, instead what is implied here is that in the limit $\omega \to 0$, $\frac{\omega}{k} \to c_0$, the speed of sound.

The parameters were as follows:

- Grid: 194×65
- ω: 0.0 ($\lambda = \infty$)
- M_0: 0.3

- α: 0.1° (flow solver) ; 10.0° (acoustic solver)
- Far–field diameter: 20 chords
- Iterations: 400 (flow solver) + 400 (acoustic solver).

Figure 1 shows the graph of $-C_p$; the dashed lines represent the nonlinear solution and the square symbols represent the linearized flow solution. Agreement is good everywhere. The largest error of $\Delta C_p = 0.016$ was at the leading edge. In terms of the total lift, the nonlinear case was 7.283×10^{-4} and the linearized case was 7.319×10^{-4} which is a difference of about 0.49%. It appears that the solver is accurate in predicting the circulation.

6.2. Sinusoidal gust (Sears problem)

The previous validation case showed that the method was valid in specifying the circulation for a steady problem. Acoustics is by definition an unsteady phenomenon, and in an unsteady lifting case there exists a singularity at the sharp edge which is relieved by the shedding of vorticity from the edge into the wake. The trailing vortex sheet induces a velocity field about the airfoil and hence also influences the forces on the airfoil. The pressure field on the airfoil, in turn, affects the radiated sound field. Since the previous steady case had no wake; the case is, by itself, insufficient for acoustic purposes.

Consider a flat plate in an incompressible flow which encounters a sinusoidal vorticity field of the form,

$$v' = e^{i(kx-\omega t)}, \qquad k = \frac{\omega}{u_0} \tag{13}$$

and u_0 is the free stream velocity. We seek to compute the time–harmonic response of the plate and its surrounding fluid perturbed by the incoming vortical gust. This sinusoidal gust problem was chosen as an ideal test case because of its simplicity, and because historically it has received a considerable amount of attention.

The given analytic solution (Bisplinghoff *et al.*, 1955) is for an incompressible flow, and hence the numerical simulation must have $M \ll 1$ and $\omega \ll 1$. The parameters were as follows:

- Grid: 250×100
- ω: 0.01

- M_0: 0.1
- Far–field diameter: 20 chords
- Iters: 9000.

A NACA00012 airfoil was used to simulate a flat plate. $\omega = 0.01$ corresponds to a reduced frequency of $\omega/2U_0 = 0.0425$ and a wavelength of 743 chords. Within a domain of 10 chords the computed solution should correspond closely to that of an incompressible approximation. Figure 2 shows the Δp along the flat plate. The lines represent the real and imaginary parts of exact Δp, and the symbols represent the corresponding computed solutions. The accuracy is good everywhere. In terms of the total lift, the analytic solution is $|\tilde{L}| = 0.3448$ and $\arg(\tilde{L}) = 7.3°$ and the computed solution is $|\tilde{L}| = 0.3496$ and $\arg(\tilde{L}) = 8.7°$. The error is approximately 1.4% in amplitude and approximately 1.4° in phase. It appears that the methodology is accurate in predicting the vortex shedding process, at least in a low frequency case where the stability of the vortex sheet is not an issue close to the flat plate.

6.3. Scattering by a NACA0012 airfoil at zero angle of attack

The diffraction and scattering of plane acoustic waves have been studied by comparing two related cases at different Mach numbers.

Acoustic diffraction occurs in a region which has a solid body of interference or in a region which allows some part of the wave front to propagate faster than another. In a homogeneous media, diffraction occurs near a solid surface as wave fronts are necessarily distorted when the travelling waves bend in order to accommodate the surface. Diffraction occurs in a inhomogeneous media by spatial changes in the local speed of sound and/or by spatial changes in the local convection velocity.

Consider a NACA0012 airfoil at a zero angle of attack in a steady $M_0 = 0.3$ flow. A plane acoustic wave of the form,

$$p_i' = e^{i(kx-\omega t)}, \qquad k = \frac{\omega}{c_0(1+M_0)} \tag{14}$$

impinges on the airfoil.

The parameters were as follows:

- Grid: 300×150
- ω: 2.602
- Far–field diameter: 20 chords
- Iterations: 2000.

$\omega = 2.602$ translates to a maximum wavelength in the downstream direction of $\lambda_{max} \approx 3.7$ chords and a minimum wavelength in the upstream direction of $\lambda_{min} \approx 2$ chords. The grid guarantees a resolution of approximately 9.5 points/λ_{min} in the azimuthal direction and approximately 30 points/λ_{min} in the radial direction, so that the current scheme should be able to resolve the relevant length scales even at the far–field. The grid also has a slight exponential packing near the airfoil to increase the accuracy on the surface.

Since the thickness of the airfoil is only 3% of the incident wavelength, the scattering effects were expected to be small. Figure 3 shows the amplitude of the scattered pressure, $\tilde{p}_s$. The blunt leading edge was responsible for most of the scattering and the leading edge emulated a compact scatterer radiating sound in all directions.

Figure 4 shows the phase of the scattered pressure. The Doppler effect is indicated by the contours which show shorter wavelengths in the upstream direction than in the downstream direction. The complexity of the phase structure close to the airfoil illustrates the diffractive effect of the accelerating fluid. The bend in the contour lines at about $\pm 45°$ illustrate that at moderate distances the airfoil behaves as two monopoles located at each of the edges. Evidently the airfoil is not very compact with respect to the incoming wavelength.

Figure 5 shows the amplitude contours of the acoustic pressure. The region ahead of the airfoil exhibits the interference fringes from the interaction of the incident wave with the scattered wave; still because of the low scattered amplitudes, the pressure field is composed almost entirely of the incident field except at the leading edge region.

A comparison case of $M_0 = 0$ was computed to illustrate the diffractive effects of a varying flow–field. The parameters for this case were chosen so that the computed values would transform to uniform flow values using simple mapping functions.

In a uniform flow the pressure perturbation $p'(x, y, t)$ is represented by a convective form of the wave equation,

$$p'_{tt} + 2u_0 p'_{xt} - (c_0^2 - u_0^2) p'_{xx} - c_0^2 p'_{yy} = 0 \tag{15}$$

The above equation can be transformed into the standard wave equation by using the following "unsteady Prandtl–Glauert" mapping variables,

$$\xi = \frac{x}{\beta}, \; \tau = \beta t + \frac{M_0 x}{c_0 \beta}$$

where $\beta = \sqrt{1 - M_0^2}$. In (ξ, y) space the equation for $\tilde{p}$ becomes,

$$c_0^2(\tilde{p}_{\xi\xi} + \tilde{p}_{yy}) = \Omega^2 \tilde{p} \qquad (16)$$

where $\Omega = \frac{\omega}{\beta}$ and $p'(\xi, y, \tau) = \Re(\tilde{p}(\xi, y)e^{-i\Omega\tau})$.

The complex amplitude $\tilde{p}(\xi, y)$ is related to $\tilde{p}(x, y)$ by the relation,

$$\tilde{p}(\xi, y) = \tilde{p}(x, y) \exp\left[\frac{iM_0\omega x}{c_0\beta^2}\right].$$

The comparison $M_0 = 0$ case was computed in (ξ, y) space. The pressure values were then mapped to (x, y) space of uniform $M_0 = 0.3$ flow. The transformed values were, in turn, subtracted from the original $M_0 = 0.3$ computed solution. The difference, $\Delta\tilde{p}$, is a measure of the diffraction in a varying flow–field. In regions where there are no flow gradients $\Delta\tilde{p}$ will be small, and thus a large $\Delta\tilde{p}$ indicates high diffraction. Figure 6 shows the amplitude contour of $\Delta\tilde{p}$; note that the amplitude is zero everywhere except in the vicinity of the leading and trailing edges of the airfoil. The edge regions close to the airfoil are regions of high diffraction.

The wave equation in (ξ, y) space can also be solved using the Green's function method. The scattered portion of the solution is,

$$\tilde{p}_s(\xi, y) = -\int_c \frac{\partial p}{\partial n}\frac{i}{4}H_0^{(1)}\left[\frac{\Omega}{c_0}\sqrt{(\xi - \xi_1)^2 + (y - y_1)^2}\right] d\xi_1 dy_1. \qquad (17)$$

The above integral is to be evaluated over the surface of the airfoil. Such a solution has been computed at the far–field using a panel–method type of a solver to assure that the current numerical method does not allow a large reflection at the far–field, and that the domain does not include significant numerical sources. Figure 7 shows the amplitude of scattered wave at the far–field for both the panel method and the current scheme. There are little noticeable differences between the two solutions.

Figure 8 shows the mean power W versus $k_{min}R$ where R is the distance measured from the surface of the airfoil and k_{min} is the minimum wavenumber in the downstream direction. W is as defined in an earlier section and was computed by numerically integrating the line integral about the airfoil at the specified location. The current case is (or should be) irrotational and isentropic everywhere, therefore the mean power radiated should be zero. The figure indicates an oscillation of mean power about zero with an amplitude of about 5×10^{-3}. Since $\lambda \sim O(1)$ and $\Delta x \sim O(0.1)$, we expect a scheme which is $O(\Delta x^4)$ accurate *locally* to be of $O(10^{-3})$ accurate *globally*. The power computation is accurate to two significant digits. A higher accuracy requires a finer grid or a higher order discretization scheme.

6.4. Scattering by a NACA0012 airfoil at an angle of attack

The diffraction and energy characteristics of plane acoustic waves scattered by an airfoil at an angle of attack have been studied. The enforcement of the Kutta condition creates an unsteady circulation about the airfoil and a vortex sheet downstream of the trailing edge.

A NACA0012 airfoil is in a steady $M_0 = 0.3$ flow at an angle of attack of 2°. A plane acoustic wave of the identical form as in the previous case impinges on the airfoil.

The parameters were as follows:

- Grid: 300×150
- ω: 2.602
- Far–field diameter: 20 chords
- Iterations: 3000.

The domain and the grid is unchanged from the previous case except for the rotation of the airfoil, and thus we expect similar resolution characteristics.

Figure 9 shows the computed amplitude of the scattered pressure $\tilde{p}_s$. Note the differences between the current case and the corresponding values from the previous case. The current values have higher amplitudes because at an angle of attack the airfoil has a higher effective cross–section imposing on the incoming acoustic wave, and the directivity pattern is asymmetric due to the acceleration of the fluid on the suction surface.

Figure 10 shows the computed phase of the scattered pressure. The plot again indicates a complex phase structure near the airfoil and the shorter wavelength in the upstream direction. The figure also illustrates three distinct bends in phase contours. They occur at approximately 50°, −40°, and along 0°. The airfoil is again not very compact, and at moderate distances the current airfoil can be thought of as two monopoles at the edges plus a dipole which accounts for the asymmetric lift.

Figure 11 shows the amplitude contours of the acoustic pressure. Note the differences between Figure 11 and Figure 5. The region ahead of the airfoil still exhibits the interference fringes from the interaction of the incident wave with the scattered wave, but the fringes are of higher amplitude and are asymmetric. Also note the contour of 1.0 along the wake; the wake is acoustically silent.

Earlier it was shown that energy transfer between the mean and the perturbation flows may occur when the source term $D_a = \rho_0 \mathbf{u}_0 \cdot (\xi' \times \mathbf{u}') + \rho' \mathbf{u}' \cdot (\xi_0 \times \mathbf{u}_0)$ is nonzero. In the current case the vorticity $\xi' \neq 0$, but D_a is still nearly zero since $\mathbf{u}_0$ is essentially parallel to $\mathbf{u}'$, at least as long as the wake remains stable. There is little or no energy transfer between the mean and the perturbation flow. Figure 12 shows the variation of the mean power versus $k_{min}R$. The plot is essentially unchanged from the previous case.

While there is little or no energy transfer between the mean and the perturbation flow, there is a conversion of energy in the perturbation field. The energy from the incoming acoustic wave is converted into hydrodynamic energy of the vortices, and since the total amount of energy is constant there should be a net decrease in acoustic energy.

The perturbation energy has been previously defined. The energy contains both the acoustic and the rotational portions, and while it is possible to separate the perturbation velocity, hence the energy, into those parts the process usually entails solving the Laplace's equation. We have chosen a less accurate but a simpler definition which becomes exact in the limit $kR \to \infty$. For isentropic flow $\rho' = p'/c_0^2$ and the acoustic portion of velocity $\mathbf{u}'_a \approx p'/\rho_0 c_0$ and thus,

$$E_{acoustic} \sim \frac{\overline{p'^2}}{\rho_0 c_0}.$$

Define a related quantity, acoustic cross section σ, as in Pierce (1981),

$$\sigma \equiv \oint_s \frac{\overline{p'^2}}{\rho_0 c_0} ds. \tag{18}$$

The choice of contour ds is not arbitrary, and one must be chosen with care. For instance, there is an infinite amount of energy associated with a plane wave. However, there is a finite amount of energy in the scattered portion of the solution, and so a comparison of energies can be made valid if ω, M_0, α and the path ds is fixed.

A comparison of σ between the computed solution in (x, y) and the transformed solution using the unsteady Prandtl–Glauert method in (ξ, y) has been made to determine the amount of energy transferred from the acoustic to the vortical. The claim here is *not* that the transformed solution corresponds to the continuous solution with no Kutta condition, but rather that the transformed solution and the solution with no Kutta condition both have an equivalent amount of energy. Without viscosity or the Kutta condition an incident acoustic wave conserves energy.

In the previous non–lifting case where $\alpha = 0$, the ratio σ_{xy}/σ_{upg} should be $\equiv 1$ since there was no energy conversion. σ_{xy} is the computed acoustic cross section in the real plane (x, y) and σ_{upg} is the computed acoustic cross section in (ξ, y) space via the unsteady Prandtl–Glauert transformation. The computed ratio of σ_{xy}/σ_{upg} was 0.999.

In the current case where there is some energy conversion at the trailing edge, the ratio σ_{xy}/σ_{upg} should be < 1. The computed ratio was 0.990. Evidently approximately 1% of the acoustic energy was converted into the vortical energy.

This suggests that if vorticity is produced by the interaction of sound with a solid surface, there could be an attenuation of the sound as proposed by, among others, Crighton (1975), and Rienstra (1981). Figure 13 shows the computed p'_{rms} versus θ at $k_{min}R = 15.0$. The line with symbols is the computed solution in (ξ, y) space, and the line without symbols is the computed solution in (x, y) space. Note that locally the sound level in (x, y) can be higher than in (ξ, y) even though globally the sound level is lower in (x, y). Note also that the peaks and troughs are not in phase due to the diffractive effect of the shed vortex sheet.

7. Conclusions

A numerical method to directly compute the acoustic pressures and bypass an acoustic analogy has been presented. Its accuracy has been validated by comparing the computed solution with a full nonlinear steady Euler solver and by comparing it versus a known analytic Sears problem solution.

The diffraction and scattering of plane acoustic waves by rigid bodies in various flow–fields were investigated. Diffraction was quantitatively shown to occur in regions of high flow–field gradients. In the case where there was an unsteady lift, some portion of the incident acoustic energy was transformed into vortical energy. The resulting energy conversion caused an attenuation of sound.

We have highlighted the numerical method's application in plane wave scattering problems. The current scheme, however, is applicable to general compressible unsteady flow cases as long as 1) the incident Mach number does not approach unity and 2) the flow–field is relatively free of dominant non–linear sources of sound (e.g. unsteady vortex pairing, turbulence, etc.).

References

Bisplinghoff, R. L., Ashley, H., and Halfman, R. L., 1957. *Aeroelasticity*, Reading, Massachusetts: Addison-Wesley.

Bussing, T. R. and Murman, E. M., 1985. "A finite volume method for the calculation of compressible chemically reacting flows," AIAA Paper 85-0331.

Colonius, T., Lele S. K., and Moin P., 1991. "Scattering of sound waves by a compressible vortex," AIAA Paper 91-0494.

Crighton, D. G., 1975. "Basic principles of aerodynamic noise generation," *Prog. Aerospace Sci.* **16**(1), pp. 31-96.

Giles, M., 1988. "Non-reflecting boundary conditions for the Euler equations," CFDL-TR-88-1, MIT Department of Aeronautics and Astronautics.

Hall, K. C., 1987. "A linearized Euler analysis of unsteady flows in turbomachinery," Sc. D. Thesis, MIT.

Hariharan, S. I., 1984. "Numerical solutions of acoustic wave propagation problems using Euler computations," NASA Contractor Report 172434, ICASE Report No. 84-39.

Huh, K. S., Agarwal R. K., and Widnall S. E., 1990. "Numerical simulation of acoustic diffraction of two–dimensional rigid bodies in arbitrary flows," AIAA Paper 90-3920.

Huh, K. S., 1992. "Computational acoustics via linearized Euler equations," Ph. D. thesis in progress, MIT Department of Aeronautics and Astronautics.

Jameson, A., Schmidt, W., and Turkel, E., 1981. "Numerical solutions of the Euler equations by finite volume methods using Runge-Kutta time-stepping schemes," AIAA Paper 81-1259.

Khan, M. M. S., Brown W. H., and Ahuja, K. K., 1986. "Computational aeroacoustics as applied to the diffraction of sound by cylindrical bodies," AIAA Paper 86-1879.

Lele, S. K., 1990. "Compact finite difference schemes with spectral-like resolution," submitted to *Journal of Computational Physics.*

Myers, M. K., 1986. "Generalization and extension of the law of acoustic energy conservation in a nonuniform flow," AIAA Paper 86-0471.

Pierce, A. D., 1981. *Acoustics, An Introduction to Its Physical Principles and Applications*, New York: McGraw-Hill.

Rienstra, S. W., 1981. "Sound diffraction at a trailing edge," *Journal of Fluid Mechanics* **108**, pp. 443-460.

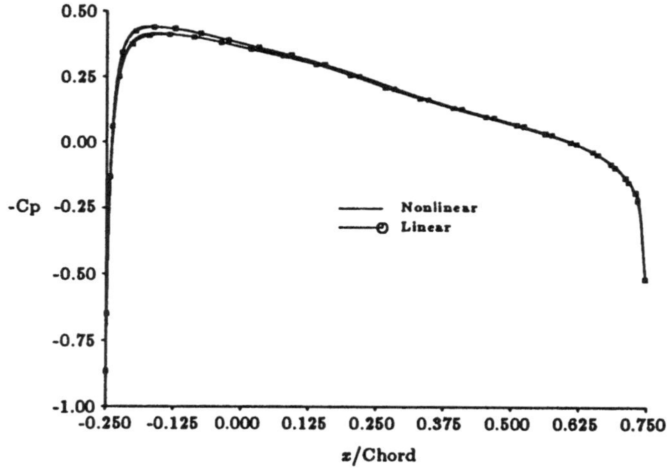

Figure 1: $-C_p$ on a NACA0012

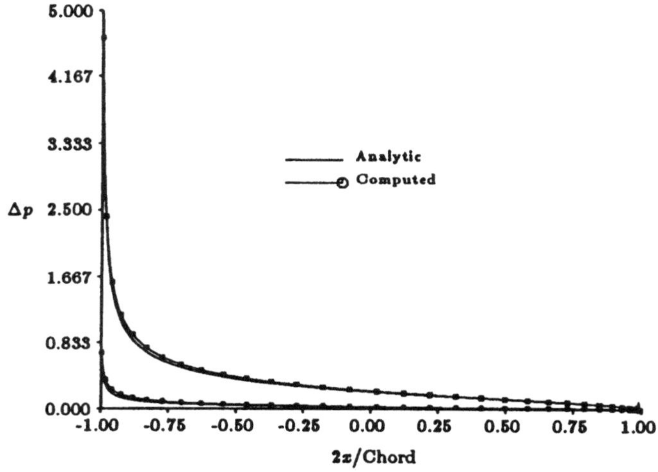

Figure 2: Δp on the Flat Plate

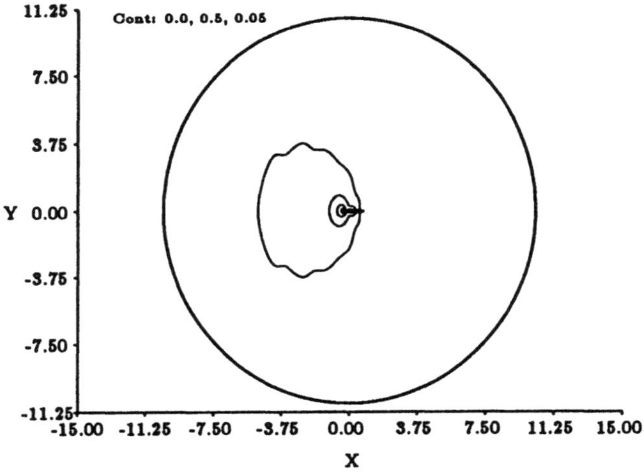

Figure 3: Amplitude of $\tilde{p}_s$

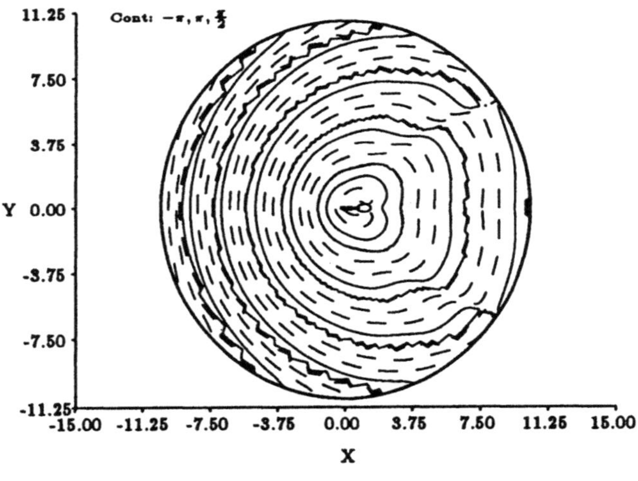

Figure 4: Phase of $\tilde{p}_s$

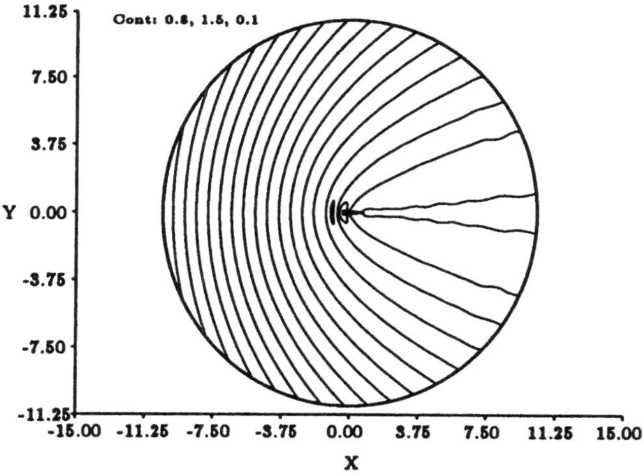

Figure 5: Amplitude of Acoustic Pressure

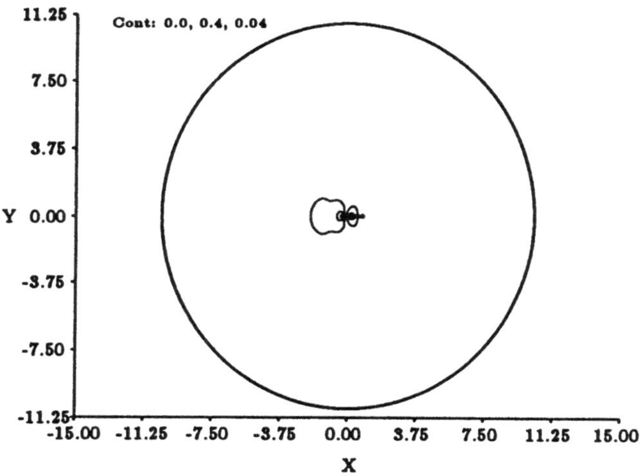

Figure 6: Amplitude of $\Delta\tilde{p}$

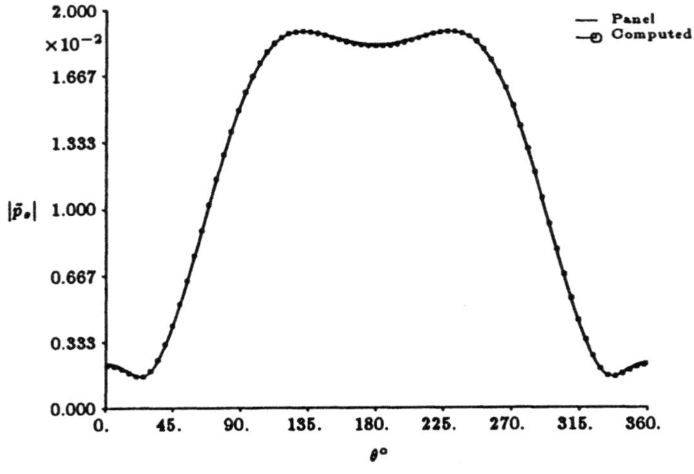

Figure 7: Comparison of $\tilde{p}_s$ at Far-Field

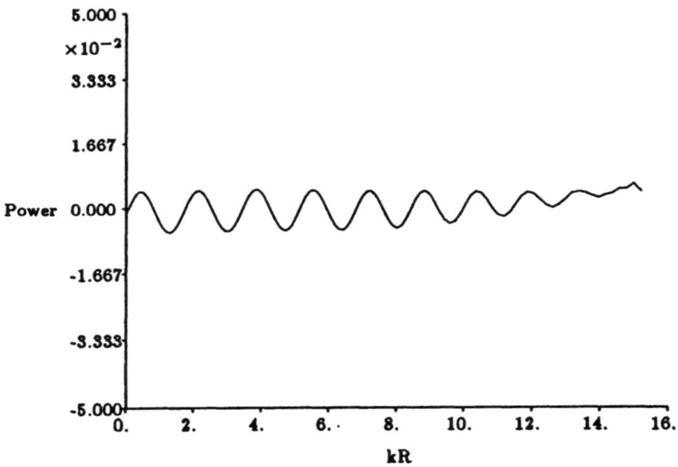

Figure 8: Mean Power versus Distance

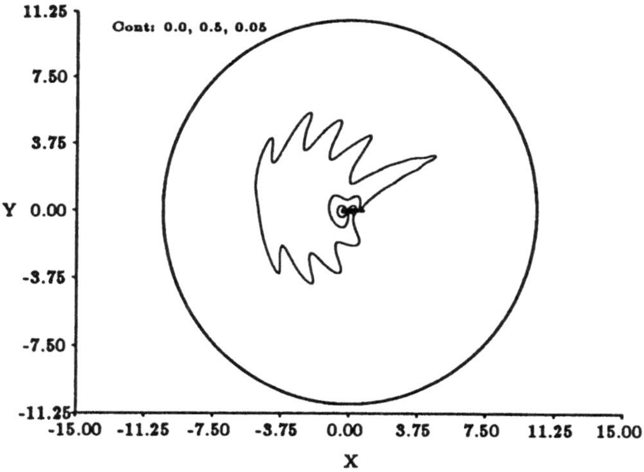

Figure 9: Amplitude of $\tilde{p}_s$

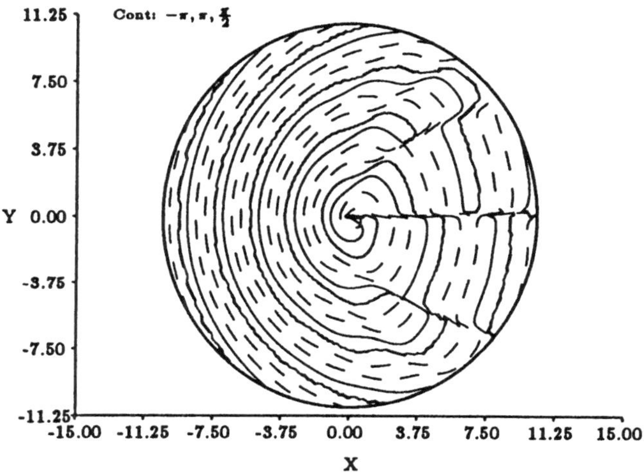

Figure 10: Phase of $\tilde{p}_s$

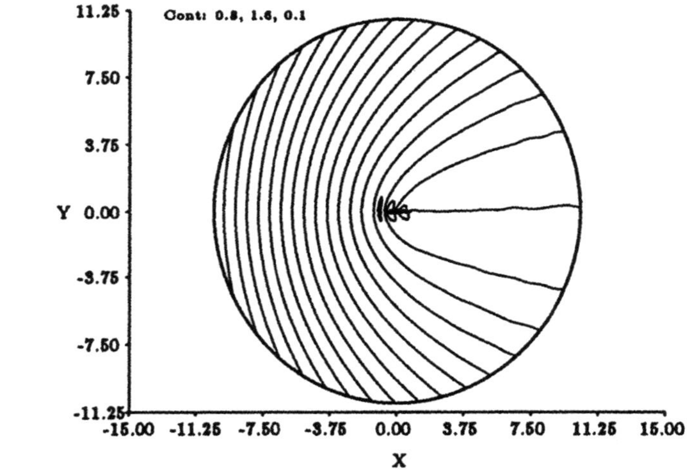

Figure 11: Amplitude of Acoustic Pressure

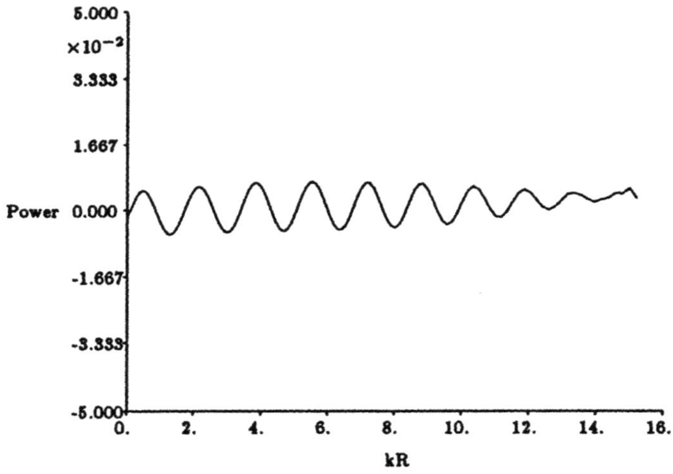

Figure 12: Mean Power versus Distance

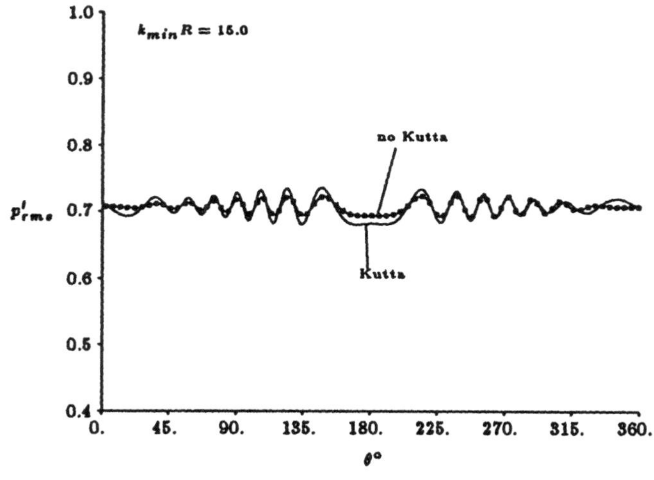

Figure 13: RMS Acoustic Pressure versus Azimuthal Angle

APPLICATION OF GEOMETRICAL ACOUSTICS TO PROPAGATION OF HIGH FREQUENCY JET NOISE

Abbas Khavaran

Sverdrup Technology, Inc.
NASA Lewis Research Center
Brook Park, Ohio 44142

ABSTRACT

High frequency geometrical acoustics based on ray-tracing methods is used to compute the spherical directivity of noise radiated from a convecting quadrupole source. The source is placed at an arbitrary position within a spreading jet. The propagation equations are solved in their general form in a rectangular coordinate system. The directivity pattern for an observer in the far field is obtained by applying an iteration scheme that finds the particular ray that connects the source point to the observer. Factors influencing the zone of silence are discussed. The formation of caustic as a result of reduction in ray tube area and the exact location where it appears is demonstrated. Finally the noise directivity due to a ring source convecting along an axisymmetric jet is obtained by a simple integration on the azimuthal directivity of compact quadrupole sources distributed on the ring.

1. Introduction

The study of jet noise starts with Lighthill's acoustic analogy. By defining the source term in his theory of aerodynamic sound generation, he opened the door to a vast scientific knowledge already developed in theoretical acoustics. Besides the generation aspects of the sound which has been extensively researched in the past (Lighthill, 1962 and Ffowcs, 1963), the propagation through moving fluids is an important aspect of jet noise computation. As far as the wave length of acoustic disturbances is concerned, within the three-way subdivision two are relatively easier. The easier cases, as noted by Lighthill (1972), correspond to situations when the emitted wave has a wave length much shorter or much longer than the characteristic length of the flow. The high frequency solution appears to provide a reasonably good approximation even at Helmholtz numbers as small as one (Tester and Morfey, 1976; Balsa, 1976; and Scott, 1979), and is of particular interest in prediction of noise directivity for high speed jets.

A number of investigators have studied the radiation field of multipole sources immersed in parallel sheared flows. Gliebe and Balsa (1976) and Goldstein (1975) studied the sound/flow interaction for round jets with arbitrary velocity profiles assuming quadrupole sources convecting along the centerline in both high and low frequency limits, respectively. The generalization to arbitrarily located sources in continuously varying monotonic profiles were derived by Balsa (1977) and Goldstein (1976a, 1976b). For a parallel flow study, Lilley's equation is considered as the starting point by most investigators studying the sound/flow interaction. For axisymmetric jets, the Green's function solution to a convected monopole of frequency ω is obtained by applying a sequence of Fourier transformations. The solution for multipole singularities are obtained by differentiating the monopole solution with respect to appropriate source coordinates. Balsa and Mani (Mani et al., 1977) provide a comprehensive analysis of the shielding effects of parallel jets when velocity and temperature profiles are functions of the radial variable only. Their derivation was used by the present author in the computation of supersonic jet mixing noise of a CD nozzle (Khavaran et al., 1992). To study the effect of asymmetry in the mean flow, Goldstein (1982) solves Lilley's equation for high frequency multipole sources in a parallel jet flow whose Mach number and temperature are functions of the cross-flow coordinates. In the high frequency limit one can recover equations similar to those governing the two-dimensional waves generated by a line source in a non-moving medium with variable index of refraction. Avila and Keller (1963) obtained the solution to this problem using matched asymptotic expansions. The corresponding sound field directivity is obtained by solving a fourth order ordinary differential equation which finds the projection of rays on a cross-flow plane. This solution technique, though successful in giving a qualitative picture of circumferential directivity of jet noise for non-axisymmetric plumes, will have limited use because of parallel flow assumption. In addition at polar angles close to the jet axis, the rays can get trapped within an envelope and go through a never-ending refraction cycle. Thus a full three-dimensional ray-tracing approach becomes necessary where one can study the refraction of acoustic rays as they emerge from the source on their way to a distant observer. This solution technique provides the tools for investigating the zone of silence as well as the caustics of geometrical acoustics.

A high frequency Green function for a convecting multipole source in a spreading jet was developed by P. A. Durbin (1983a) and was applied to predict the directivity of noise for a source convecting along the center-axis of an axisymmetric jet. The symmetric nature of this problem, confines the acoustic rays to a constant azimuthal plane and thus reduces the number of propagation equations from six to two. Following the methodology described in Durbin (1983a, 1983b), an attempt has been made here to predict the spherical directivity of a convecting quadrupole source in an arbitrary flow. The six propagation equations are solved in a rectangular coordinate system. As we shall describe under numerical results, the solution to the corresponding boundary value problem obtained based on specified source and observer locations, requires an iteration on the initial ray angles. The jet spreading as well as plume profile strongly influence the size of zone of silence. The jet spreading does not necessarily remove the zone of silence, rather velocity gradients near the center-axis have a dominant role in defining the boundary of zone of silence. The nature of caustic and its exact location is described by studying the variation in ray tube area obtained from transport equation. It should be emphasized that the flow gradients should be small for the geometric ray theory to work. For cases when shocks are present, special treatment will be necessary to properly trace the rays across a discontinuity. In addition, no attempt has been made here to derive expressions valid near the caustics, although solutions similar to those obtained for a still medium can be expected (Kay and Keller, 1954; Ludwig, 1966; and Zauderer, 1970).

2. Governing Equations

The relevant equations are the linearized gas-dynamic equations governing the propagation of small disturbances through a steady mean flow. For a source with time harmonic factor $e^{-i\omega t}$ the inhomogeneous equations of continuity and momentum are

$$-i\omega\rho' + \nabla.(\mathbf{U}\rho' + \mathbf{u}\rho) = a^2 Q_d$$

$$-i\omega\mathbf{u} + \mathbf{U}.\nabla\mathbf{u} + \mathbf{u}.\nabla\mathbf{U} + \nabla(a^2\rho'/\rho) = -(a^2/\rho)Q_m \qquad (1)$$

where ρ' and $\mathbf{u}$ are the density and velocity of the acoustic fluctuations and ρ and $\mathbf{U}$ are the corresponding quantities for the mean flow, a is the mean flow sound speed and Q_d and Q_m represent the source terms. In a region not close to the source, the amplitude variation of the acoustic disturbances is slow and a high frequency solution to the homogeneous form of equations (1) can be obtained by substituting the classical expansions in inverse powers of the wave number $k = \omega/a_r$:

$$\rho' = \{\varrho + (1/ik)\varrho_1 + ...\}\, e^{ikL(\mathbf{X})}$$
$$\mathbf{u} = \{\mathbf{v} + (1/ik)\mathbf{v}_1 + ...\}\, e^{ikL(\mathbf{X})} \tag{2}$$

where a_r is a typical reference sound speed. Factoring the coefficients of equal powers of k and setting each factor equal to zero independently, one can obtain a series of equations describing the behavior of wave fronts of constant phase and equations for transport of energy. The first of such equations is eikonal equation. It describes the change in normal to the phase front due to refraction caused by nonuniformities in the flow velocity and temperature. Introducing normal to the phase front as $\mathbf{p} = \nabla L$, and Mach number $\mathbf{M} = \mathbf{U}/a$ and dimensionless sound speed $C(\mathbf{X}) = a/a_r$, the eikonal equation is expressed as

$$(M_i p_i - 1/C)^2 = p_i p_i. \tag{3}$$

The rays generated by the eikonal equation can be obtained by the method of characteristics, thus along the ray $\mathbf{X}(t)$ and $\mathbf{p}(t)$ can be calculated by solving a system of six ordinary differential equations (Durbin, 1983a and Jones, 1977):

$$\dot{X}_i = T_{ij}p_j + M_i/C \tag{4}$$

$$\dot{p}_i = -\frac{1}{2}p_j\frac{\partial T_{jk}}{\partial X_i}p_k - p_j\frac{\partial}{\partial X_i}\left(\frac{M_j}{C}\right) + \frac{1}{2}\frac{\partial}{\partial X_i}(C^{-2}). \tag{5}$$

Tensor T_{ij} is related to Mach number $\mathbf{M}$; ($T_{ij} = \delta_{ij} - M_iM_j$). The initial conditions for (4) and (5) are the source location $\mathbf{X} = \dot{\mathbf{X}}_o$ and the direction at which the ray leaves the source. When the source is placed at the apex of a cone of angle μ (referred to as cone of emission), each ray can be defined along the generator of the cone through angle δ (Figure 1). Along such a ray the initial velocity can be expressed through unit vector $\hat{\mathbf{X}}_o = (\cos\mu,\ \sin\mu\cos\delta,\ \sin\mu\sin\delta)$ and ray speed $\lambda = |\dot{\mathbf{X}}|$ as $\dot{\mathbf{X}}_o = \lambda_o\hat{\mathbf{X}}_o$. Substituting $\dot{\mathbf{X}}_o$ in (4) and solving (3) and (4) simultaneously, one can find the initial condition on $\mathbf{p}$ as

$$\mathbf{p}_o = \lambda_o\hat{\mathbf{X}}_o + (\lambda_o\hat{\mathbf{X}}_o.\mathbf{M}_o - \frac{1}{C_o})\frac{\mathbf{M}_o}{{\beta_o}^2} \tag{6}$$

$$\lambda_o = [\beta_o{}^2 + (\hat{\mathbf{X}}_o.\mathbf{M}_o)^2]^{-\frac{1}{2}}/C_o \tag{7}$$

where subscript o applies to the source location and $\beta^2 = 1 - |\mathbf{M}|^2$. Limiting the analysis to the case when $\beta_o{}^2 \neq 0$, it can be concluded from (7) that for the ray speed λ_o to be real, the direction of emission $\hat{\mathbf{X}}_o$ for a supersonic flow, $|\mathbf{M}| > 1$, should satisfy the condition

$$1 - |\mathbf{M}_o|^2 + (\hat{\mathbf{X}}_o.\mathbf{M}_o)^2 > 0. \tag{8}$$

The transport equation, obtained from the next order of k in the governing equations, can be simplified as (Jones, 1977):

$$\nabla.(\rho\dot{\mathbf{X}}\Gamma^2) = 0, \qquad \Gamma = \frac{(\varrho/\rho)}{\frac{1}{C} - \mathbf{M}.\mathbf{p}}. \tag{9}$$

Equation (9) leads to the wave action conservation equation (Candel, 1977) which states that the action contained in a wave packet is invariant along each characteristic line

$$\lambda\Gamma^2\rho J = constant \tag{10}$$

where J measures the variation of the ray tube cross-section along each ray. The variation in the ray tube area can be expressed in terms of a transformation which maps the initial ray angles $(\mu,\, \delta)$ onto the final direction at the observer location $(r,\, \theta_\infty,\, \phi_\infty)$. Assuming that in the far field the rays become straight, it can easily be shown that

$$J = r^2\, sin\,\theta_\infty \left| \frac{\partial(\theta_\infty,\, \phi_\infty)}{\partial(\mu,\, \delta)} \right|. \tag{11}$$

Thus the field amplitude ϱ can be obtained from (10) once the elemental wave front area J has been determined. To find the constant in (10), an appropriate near source solution satisfying (1) is asymptotically matched with the outer solution. The details of the analysis are spelled out in Durbin (1983a) and are not repeated here. If a_r is taken to be the stagnation sound speed a_∞, the first term in the high frequency approximation for (2) is

$$\begin{aligned} \rho' \propto \varrho\, e^{ikL} \;=\; & \left(\frac{1}{4\pi r}\right)\left(\frac{C_o}{C}\right)\frac{1 - \frac{\mathbf{U}}{a_\infty}.\mathbf{p}}{1 - \frac{\mathbf{U_o}}{a_\infty}.\mathbf{p_o}} \\ & \left\{\frac{\rho}{\rho_o}\frac{\lambda_o^3}{\lambda}\frac{sin\,\mu}{sin\,\theta_\infty}\frac{C_o{}^2}{\left|\frac{\partial(\theta_\infty,\,\phi_\infty)}{\partial(\mu,\,\delta)}\right|}\right\}^{1/2} e^{ikL}. \end{aligned} \tag{12}$$

This Green function can be used to derive an expression for the mean square pressure for a convecting quadrupole source (Goldstein, 1982 and Durbin, 1983a). In the high frequency approximation the source frequency ω_o can be related to the observer frequency ω through the Doppler factor

$$\omega = \omega_o/(1 - \frac{\mathbf{U_c}}{a_\infty}.\mathbf{p_o}) \tag{13}$$

where $\mathbf{U_c}$ is the source convection velocity.

If the spectral power intensity for the source is represented as $\overline{|\hat{Q}(\omega_o)|^2}$, the source power within a narrow-band $\Delta\omega_o$ is $Q_o{}^2 = \overline{|\hat{Q}(\omega_o)|^2}\Delta\omega_o$. The resulting mean-square pressure directivity for a convecting isotropic quadrupole is given as

$$\overline{P^2}(\omega, \mathbf{X}) \propto (\frac{1}{4\pi r})^2 \omega_o{}^4 Q_o^2 (\frac{\rho}{\rho_o})(\frac{\lambda_o^3}{\lambda})(\frac{\sin\mu}{\sin\theta_\infty})\frac{(1-\frac{\mathbf{U_o}}{a_\infty}.\mathbf{p_o})^2(1-\frac{\mathbf{U}}{a_\infty}.\mathbf{p})^2}{(1-\frac{\mathbf{U_c}}{a_\infty}.\mathbf{p_o})^5 \mid \frac{\partial(\theta_\infty,\phi_\infty)}{\partial(\mu,\delta)} \mid}. \tag{14}$$

It can be seen from (13) that the source frequency should be different from point to point in order for a fixed observer to hear the same frequency. This is essentially due to refraction phenomena resulting in variation of $\mathbf{p_o}$ for each ray reaching a given observer and will be explored later by computing the Doppler factor for source points distributed around the axis of an axisymmetric jet. It is also important to remember that for the purpose of numerical computation of equation (14), no attempt has been made to specify the source strength. Determination of source correlation term and its spectrum, by itself, is a major area of research in computational aeroacoustics. An approach based on computation of turbulence kinetic energy and its dissipation rate using CFD is described in Khavaran et al. (1992).

3. Numerical Results and Discussion

The spherical directivity pattern corresponding to equation (14) can be written as

$$\overline{P^2}(\mathbf{X}) \propto (\frac{\lambda_o^3}{\lambda})(\frac{\sin\mu}{\sin\theta_\infty})\left\{\frac{(1-\frac{\mathbf{U_o}}{a_\infty}.\mathbf{p_o})^2(1-\frac{\mathbf{U}}{a_\infty}.\mathbf{p})^2}{(1-\frac{\mathbf{U_c}}{a_\infty}.\mathbf{p_o})^5 \mid \frac{\partial(\theta_\infty,\phi_\infty)}{\partial(\mu,\delta)} \mid}\right\} \tag{15}$$

where $\mathbf{U_o}$ is the mean flow velocity at the source location and in general is different from the source convection velocity $\mathbf{U_c}$. For a parallel jet, the directivity pattern outside the zone of silence is commonly accepted to follow the Doppler factor to -3 power. For a spreading jet, this remains valid, that is $\overline{P^2} \propto (1 - \mathbf{U_o}.\mathbf{p_o}/a_\infty)^2/(1 - \mathbf{U_c}.\mathbf{p_o}/a_\infty)^5$, as long as the source is convecting along the center-axis of an axisymmetric jet.

As the source is placed off-axis, the non-axisymmetric nature of the problem creates a directivity pattern that will not follow the above simple rule; rather the source eccentricity as well as the relative azimuthal angle between the source and observer will influence the directivity pattern. However, the directivity due to a ring source, obtained from the circumferential integral of the noise produced by compact quadrupole sources distributed on the ring will be shown to, more or less, follow the Doppler factor to -3 power outside the quieting zone. Here for numerical computations, the source convection velocity is taken to be a weighted average of the mean flow velocity at the jet exit and the source location

$$\frac{\mathbf{U_c}}{a_\infty} = .5(\frac{\mathbf{U_o}}{a_\infty}) + \beta_c(\frac{\mathbf{U_j}}{a_\infty}). \tag{16}$$

In order to integrate the propagation equations, the jet velocity and sound speed are expressed in a closed form to speed up the computation of derivatives. For a jet spreading at angle α, a self-similar velocity profile with appropriate decay characteristics can be expressed as

$$\mathbf{U}(r,\theta)/a = (\mathbf{U}/a)_{CL}\ e^{-(\frac{\theta}{\alpha})^n} \tag{17}$$

where $(\mathbf{U}/a)_{CL}$ is the Mach number on the centerline and θ is the polar angle defined earlier. The velocity decay on the center-axis of the jet has been measured experimentally for different nozzle geometries and exit conditions. For a convergent nozzle, the peak Mach can be expressed as a function of axial distance parameter L_x (Groesbeck, Huff, and von Glahan, 1977)

$$(\mathbf{U}/a_\infty)_{CL} = (\mathbf{U_j}/a_\infty)\left\{1 + (.15L_x)^4\right\}^{-.25}, \quad L_x = \frac{X/D}{C_n\sqrt{1+M_j}}. \tag{18}$$

The jet exit Mach number $M_j = U_j/a_\infty$ can vary in a relatively wide range as indicated in Figure 2. D is the exit diameter and X is measured from exit plane (Figure 1). Using the adiabatic flow relation combined with (17) and (18), the Mach number and sound speed are given as

$$M = \frac{U}{a} = \left\{ \frac{\sqrt{1 + (.15L_x)^4}}{(U_j/a_\infty)^2} - \frac{\gamma - 1}{2} \right\}^{-1/2} e^{-(\frac{\theta}{\alpha})^n}$$

$$C = \frac{a}{a_\infty} = \left\{ 1 + \frac{\gamma - 1}{2} (\frac{U}{a})^2 \right\}^{-1/2} . \tag{19}$$

A large value for exponent n corresponds to a slug flow profile which is the characteristic of the exit. By allowing n to gradually decrease in flow direction, one can recover a Gaussian distribution for the downstream profile. To illustrate the refraction effects, two plume models have been considered. In model one n is constant ($n = 4$), whereas in model two n starts from near 6 at exit and decays to 2 at about 12 diameters from the exit according to

$$n = \frac{5b}{(X/D + .1)^{1.25} + b} + 1, \quad b = 5. \tag{20}$$

The velocity profiles for the two models are illustrated in Figure 3 for a jet spreading at $\alpha = 10°$. The location of the exit plane is defined as the section at which the jet diameter is one. All numerical computations are done for a jet exit Mach of $U_j/a_\infty = .99$ and at seven diameters from the exit where $U_o/a_\infty = .87$ on the centerline. The source location at any axial section can be defined through angles θ_o and ϕ_o. For a sample pair of initial ray angles (μ, δ), a typical ray has been traced using model two and its trajectory is projected on three orthogonal planes (Figure 4). The broken lines show the outer boundary of the jet mixing region. A full ray-tracing analysis for an off-axis source is given in Figure 5. The final angles of the ray $(\theta_\infty, \phi_\infty)$ are plotted vs. the initial angles (μ, δ) for model two. It can be seen that for small angles of cone of emission ($\mu \leq 10°$), the rays are limited to a narrow envelope in the azimuthal plane ($| \phi_\infty | < 48.4°$). For example for $\mu = 10°$, even those rays emitted at $\delta \to 180°$ will not reach $\phi_\infty = 180°$, rather they go through a complete refraction and reach back at $\phi_\infty = 0$. Such rays can penetrate polar angle $\theta_\infty \to 46.8°$, shallower than $\theta_\infty = 54.7°$ set by $\mu = 0$. As the cone of emission expands, the rays start to fill the entire azimuthal plane. For $\mu > 50°$, the rays begin to penetrate the upstream angles. As the cone in expanded further, for $\mu = 70°$, one can detect rays that travel more than 180° in circumferential direction to reach the upstream directions. It should be emphasized that even for such cases, there is only one ray reaching an observer location. The only exception resulting in multi-arrival is in the vicinity of caustics. This case will be discussed later.

To investigate the boundary of zone of silence, the source was moved toward the centerline by changing θ_o while X/D is constant. For a ray beaming straight downstream ($\mu = 0$), the final polar angle $\theta^\star$ is plotted vs. the source location. Figure 6a, obtained for model one, shows how the zone of silence (defined through angle $\theta^\star$) is removed by allowing a small jet spreading angle. For model two, however, Figure 6b shows that the zone of silence is still present even at spreading angles as large as $\alpha = 20°$. This comparison clearly demonstrates the significance of velocity gradients in bending the rays away from the center axis. Rays that are emitted near the centerline should travel a relatively longer distance before they get refracted and as a result they run into high gradient regions of the plume (similar to the Gaussian profile in model two) and a zone of silence is formed. Another significant observation that can be made here is related to what is termed as the "quieting zone". As the source moves away from the center, there are rays emitted on a cone of emission with $\mu > 0$ that can go through complete refraction and reach values of θ_∞ smaller than $\theta^\star$ (an example was given in Figure 5). therefore one can conclude that within the zone of silence, the sound pressure level (SPL) does not drop to zero at once, rather there is a gradual decay in noise level. As the jet spreading angle approaches zero, the size of zone of silence in both 6a and 6b approach the parallel flow solution (Durbin, 1983b) given as

$$\theta^\star \ < \ cos^{-1}\frac{1}{C_o(1+M_o)} \tag{21}$$

where $C_o = .92$ and $M_o = .94$ at $X/D = 7$.

In all remaining discussions the jet spreading angle is held at $\alpha = 10°$ and the convection velocity follows (16) with $\beta_c = .3$. For a source on the axis, the sound field is axisymmetric and results similar to those in Durbin (1983b) are obtained. In Figure 7 the slope of the curve near the downstream axis changes with the size of zone of silence. By selecting a constant exponential factor n, Durbin (1983b) discusses the directivity pattern of a similar source for a set of constant values for n as well as the jet spreading angle. As we have seen earlier in the context of Figure 6, a constant n can artificially suppress refraction. To compare the present prediction with data, Stone's correlation (Stone et al., 1981) was employed. The directivity pattern $\overline{P^2}(1 - M_c \cos\theta)^3$ obtained from these correlations is identically equal to one outside the quieting zone; within the quieting zone the slope gradually increases with frequency. For a Helmholtz number of $He = fD/a_\infty = 31.6$, the data correlation shows a slope of 19° per octave compared with 10.5° per octave calculated from the high frequency geometrical theory. As we shall demonstrate later on, by using a ring source, this difference in slopes narrows down substantially. In applying Stone's correlation, the convection Mach number for the source was substituted according to (16).

When the source is placed off-axis, the numerical work in solving propagation equations is greatly increased. In general an iteration process should be applied to find a ray that propagates through a specified observer location. Figure 8a shows the variation in the shielding effect as a function of source eccentricity. The source is placed at three locations defined through $\theta_o = (0, 2°, 5°)$ with $\phi_o = 0$ and $X/D = 7$ while the azimuthal angle for the observer is held at $\phi_\infty = 0$. Notice that as the source is brought closer to the observer, the shielding effect of the mean flow decreases in downstream directions resulting in an increase in SPL. At $\theta_\infty = 90°$ all these sources produce the same sound field while in the upstream angles an opposite effect is observed. Figure 8b shows that as the source moves further away from the axis, the SPL does not scale with power -3 of the Doppler factor.

Of particular interest is the spherical directivity of an off-axis convecting quadrupole. This is demonstrated in Figures 9 and 10 for a source at $\theta_o = 5°$ and $\phi_o = 0$. In Figure 9, ϕ_∞ is held as a parameter and directivity pattern is plotted against θ_∞, while the same results are plotted in Figure 10 by exchanging the roles of θ_∞ and ϕ_∞. In general one can observe an increase in azimuthal variation of sound further downstream. At $\theta_\infty = 90°$, there is no azimuthal directivity and in upstream direction a change in the trend of these curves is observed. For $\theta_\infty > 110°$, an increase in SPL is predicted as the observer moves azimuthally to the opposite of the source (Figure 9b). In fact if the computation is carried out for $\phi_\infty > 160°$, a singularity starts to develop at $\phi_\infty \rightarrow 180°$ and $\theta_\infty \rightarrow 141°$ as the Jacobian in equation (15) approaches zero. This is further clarified in Figure 11 by looking more closely at the acoustic rays reaching this neighborhood. For a cone of emission with $\mu = 67°$, the rays that leave the source at a range of δ say from 150° to 180°, will all end up at the same observer angles. This indicates a reduction in ray tube area resulting in focusing or concentration of rays in a neighborhood which is the caustic of geometrical acoustics. The sound field obtained according to (12) is essentially the first term in asymptotic expansion in inverse powers of wave number. As was pointed out by J. Keller (Kay and Keller, 1954 and Keller, 1958), the zero-order term in this expansion can become infinite even when the field is perfectly regular. A change in the form of the asymptotic expansion is necessary such that the first term contains factor k raised to a fractional power. A uniform asymptotic expansion at caustics for a still medium was suggested by Ludwig (1966) in terms of Airy function and its derivatives. A more rigorous analysis was suggested by Zauderer (1970) by looking into the modified forms of asymptotic expansions of the reduced wave equation in transition regions. One might expect similar expansions for the case of a moving medium. No attempt has been made here to replace (12) by a solution uniformly valid near the caustic, rather a simple extrapolation of the solution given in Figure 10 was applied to predict the field at $180° \geq \phi_\infty \geq 160°$. Assuming that a ring source is obtained from the superposition of independent correlation volume elements distributed around the center axis, for an axisymmetric jet one can simply integrate the area under each curve in Figure 10 and obtain the sound field due to a convecting ring source. This idea can further be extended to an entire jet by subdividing the plume into axial slices that are formed from ring elements.

The most important result of the present theoretical prediction is summarized in Figures 12, 13 and 14. The noise directivity for a ring source is shown in Figure 12. The presence of caustic manifests itself as a change in slope around 120 degrees. Such a change in slope can also be observed in most experimental reports on the directivity of jet mixing noise (Stone et al., 1981). Figure 13 presents the Doppler factor as calculated from the ray theory normalized with respect to the conventional definition based on line-of-sight method. At any given polar angle θ_∞ (used as parameter here), each curve represents the change in the Doppler factor as the source is allowed to navigate on a ring relative to a fixed observer. This can clearly be attributed to the required change in the initial direction of the ray in order to make it propagate through the observer point. Finally, the scaling of directivity with power -3 of the Doppler factor is investigated in Figure 14. Within the quieting zone, the prediction gives a slope of 17° per octave compared with 19.3° per octave obtained from Stone's correlation with a Helmholtz number of 31.6. This is a significant improvement over the earlier comparisons for a point source. It should also be noted that even at a Helmholtz number of 10, the data correlations give a slope of 23° per octave, not too far from our high frequency prediction.

4. Suggestions for Future Work

It has been shown, through a relatively simple jet model, that geometrical ray theory can be used to study the propagation of high frequency jet noise. A more complete study, perhaps, should employ an actual plume where the spreading angle can also change along the jet. Such an application will require appropriate spline interpolations on data and numerical computation of derivatives that appear in propagation equations. In addition ray-tracing across a discontinuity will also be of interest for cases when shock cells are present. Finally, an asymptotic solution uniformly valid near the transition points can be considered a good exercise in applied mathematics.

Acknowledgements

The author would like to thank M. E. Goldstein of NASA Lewis Research Center for suggesting the high frequency problem of geometrical acoustics.

References

Avila, G. S. S., and Keller, J. B., 1963. "The high frequency asymptotic field of a point source in an inhomogeneous medium," *Comm. Pure Appl. Math.* **16**, pp. 363-381.

Balsa, T. F., 1976. "The far field of high frequency convected singularities in sheared flows, with application to jet-noise prediction," *JFM* **74**, pp. 193-208.

Balsa, T. F., 1977. "The acoustic field of sources in shear flow with application to jet noise: convective amplification," *JFM* **79**, pp. 33-47.

Candel, S. M., 1977. "Numerical solution of conservation equations arising in linear wave theory: Application to aeroacoustics," *JFM* **83**, pp. 465-493.

Durbin, P. A., 1983a. "High frequency Green function for aerodynamic noise in moving media, Part I: General theory," *J. Sound Vibration* **91**(4), pp. 519-525.

Durbin, P. A., 1983b. "High frequency Green function for aerodynamic noise in moving media, Part II: Noise from a spreading jet," *J. Sound Vibration* **91**(4), pp. 527-538.

Ffowcs Williams, J. E., 1963. "The noise from turbulence convected at high speed," *Philos. Trans. Roy. Soc. London* **A255**, pp. 469-503.

Gliebe, P. R. and Balsa, T. F., 1976. "The aerodynamics and acoustics of coaxial jet noise," AIAA Paper 76-492.

Goldstein, M. E., 1975. "The low frequency sound from multipole sources in axisymmetric shear flows," *JFM* **70**, pp. 595-604.

Goldstein, M. E., 1976a. *Aeroacoustics*, McGraw-Hill.

Goldstein, M. E., 1976b. "The low frequency sound from multipole sources in axisymmetric shear flows - Part II," *JFM* **75**, pp. 17-28.

Goldstein, M. E., 1982. "High frequency sound emission from moving point multipole sources embedded in arbitrary transversely sheared mean flows," *J. Sound Vibration* **80**(4), pp. 499-522.

Groesbeck, D. E., Huff, R. G., and von Glahan, U. H., 1977. "Comparison of jet Mach number decay data with a correlation and jet spreading contours for a large variety of nozzles," NASA TN D-8423.

Jones, D. S., 1977. "The mathematical theory of noise shielding," *Progress in Aerospace Science* **17**, pp. 149-229.

Kay, I. and Keller, J. B., 1954. "Asymptotic evaluation of the field at a caustic," *J. Applied Physics*, **25**, pp. 876-883.

Keller, J. B., 1958. *Geometrical Theory of Diffraction, Calculus of Variations and its Applications, Proc. Symposia Appl. Math.*, Vol. 8, pp. 27-52, McGraw-Hill, NY.

Khavaran, A., Krejsa, E. A., and Kim, C. M., 1992. "Computation of supersonic jet mixing noise for an axisymmetric CD nozzle using $k\epsilon$ turbulence model," AIAA Paper 92-0500, Also NASA TM-105338.

Lighthill, M. J., 1962. "Sound generated aerodynamically," *Proc. Roy. Soc.*, **A267**, pp. 147-182.

Lighthill, M. J., 1972. "The fourth annual fairey lecture: The propagation of sound through moving fluids," *J. Sound Vibration* **24**(4), pp. 471-479.

Ludwig, D., 1966. "Uniform asymptotic expansions at a caustic," *Comm. Pure Appl. Math.* **19**, pp. 215-250.

Mani, R., et al., 1977. "High velocity jet noise source location and reduction," Task 2, FAA-RD-76-79-II.

Scott, J. N., 1979. "Propagation of sound waves through linear shear layer," *AIAA Journal*, **17**, pp. 237-244.

Stone, J. R., Groesbeck, D. E., and Zola, C. L., 1981. "An improved prediction method for noise generated by conventional profile coaxial jets," AIAA Paper 81-1991, Also NASA TM-82712.

Tester, B. J. and Morfey, C. L., 1976. "Developments in jet noise modeling - theoretical predictions and comparisons with measured data," *J. Sound Vibration* **46**(1), pp. 79-103.

Zauderer, E., 1970. "Uniform asymptotic solutions of the reduced wave equation," *J. Math. Anal. Appl.* **30**, pp. 157-171.

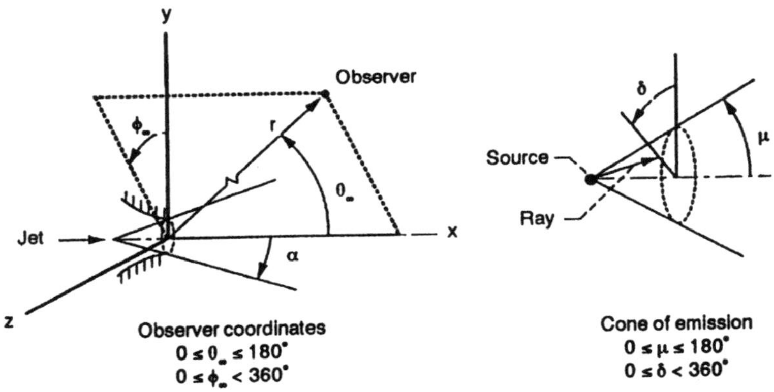

Figure 1. **Source and observer coordinates.**

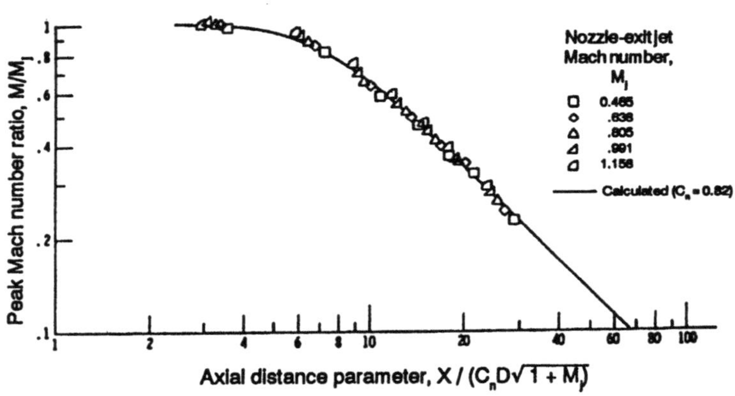

Figure 2. **Peak axial Mach number decay of 3.58-centimeter-diameter convergent nozzle (Groesbeck 1977).**

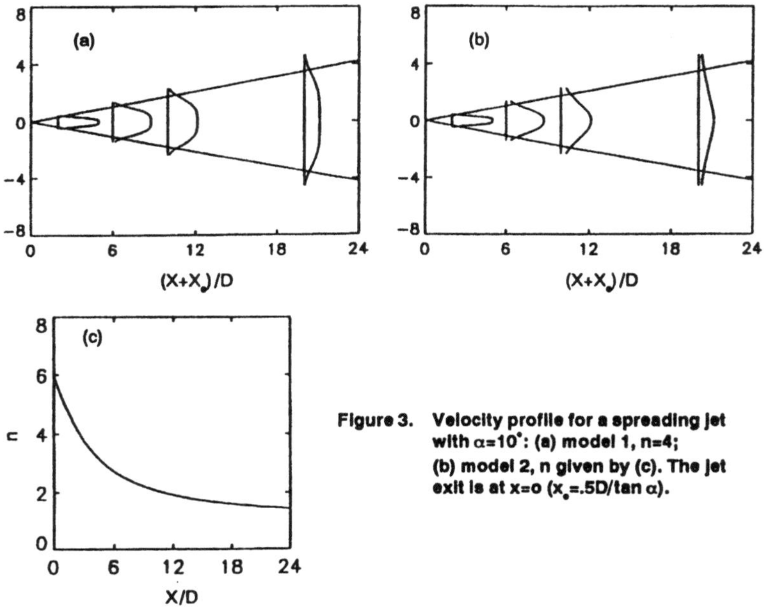

Figure 3. Velocity profile for a spreading jet with $\alpha=10^{\circ}$: (a) model 1, n=4; (b) model 2, n given by (c). The jet exit is at x=o (x_e=.5D/tan α).

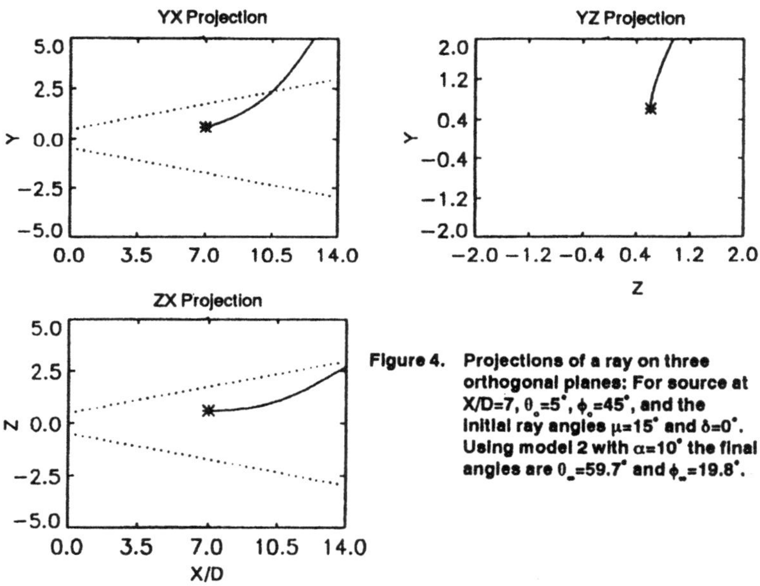

Figure 4. Projections of a ray on three orthogonal planes: For source at X/D=7, $\theta_o=5^{\circ}$, $\phi_o=45^{\circ}$, and the initial ray angles $\mu=15^{\circ}$ and $\delta=0^{\circ}$. Using model 2 with $\alpha=10^{\circ}$ the final angles are $\theta_{\infty}=59.7^{\circ}$ and $\phi_{\infty}=19.8^{\circ}$.

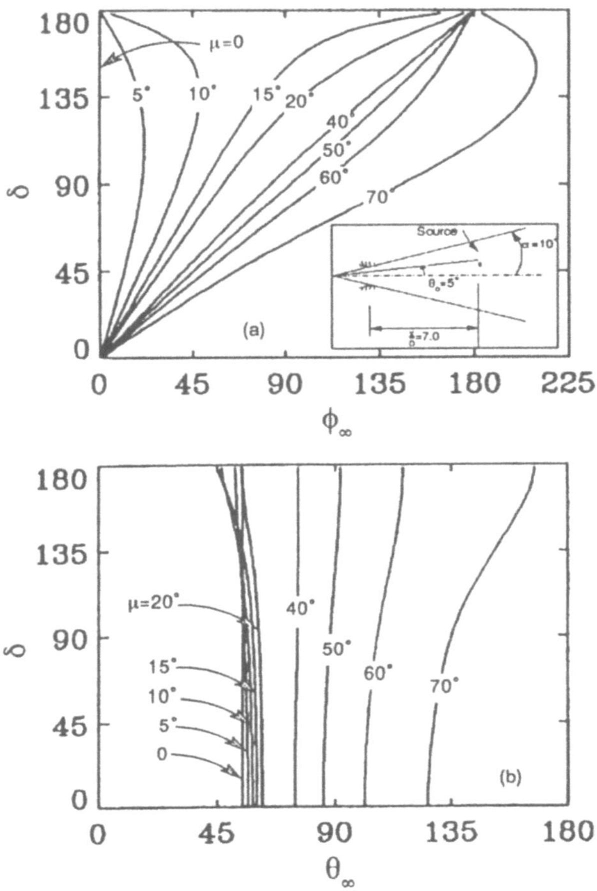

Figure 5. A complete ray-tracing for a source at X/D=7, θ_o=5°, ϕ_o=0°, using model 2 with α=10°.

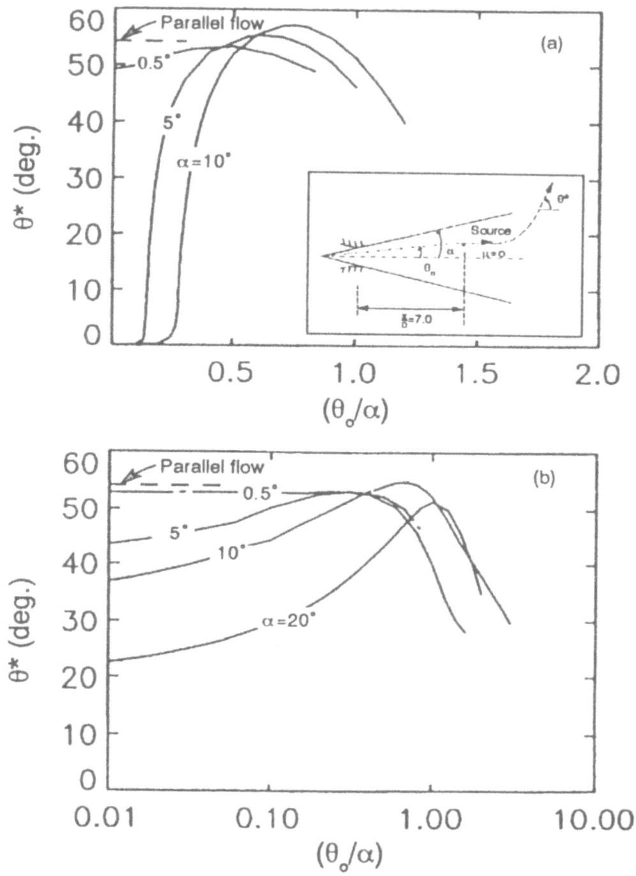

Figure 6. Boundary of zone of silence for a source at x/D=7. (a) Model 1; (b) Model 2. The jet spreading angle α is indicated as a parameter.

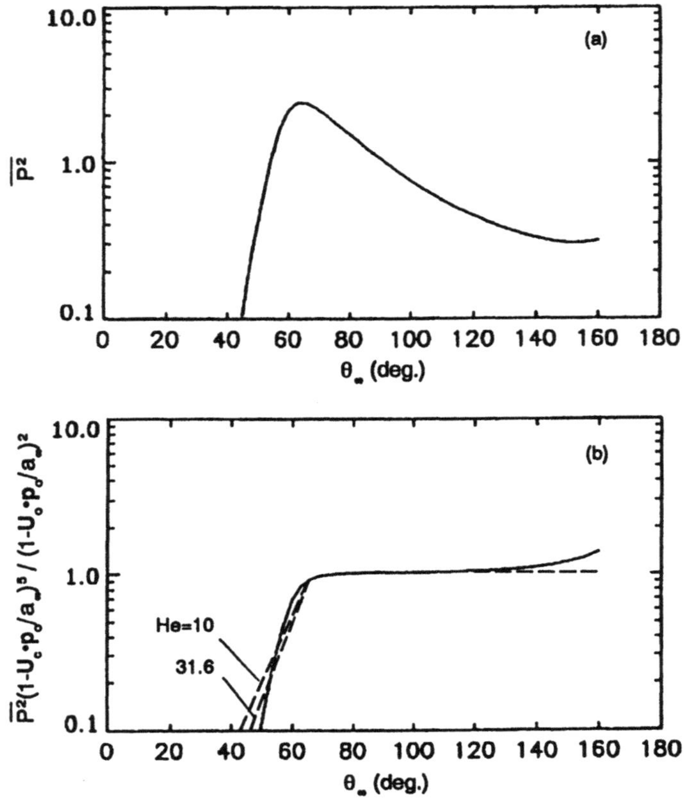

Figure 7. **Noise directivity for a source convecting along the center axis at X/D=7 for $\alpha=10°$ using model 2. In 7(b) (—) is the high frequency computation and (– –) is $\overline{P^2}(1-U_c \cos\theta_\infty/a_\infty)^3$ obtained from data (Stone 1981) for indicated Helmholtz numbers.**

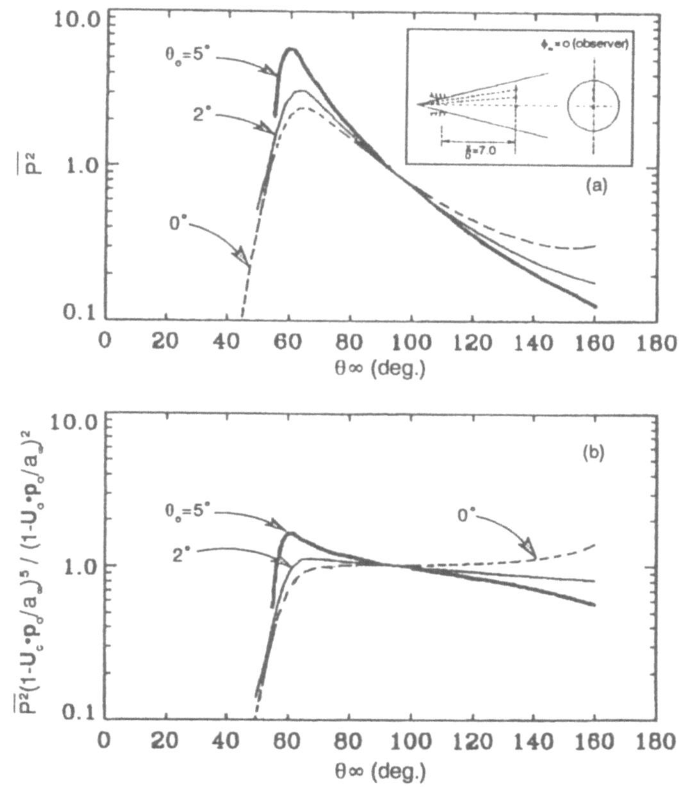

Figure 8. The effect of source location on directivity pattern; the sources are at X/D=7, ϕ_o=0° and θ_o=0°, 2° and 5°. The observer is at ϕ_∞=0°.

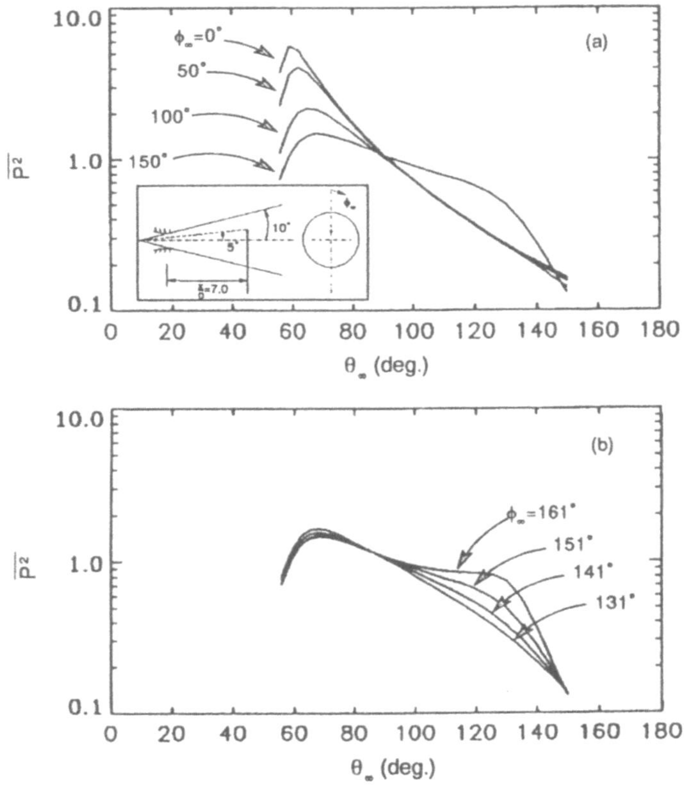

Figure 9. **Spherical directivity pattern for an off-axis source. The source is located at X/D=7, ϕ_o=0 and θ_o=5°. The formation of singularity near the caustic is demonstrated in 9(b).**

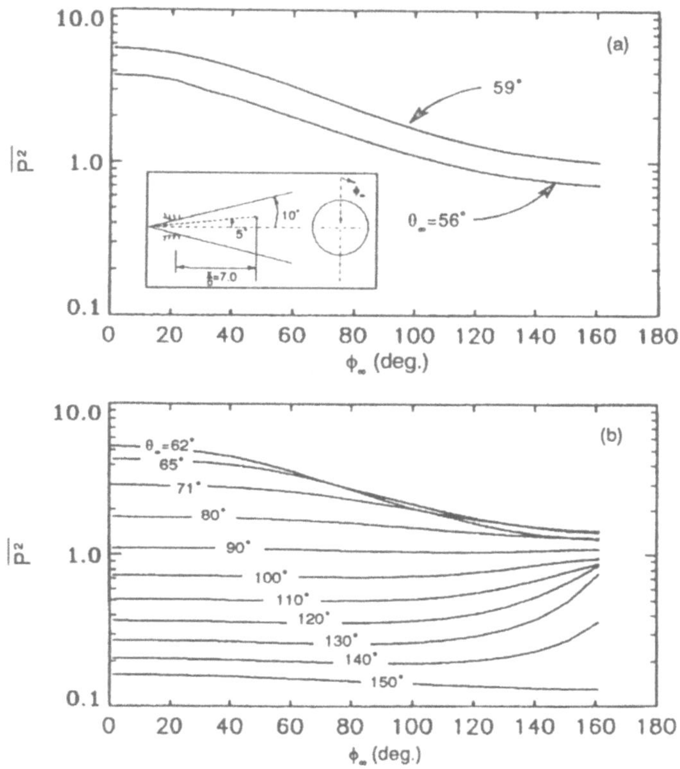

Figure 10. Spherical directivity vs the azimuthal observation angle (polar angle is indicated as a parameter). The source location is the same as Figure 9. Figure 10(a) shows the sharp drop in noise level within the quieting zone.

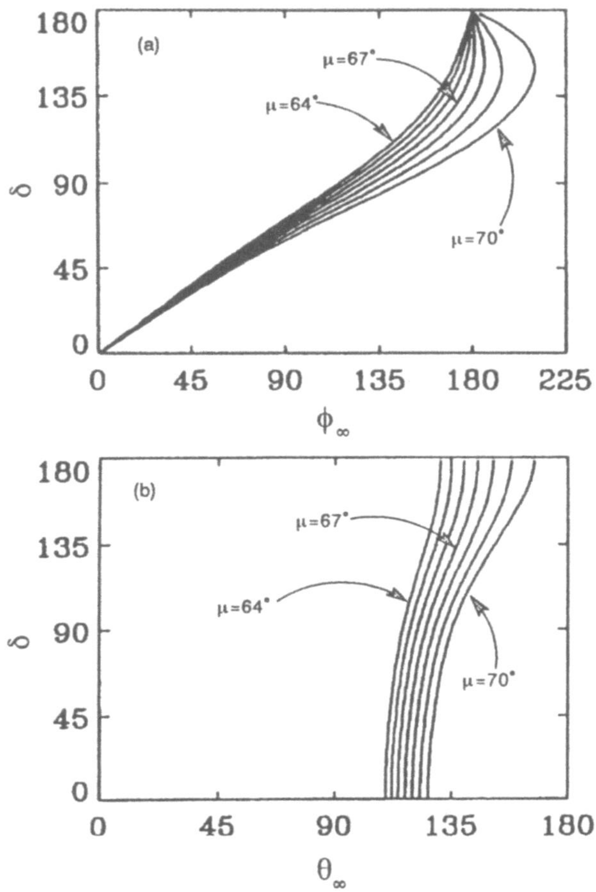

Figure 11. Ray-tracing showing the reduction in the area of ray tube in the neighborhood of μ=67°. This results in a caustic at $\theta_\infty \to 141°$ and $\phi_\infty \to 180°$.

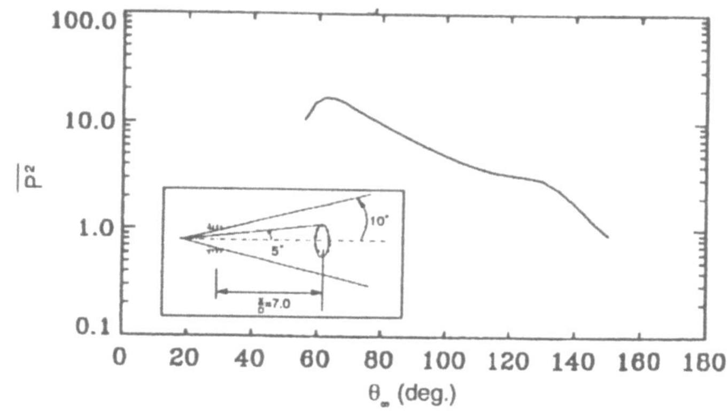

Figure 12. Directivity pattern for a ring source.

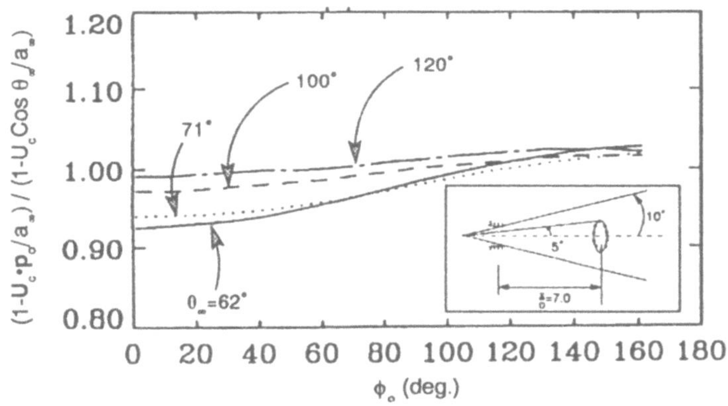

Figure 13. Variations in the Doppler effect due to change in source location. The observer is fixed at ϕ_∞=0; while the source moves on a ring.

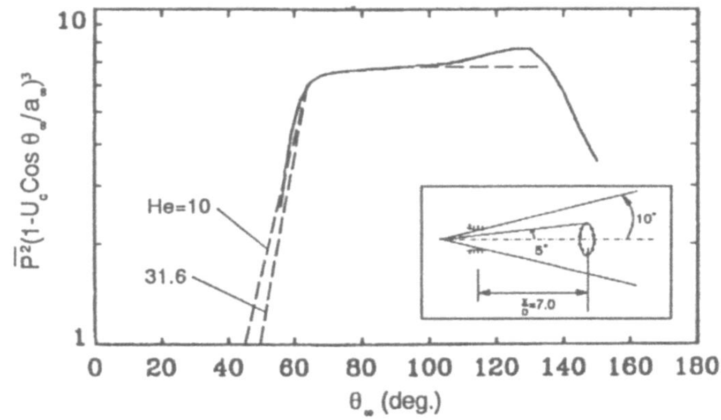

Figure 14. Comparison between data (Stone 1981) and high frequency approximation. The solid line (—) is the computation for a ring source. The data correlation is given as (– –) for indicated Helmholtz numbers.

APPLICATION OF A CFD CODE FOR UNSTEADY TRANSONIC AERODYNAMICS TO PROBLEMS IN AEROACOUSTICS

Gautam SenGupta

The Boeing Company
Seattle, Washington 98124-2207

ABSTRACT

Prediction of unsteady aerodynamic pressure distributions is an important step towards analyzing the flutter characteristics of an airplane. For this purpose, a Computational Fluid Dynamics code, originally developed for predicting steady transonic flow past bodies of complex geometry, was extended to handle unsteady aerodynamic problems, in which the unsteady solution could be viewed as a harmonic perturbation to the steady transonic flow. The predicted unsteady flow effects were compared with linear solutions for subsonic flow problems. The predicted results were also compared with available experimental data on unsteady pressure distributions on an oscillating wing in *transonic* flow. In all these cases, excellent agreements were obtained.

Scattering and diffraction play important roles in determining noise generated by turbulent sources near solid bodies. Scattering of engine noise by the fuselage, wings and the empennage influences community and interior noise levels in an aircraft. There are many areas in aeroacoustics where application of advanced Computational Fluid Dynamics (CFD) tools can provide valuable insight into the physics of the problem. In this paper, applications of the above unsteady CFD code to several specific areas are illustrated.

1. Introduction

Prediction of unsteady pressure distributions is an important step toward analyzing the aeroelastic response of an airplane. Past methods for predicting the unsteady aerodynamic characteristics were based either on linearized theory (Dusto and Epton, 1980) or on small disturbance transonic flow theory (Borland, 1986 and Bennet *et al.*, 1991). Recently, a CFD code called TRANAIR has been developed and it simulates all steady nonlinear flow effects (in the context of the full potential theory). It then develops the unsteady solution as a time harmonic linear perturbation to that steady flow.

Application of CFD codes developed for unsteady aerodynamic problems can also provide valuable insight into scattering and diffraction problems in aeroacoustics, since noise generated by turbulent sources near solid bodies is significantly influenced by scattering and diffraction phenomena. In this paper, applications to several specific areas are shown. One is the subject of trailing edge noise. Another is the subject of scattering of an incident plane wave by a flat plate. Mathematically, the problem is similar to that of sound generation by an incident vortical field interacting with an airfoil, in which the incident field and the generated acoustic field are coupled through the boundary condition of zero normal velocity on the surface of the airfoil. For this reason, the analogous problem of scattering of an incident plane wave by a flat plate was considered. Next we examined propagation of sound through ducts, where there is an abrupt change in duct cross sections. It is observed that the plane wave model typically used in these problems is adequate away from the area of cross sectional discontinuity. However, in the vicinity of the discontinuity there can be significant cross sectional variations in the sound pressure level. Another problem we considered is scattering of propeller noise by an aircraft. For this purpose, propeller noise was simulated by using a distribution of axial and circumferential dipoles. Scattering of these sources by an aircraft fuselage was then simulated numerically. Although further work remains to be done, preliminary predictions showed a significant degree of scattering, qualitatively similar to some of the wind tunnel test data. The final problem we considered is propagation of acoustic disturbances near the inlet of an engine in the presence of transonic flow.

These are just a few examples which illustrate that application of advanced CFD codes developed for unsteady aerodynamic problems can provide valuable insight into scattering and diffraction problems in aeroacoustics.

1.1. Theoretical approach

The TRANAIR method is based on the observation that the equations for potential fluid flow may be derived from a variational principle. This method is described in detail in references (Johnson *et al.*, 1989, Burkhart *et al.*, 1989 and SenGupta *et al.*, 1991a) and only the main points will be summarized in this section. The variational principle guides the discretization of the continuous problem in a manner which ensures that the discrete problem is an accurate approximation to the continuous one. The key to the ability to handle complex configurations is the use of a rectangular rather than a surface fitted grid. The variational problem is discretized on a rectangular grid which extends over all of space. Computations are restricted to a finite domain through the use of a source approach employed in integral equation methods. That is, the fundamental unknowns are sources for the constant coefficient differential operator governing the far field, rather than the velocity potential at grid point. In theory, the grid and the field solution extend to infinity. However, the sources are confined to the finite subset of the grid (computational domain) containing flow not governed by the far field form of the differential equation. Even within the computational domain, the sources are generally weak except near boundaries, shocks and wakes. Hence the problem size (measured by the number of significant unknowns in the equations) is effectively reduced with a corresponding reduction in solution cost.

Since the differential equation involves potential, it is important that the potential induced by specified sources be computed efficiently. The potential is given by a convolution integral of the sources with a discrete exterior Green's function. Because a rectangular grid is used the convolution can be performed very efficiently with Fast Fourier Transforms. Use of a discrete Green's function extending to infinity allows one to look at steady and unsteady flow variables beyond the boundaries of the computational domain, without being concerned about spurious reflections from the boundaries of that domain.

2. Results

2.1. Application to unsteady aerodynamic problems

The details of our applications to unsteady aerodynamic problems can be seen in SenGupta *et al.* (1991a), and only the key points are being summarized here. The unsteady C_p values predicted by a linear version of the code was compared with another (linear) code based on the panel method for a wing in pitching oscillation, and excellent agreement was obtained. For validation in the transonic flow regime, we compared the predicted levels against available experimental data for a pitching wing. The experimental data available in Tijdeman *et al.* (1979) was used for this comparison. The details of the comparison with experimental data are given in SenGupta *et al.* (1991a). In general, the predicted steady C_p values were seen to be in general agreement with the measured data at $M = .9$ and .95. Agreement among the experimental and numerical unsteady C_p values was also excellent at all span stations.

2.2. Application to problems in aeroacoustics

2.2.1. Comparison with analytical solutions and laboratory test data

For initial verification, we considered the simplest possible case, i.e., the unsteady pressure distribution on a sphere pulsating radially in a stationary medium. Comparing predicted results with analytical solutions at a given frequency allowed us to check the mass matrix that was being generated by the code and make sure that the boundary conditions on the sphere were being handled correctly.

The next problem considered was scattering of an acoustic plane wave by a sphere, since the analytical solution for this problem is well known (Morse, 1948). Fig. 1 shows the predicted Acoustic Scattering Factor (ASF) on the surface of the sphere for $ka = \omega a/c = 3$, where a is the radius of the sphere and c is the speed of sound in a stationary medium. The ASF is defined as the difference (in dB) between the total (i.e. incident plus scattered) and the incident fields. The maximum and minimum analytical values are also shown in parenthesis. As can be seen, the numerical solutions are very close to the analytical values.

In addition, we also verified the numerical predictions against laboratory test data on scattering of acoustic signals from a compact source by a cylinder. The details of that comparison are given in SenGupta (1987) and will not be repeated here.

2.2.2. Application to trailing edge noise

Trailing edge noise is a significant component of airframe and jet-airframe interaction noise, and is caused by interaction of turbulent fluid pressure fluctuations with the trailing edge of an airfoil. Fig. 2

shows far field directivity patterns generated by a dipole source located near the trailing edge of a NACA 0012 airfoil at three different frequencies corresponding to three different values of λ/C_l. It is seen that for $\lambda/C_l = 10$, the far field directivity pattern resembles that of a classical dipole source associated with an acoustically compact airfoil. On the other hand, for $\lambda/C_l = .025$, the directivity pattern approaches that of a dipole source in the presence of a semi-infinite surface, except for the cancellation at $\phi = 0$, due to the finite chord length. In addition, there are some minor lobes due to interference between diffracted waves from the leading and trailing edges. At an intermediate frequency with $\lambda/C_l = 1$, the directivity pattern looks different from any of the above patterns. Similar trends have been predicted for flat plates of finite length (Tam and Yu, 1975, and Amiet, 1976) and observed in experimental data (Miller, 1983).

2.2.3. Scattering of a plane wave by a flat plate

In this example, we consider scattering of an acoustic plane wave by a flat plate of finite dimensions. Mathematically, this problem is similar to the problem of prediction of dipole sound due to a vortical field incident on a thin airfoil. According to Goldstein (1976), the incident field and the generated acoustic field are coupled through the boundary condition of zero normal velocity on the surface of the airfoil. For this reason, the analogous problem of scattering of an incident plane wave by a flat plate was considered.

Fig. 3 shows the results predicted by TRANAIR. The ratio of the wavelength to the width of the plate (λ/w) was 0.98. The angle of incidence measured from the normal to the plate was 50 degrees. For an infinite flat plate, the magnitude of the Acoustic Scattering Factor is 6 dB. An interesting point to note is that on the incident side of the finite plate, the maximum value of ASF was more than 6 dB. Examination of the variation of the phase of the total field showed that the presence of the flat plate also causes considerable changes to the sound field near the plate (SenGupta, 1990).

2.2.4. Propagation of acoustic waves through ducts of varying cross sections

In many acoustic problems dealing with propagation of acoustic signals through ducts, it is assumed that the acoustic waves can be modeled as a plane wave, with no variations in a direction perpendicular to the direction of propagation. The plane wave model needs to be examined when there is an abrupt change in cross section, or when there is a bend in the duct.

For this reason, a problem was simulated where the duct diameter was changed from 7 in. to 21 in. and then back to 7 in. There was a piston vibrating at the left end of the duct and the other end was closed. As can be seen from Fig. 4, there is a considerable variation in the pressure magnitude near the junction between the two ducts, and the sound level near the junction differs significantly from that predicted by the plane wave model. In the vicinity of the expansion, the acoustic wave attempts to spread spherically, but because of the constraints of the system, it becomes essentially a plane wave a short distance downstream. The second discontinuity near the right end apparently does not cause any significant change to the acoustic wave. The incident and the reflected waves interfere with each other, causing a dip in the level near the first discontinuity.

2.2.5. Scattering of propeller noise by an aircraft fuselage

The details of our work in this area can be seen in SenGupta, *et al.* (1990) and only the key points are summarized here. The propeller model chosen for this study was the ten bladed NASA SR-6 propeller. For the present study, predictions were made only for the first harmonic. The operating conditions for the 1/5th scale model propeller corresponded to a free stream Mach number of 0.26, and RPM of 7650.

The propeller noise sources have been simulated by a distribution of acoustic dipoles oriented in the axial and in the circumferential directions in accordance with the Gutin theory (Gutin, 1948). The original Gutin theory was for propellers with straight blades, whereas the SR-6 blades are swept. The effect of the blade sweep was included in accordance to the procedure described by Hanson (1979).

The aircraft fuselage geometry was simulated by using a panel description. The predicted scattering effect is also shown in Fig. 5 which shows the angular variation of the acoustic scattering factor (ASF). As can be seen, the scattering effects were significant in the aft region. Qualitatively, these results are similar to what was observed in wind tunnel test data presented in Woodward and Hughes (1989), and further work remains to be done in this area.

2.2.6. Acoustic propagation near an engine inlet

In this example, we consider propagation of acoustic signals near an engine inlet. For a freestream Mach number $M_\infty = .26$ typical of a take-off situation, there can be considerable density and Mach number variations near an engine inlet, as shown in Fig. 6, based on TRANAIR simulation of the steady flow field near the inlet. (The density values are normalized with respect to ρ_∞, the density of the freestream.) Understanding of propagation of acoustic signals associated with the fan through this inhomogeneous medium is an important issue in the design of acoustic lining. To illustrate this, we simulated unsteady velocity fluctuations at the plane of the fan, and looked at the acoustic propagation near the engine inlet. Fig. 7 shows the variation of the magnitude and phase of the acoustic field at a frequency of $kD = 6.4$, where D is the diameter of the fan. It should be noted that this frequency was chosen arbitrarily for illustrative purposes only and does not represent the upper frequency limit of the code. One can think of many uses of the code for conducting parametric studies for problems of this type.

3. Conclusions and Recommendations

A CFD code called TRANAIR has been developed and it simulates all steady nonlinear flow effects (in the context of full potential theory). It then takes into account time harmonic linear perturbations to that steady flow. Predicted results agreed well with analytical solutions, numerical solutions and experimental data. The TRANAIR method is particularly well-suited for handling steady transonic flow past complex bodies of arbitrary shapes.

This paper describes a few applications of the TRANAIR code to problems in aeroacoustics related to scattering, diffraction and propagation effects. Several examples have been shown. In all these cases, application of the numerical method provided valuable insight into the physics of the problems. In the past, computational capabilities were limited, and there was no alternative to relying on experimental data and analytical solutions for simple geometries like a cylinder. With today's supercomputers, and with the advances being made in the field of numerical simulation in all branches of science, the field of aeroacoustics need not lag behind, at least for the types of problems considered in this paper.

It may also be noted that another spin-off of this work has been in the area of simulation of acoustic scattering from submarines (Sen-Gupta *et al.*, 1991b).

Acknowledgement

The cooperation received from the TRANAIR code development team during the course of this work is gratefully acknowledged.

References

Amiet R. K., 1976. "Noise due to flow past a trailing edge," *J. Sound & Vib.* **47**(3), pp. 387-393.

Bennet, R. M., Bland, S. R., Batina, J., Gibbons, M. D., and Mabey, D. G., 1991. "Calculation of steady and unsteady pressures on wings at supersonic speeds with a transonic small-disturbance code," *J. Aircraft* **28**(3), pp. 175-180.

Borland, C. J., 1986. "XTRAN3 - Transonic steady and unsteady aerodynamics for aeroelastic applications - Vol. 1," AFWAL-TR-85-3124.

Burkhart R. H., SenGupta, G. *et al.*, 1989. "Numerical prediction of scattering of acoustic waves by an aircraft," Paper presented at the Second IMACS Symposium on Computational Acoustics, Princeton University, NJ.

Dusto, A. R. and Epton, M. A., 1980. "An advanced panel method for analysis of arbitrary configurations in unsteady subsonic flow," NASA CR-152323.

Goldstein, M. E., 1976. *Aeroacoustics*, McGraw-Hill, Inc.

Gutin, L., 1948. "On the sound field of a rotating propeller," NACA Tech. Memo. 1195.

Hanson, D. B., 1979. "Aeroacoustics of advanced turbopropellers," in *Mechanics of sound generation in flows*, Springer Verlag, New York, NY, pp. 282-293.

Johnson, F. T. *et al.*, 1989. "TRANAIR Computer Code (Theory Document and Users' Manual)," Final Report, NASA Contract NAS2-12513.

Miller W. R., 1983. "Flight effects for jet-airframe interaction noise," AIAA Paper 83-784, 8th Aeroacoustics Conference, Atlanta, GA.

Morse, P. M., 1948. *Vibration and Sound*, McGraw-Hill Book Co.

SenGupta, G., 1987. "Computation of aeroacoustic scattering effects," AIAA Paper 87-2669, 11th AIAA Aeroacoustics Conference in Sunnyvale, CA.

SenGupta, G. *et al.*, 1990. "Application of computational methods in aeroacoustics," AIAA Paper 90-3917, 13th AIAA Aeroacoustics Conference, Tallahassee, FL.

SenGupta, G. *et al.*, 1991a. "Analysis of unsteady aerodynamic and flutter characteristics of an aeroelastic model in transonic flow," Paper presented at the AGARD Specialists' Meeting on *Transonic Unsteady Aerodynamics and Aeroelasticity*, in San Diego, CA, October 9-11.

SenGupta, G. *et al.*, 1991b. "Numerical simulation of acoustic scattering from submarines," Paper presented at the Third IMACS Symposium on Computational Acoustics, Harvard University, Cambridge, MA, June 26-28.

Tam, C. K. W. and Yu, J. C., 1975. "Trailing edge noise," AIAA Paper 75-489, 2nd Aeroacoustics Conference, Hampton, VA.

Tijdeman J. *et al.*, 1979. "Transonic wind tunnel tests on an oscillating wing with external stores; Part II - The clean wing," AFFDL-TR-78-194, Part II.

Woodward, R. P. and Hughes, C. E., 1989. "Noise of a counter-rotating propeller with simulated fuselage and support pylon at take-off/approach conditions," AIAA Paper 89-1143, Presented at the AIAA 12th Aeroacoustics Conference, San Antonio, Texas.

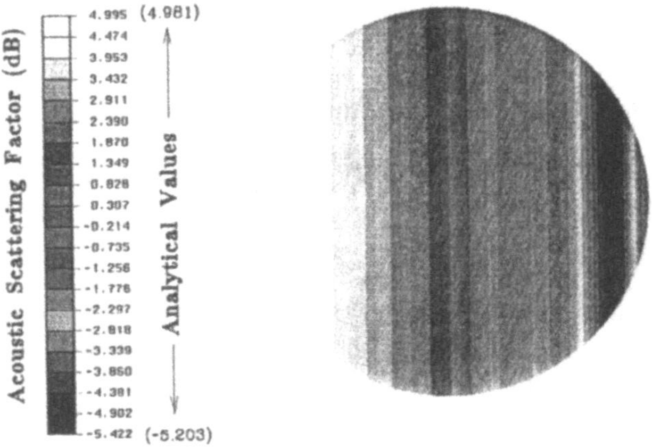

Figure 1. Scattering of an acoustic plane wave by a sphere (ka = 3.0)

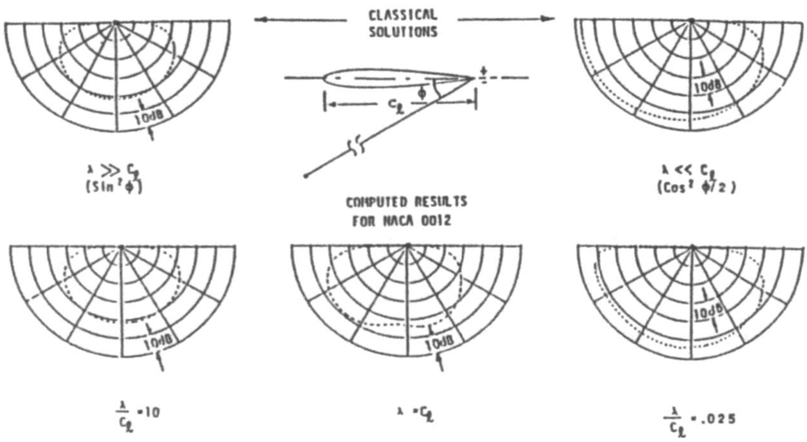

Figure 2. Application to trailing edge noise

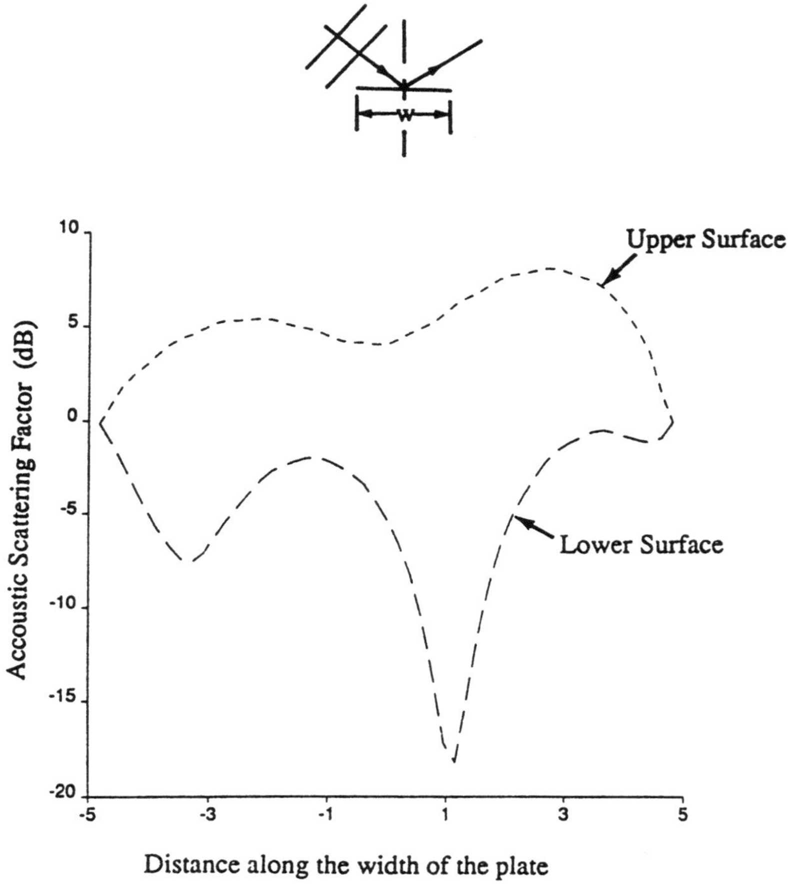

Figure 3. Scattering of a plane wave by a flat plate

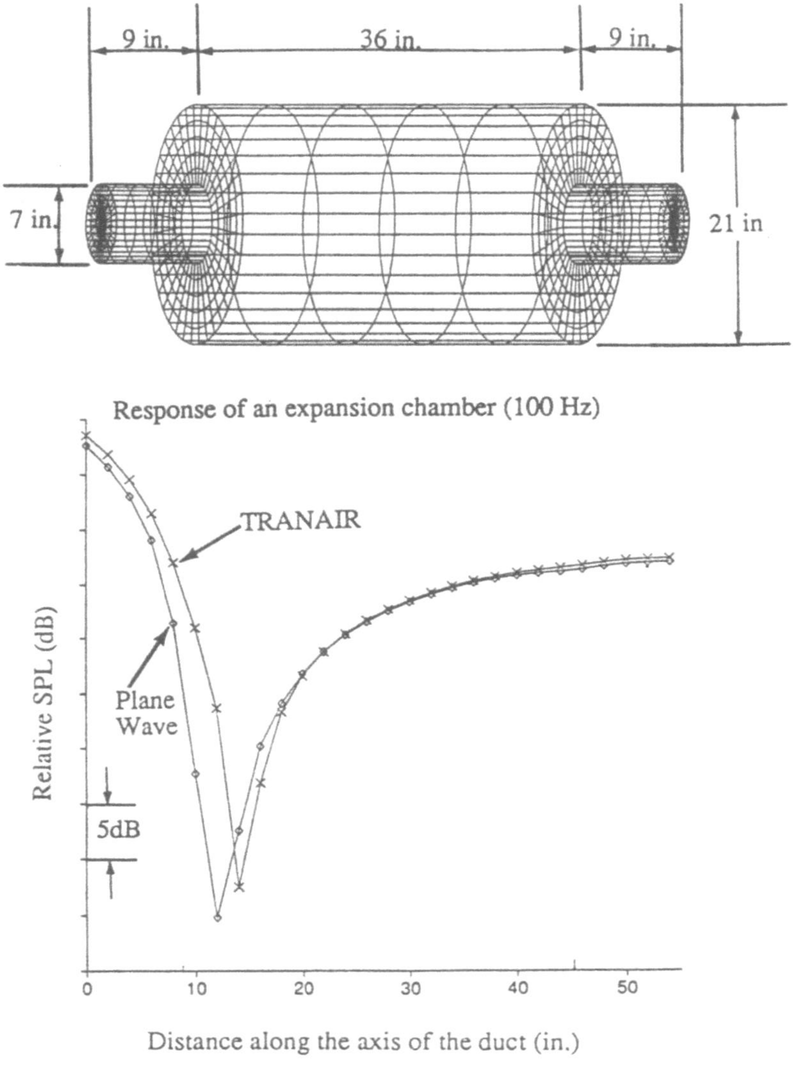

Figure 4. Acoustic propagation in a duct

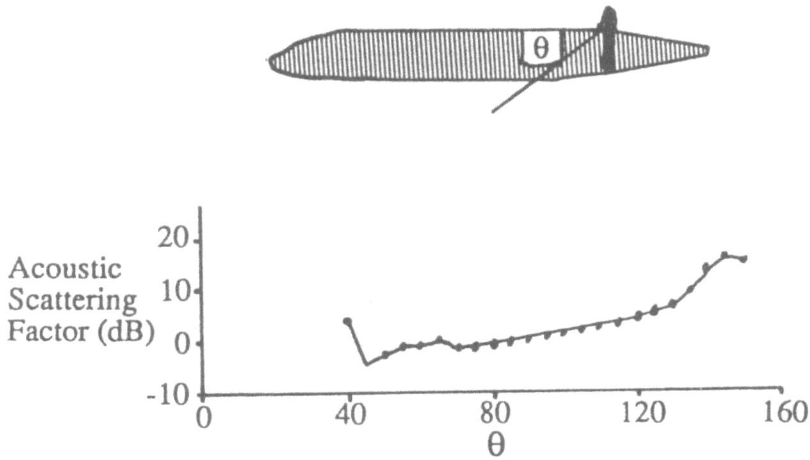

Figure 5. Prediction of scattering of propeller noise by an aircraft fuselage (255 Hz)

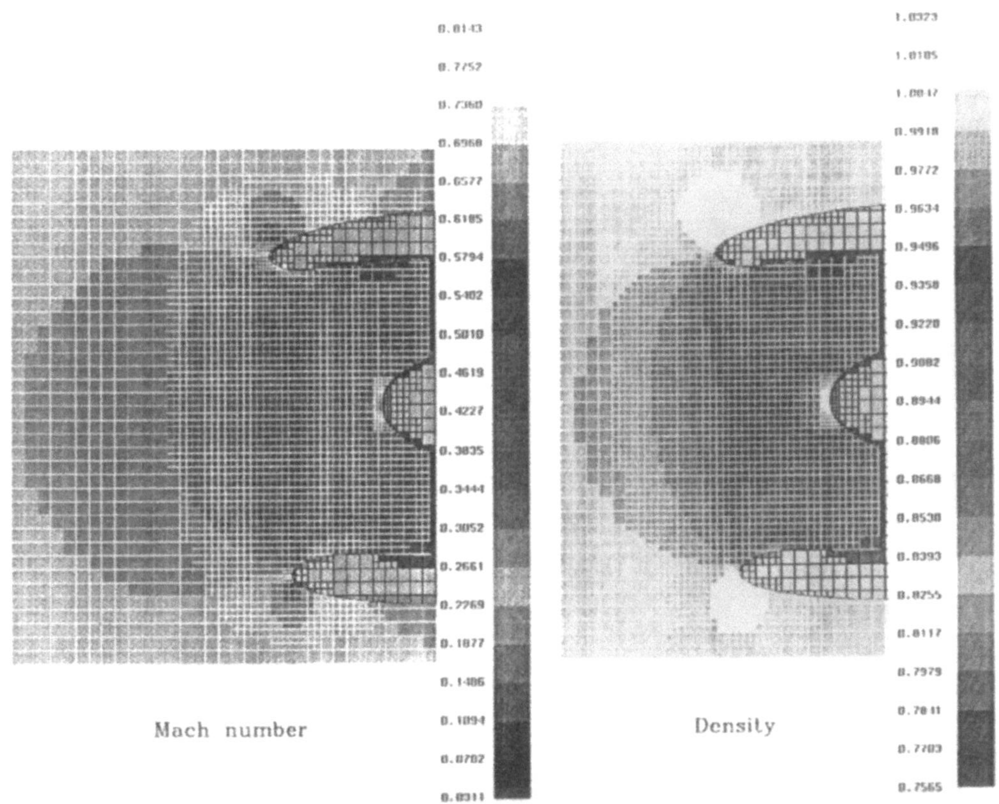

Figure 6. Variation of Mach number and density near an engine inlet ($M_\infty = .26$)

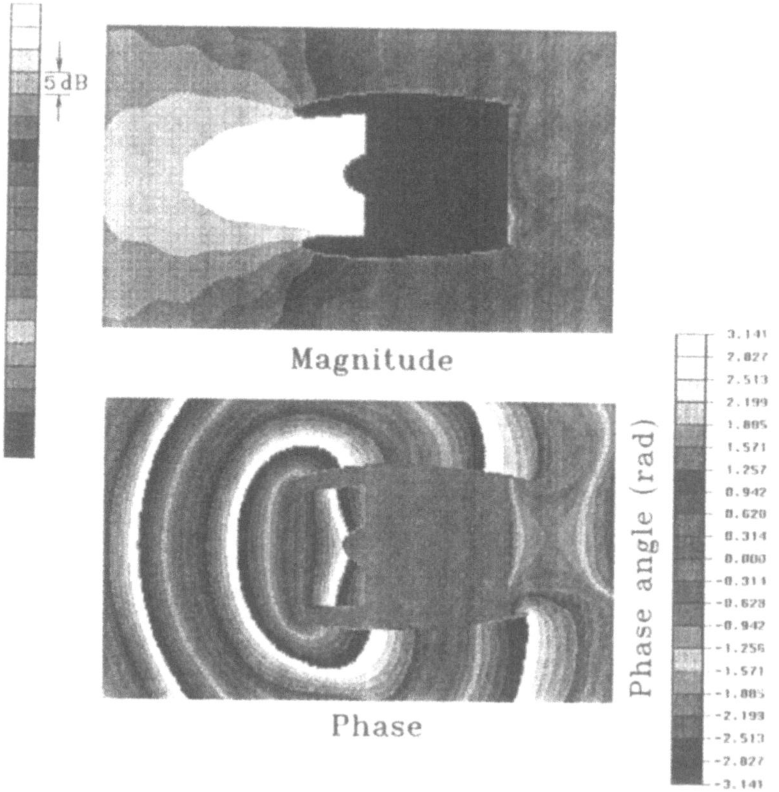

Figure 7. Variation of acoustic field near an engine inlet in the presence of flow ($M_\infty = .26$)

REPORT ON FINAL PANEL DISCUSSION

THE FINAL PANEL DISCUSSION

James Lighthill

Department of Mathematics
University College London
Gower Street, London WCIE 6BT
UNITED KINGDOM

Abstract

This paper by the Panel Chairman summarises some important conclusions about future prospects for aeroacoustics in general, and for computational aeroacoustics in particular, that were reached in the course of the Final Panel Discussion of the Workshop on Computational Aeroacoustics held from 6 to 9 April 1992 by ICASE and NASA Langley Research Center.

1. Aeroacoustics at the Start of Its Second Golden Age

In 1992 aviation faces vast new opportunities. Simultaneously, it recognizes environmental responsibilities that are even more exacting than in the past. A new, determined effort to expand aeroacoustics still further is now essential to reconcile shining aviation opportunities with enhanced environmental responsibilities.

For example, the U.S. High Speed Transport Project (HST) aims at achieving big improvements in the economics of supersonic civil transport, which will bring to a wide public the benefits of greatly reduced travel times. Yet its airframe design principles brilliantly minimize levels of supersonic-boom annoyance. Now, the approach to its engine design must be guided by massive aeroacoustic innovation in order to allow HST engine-noise levels to meet the Federal Aviation Agency's "FAR Stage III" goals.

Equally exacting challenges face the developers of new subsonic aircraft (which will, of course, continue to fulfill most of civil aviation's needs) as they plan methods for reducing noise levels of such aircraft to meet "Stage III" requirements, even though thrusts are likely to reach values twice as great as those attained by engines currently in service. If acousticians can rise to this exacting challenge, communities all over the world will benefit from huge shrinkages in aerodynamic-noise "footprints."

The new opportunities and challenges recall to many of us the excitements of the first golden age of aeroacoustics. A brief summary of its achievements, which now follows, may indeed be found instructive.

2. Achievements of the First Golden Age of Aeroacoustics

Fundamental changes in the human condition have resulted from the wide-spread availability of flight at the speeds of jet aircraft. Such aircraft can travel to the farthest place on the earth's surface in less than the time taken for the earth to rotate on its axis. International understanding and cooperation have as result increased immeasurably; for example, current levels of cooperation in Europe would have been impossible if traditional obstacles to such cooperation had not been overcome by airborne vehicles making the Rome-London journey (say) in only two hours.

Yet no such developments would have been possible without the achievements of the first golden age of aeroacoustics. This began when, in the late 1940s, far-seeing sponsors of aeronautical research recognized that the terrifyingly high noise levels of the current small military jet aircraft would need to be greatly reduced if there were to be any chance that the use of much bigger jet aircraft for civil transport might be publicly tolerated. This led in England to an organized research effort by several university groups (at Manchester, Cranfield and Southampton) in cooperation with Rolls-Royce and in correspondence with parallel U.S. activity based on a group at (the then) Langley Field.

The initial work, involving very close experimental/theoretical collaboration, was directed at understanding the basic science of what we now call "jet noise proper": the sound radiated from jets in the absence of any interaction of jet turbulence with solid boundaries. A fundamental conclusion of this work was that, for reasons associated with the quadrupole character of such radiation, the acoustic power output varies as $U^8\ell^2$ for subsonic jets of exit velocity U and diameter ℓ. Yet propulsive power varies as $U^3\ell^2$; accordingly, it has become possible for jet engines needed in civil aircraft to combine large gains in propulsive power with greatly reduced noise radiation by means of a progressive move towards wide jets of high bypass ratio and relatively low exit velocity[1].

Later, when jet noise proper had by these means been enormously diminished, a similar effort needed to be put into reduction of aircraft noise from other sources which had by then become relatively more important. These included rotor noise (fan noise from the front of the engine and turbine noise from the rear) and airframe noise from the interaction of boundary-layer turbulence with flexible surfaces, and with control surfaces and flow over trailing edges. Such noise sources are characterized by aeroacoustic theory as combinations of monopoles, dipoles and quadrupoles of well defined strengths, and this analysis proved important for their reduction.

In the meantime, that change in the character of jet noise which is observed as exit speeds rise substantially above the atmospheric sound speed had become understood as a consequence of high-speed convection of aeroacoustic sources producing an effective loss of source "compactness." Essentially, the radiation changes progressively to one of monopole character; and, thereafter, the acoustic power output varies as $U^3\ell^2$. It amounts to almost 1% of propulsive power and, at supersonic convection speeds, takes the form of "shocklet" emissions (analogous to supersonic booms) in the Mach direction. Such fundamental understanding was used in the first steps towards reduction of noise from supersonic transport aircraft.

3. The Challenge of Exploiting CFD Advances to Meet Today's More Exacting Goals

Those enormously more exacting aeroacoustic goals which face the aviation community in 1992 (see §1) call urgently for yet another massively concerted effort. This time, it is the possibility of fully utilizing great recent advances in CFD capability which offers us a realistic hope that such an effort may achieve the level of improvement that is demanded both for supersonic and for subsonic aircraft.

Indeed, such approaches are needed even for analyzing how diffraction around a complicated aircraft shape modifies the sound field

[1]Initially – that is, before such radically new aero-engine designs could be developed – a similar effect was in part achieved by fitting jet orifices with so-called "silencers" whose essential noise-reducing effect was to promote a massive increase in the rate of entrainment of air into the jet (so as artificially to induce an enhanced bypass ratio).

from aeroacoustic sources; as well as for the refined investigation of noise from those sources themselves whether they be rotors, jets, or airframe boundary layers. We also advocate continued work on "model problems" where methods can be tested and validated through rigorous comparisons with experiment.

For all of the required activity, indeed, it will be essential that, just as the theoreticians of the first golden age of aeroacoustics worked in the closest possible collaboration with experimental scientists, so also the theoreticians of the coming second golden age – notwithstanding their immensely powerful support from modern CFD techniques – should subject their methods to the essential test of experimental validation. All of our recommended programs in Computational Aeroacoustics will need to be pursued, then, in cooperation with meticulous noise measurement programs.

We see two main classes of method as available to the practitioners of Computational Aeroacoustics. The first of these utilizes the characterization of aeroacoustic sources that was developed early in the history of the subject and which is usually described as the Acoustic Analogy; in this method, CFD would be used within the flow itself to evaluate the source strengths, after which simple integrations over the flow field and its boundaries would suffice to determine the acoustic far field. The second class of available methods seeks to apply CFD techniques *comprehensively*; that is, over a much wider region which extends beyond the flow field as such to include at least "the beginnings" of the acoustic far field. We strongly recommend that both classes of method be exhaustively developed, and we see good reasons for expecting that each will be found especially appropriate for certain groups of aeroacoustic problems and relatively less so for other groups.

The original form of the Acoustic Analogy was developed for the purpose of studying the sound radiated from subsonic jets where, as already noted, the aeroacoustic sources are of quadrupole character with enormous disparities in energy level between near-field pressure fluctuations and the very much smaller far-field sound radiation; most of which, moreover, involves wavelengths large compared with typical length-scales in the flow. All of these facts seem to make subsonic jet noise the aeroacoustic problem which is least suitable for treatment by a comprehensive application of CFD throughout the field. Not only would scale separations and energy-level disparities create obvious difficulties in its use but any numerical errors that might effectively introduce spurious sources of monopole or dipole character could seriously distort the inherently less powerful quadrupole radiation.

For subsonic jet noise studies, then, we specifically recommend continued use of the Acoustic Analogy in one or more of its many available forms. Each of these employs a particular expression for the source strength, preferably chosen so that its values are insignificantly small except in the flow. The classical form T_{ij} of the quadrupole strength satisfies this criterion and, in subsonic jets, has the advantage that its statistics involve length scales comparable with those of the main energy-containing turbulent motions. Alternative forms involving the vorticity are valuable for many aeroacoustical purposes; while, on the other hand, raising some difficulties in jet turbulence because their statistics reflect smaller length scales associated with energy-dissipating motions. In §10 we propose the application of modern CFD methods (including Large Eddy Simulation) to determining the required statistical behavior of T_{ij} in jets. Also, we review the various appropriate types of Green's function (some based on "the wave equation" and some on certain available alternative forms of linear partial differential equation) by means of which the far-field radiation from those sources may be calculated.

Here, we offer one additional remark that may indeed prove relevant in a wider range of aeroacoustic problems. It is simply that, whenever an acoustic far field has been determined on linear theory, then classical methods are available for "immediately writing down" expressions that correctly describe those gradual modifications to the waveform which result from "nonlinear-acoustics" effects. We strongly recommend that such modifications be routinely calculated; especially, because they involve energy shifts to higher frequency such as may influence "perceived" noise levels (EPNdB).

Out of the above-mentioned group of three serious difficulties for application of CFD techniques "right out to the far field" in estimating subsonic jet noise, all are largely absent in the other principal problems of aeroacoustics; since, for example, the sources assume (§2) a monopole form in supersonic jets while rotor noise involves a mix of monopole, dipole and quadrupole sources. In all of the problems, therefore, we recommend vigorous action to develop effective techniques in *comprehensively* Computational Aeroacoustics, using numerical-analysis philosophies which we now proceed (§4) to outline. Most of the problems, on the other hand, are additionally suited[2] to the use of advanced forms of the Acoustic Analogy, and we also strongly recommend in these cases that CFD be actively applied to the improvement of knowledge of aeroacoustic source strengths. The challenging nature of the new demands on aeroacousticians forces us indeed to conclude that they need to be equipped with more than one powerful approach towards the estimation of aerodynamic noise.

4. Numerical-Analysis Philosophies for Comprehensively Computational Aeroacoustics

Before sketching specific numerical-analysis principles which are needed in this field of application, we may appropriately emphasize one principle which, in fact, ICASE has consistently espoused in all its work on Computer Applications in Science and Engineering. This is the principle that success in such Applications demands deep, well coordinated thought by human brains working with one or more computers in a very close and effective "symbiosis."

There can, in short, be nothing "mechanical" in the application of computational techniques to difficult problems like those in aeroacoustics. On the contrary, it is essential that powerful intellectual processes (commonly, processes that utilize a vast amount of available analytical information about the anticipated behavior of solutions to problems) be applied in parallel with the numerical analysis, with the computer programming and with the study of computer output.

We shall not repeat this sufficiently obvious and well accepted maxim after thus giving it prominence at the opening of this section. We emphasize however that such a maxim about utilizing to the full intellectual processes based on analytical theories, which may include Acoustic Analogy studies, implies a close coordination between the two "prongs" of that bimodal attack on aeroacoustical problems which we strongly advocate.

It is of course on the solid foundation created by remarkable successes in aeronautical applications of CFD that the relatively new art of comprehensively Computational Aeroacoustics must be firmly based. In particular, some of the necessary techniques can be taken

---

[2] An exception, perhaps, is the problem of diffraction of aeroacoustic sound fields around complicated aircraft shapes.

over directly and we stress that this includes the proper handling of boundary conditions at solid surfaces.

The new subject needs, however, to apply a reliable boundary condition at an outer boundary situated (not *too* distantly) within the acoustic far field. Great care is needed to ensure that this is truly a "non-reflecting" boundary condition, and we note that methods derived from the analytic theory of hyperbolic equations (including the mathematical properties of characteristics) can be very effective in achieving this aim.

A feature which even more strongly differentiates Computational Aeroacoustics from classical areas of CFD is, however, the need for a faithful representation of (linear and nonlinear) wave-propagation processes. This becomes increasingly more difficult at shorter and shorter wavelengths, and a realistic approach to the computational problems must, on any given grid, place a lower limit on the wavelengths which the program seeks to resolve. Unless this is done, severe problems including those frequently described as "numerical dispersion" are unavoidable.

A consensus from the Workshop is that such difficulties are best avoided if carefully chosen "high-order" numerical schemes are applied. Such schemes can avoid numerical dispersion at wavelengths over four times the grid spacing, and they need to be combined with program features that damp out any waves shorter than this.

There is one essentially local exception to the above rule. Long experience with effective CFD codes has shown that well designed codes can reliably locate and characterize shock waves in a flow, but that this is possible only when numerical schemes of very low order are used. This poses the problem of how Computational Aeroacoustics can best handle sound radiation from flows incorporating shock waves.

The answer, as already indicated, is to make a "local exception" in the general neighborhood of any shock waves. Essentially, this means that "high-order" numerical schemes are applied in almost all parts of the flow field, but are caused to give way to schemes of very low order in the neighborhood of shock waves. Experience has shown how carefully compiled codes can successfully achieve this dual objective.

5. Diffraction of Aeroacoustic Radiation around Aircraft Shapes

Because well-established CFD codes used by aircraft companies have, of necessity, acquired impressive capabilities for accurately applying Euler-flow boundary conditions all around a complicated aircraft shape, they form an excellent foundation for codes aimed at solving those diffraction problems that are important in aeroacoustics. These include, for example, the distortion of fan-noise radiation patterns produced by diffraction in the presence of the aircraft shape of the flow around it.

We acclaim the concept of applying (where possible) all the effort that has gone into compiling complicated but effective CFD codes to an important aeroacoustic objective. We confirm, furthermore, that the diffraction problem is just such an objective which, as already noted (§3), cannot in practice be tackled by other methods based on relatively simple Green's functions.

6. Rotor Noise

In their applications to the rotor-noise problem as such, both of the main approaches (§3) to Computational Aeroacoustics are already flourishing, and firms in both the U.S. and Europe have expressed strong appreciation of what has so far been achieved with these approaches. The field moreover is one where we can predict further exciting and important developments in both methods.

In the Acoustic Analogy approach, precise forms for the surface distribution of monopole and dipole sources associated with moving rotor blades of given shape with specified thickness and loading distributions are well established. The associated "spinning acoustic field" is readily derived therefrom by surface-integral computations. For high-speed rotors this needs to be supplemented by the field of quadrupole sources associated with important flow features which may include (i) coherent features conveniently estimated in e.g. cascade-type calculations, (ii) shock waves attached to the blades, (iii) incoherent features such as wake turbulence and (iv) effects of blade/vortex interaction. Not all of these features have yet been satisfactorily incorporated in the theories and we recommend an intensification of research aimed at achieving this, research which, needless to say, should use CFD wherever appropriate.

We draw attention also to theoretical approaches utilizing high-blade-number asymptotics. These are important, not because exact numerical evaluation of the necessary integrals poses any severe computational difficulty, but because the asymptotic analysis demonstrates how just a very limited part of the complete domain of integrations generates almost all the radiated sound. This, then, is the region where special effort to estimate flow quantities accurately needs to be made.

Acoustic emission from rotors has also been investigated very successfully by comprehensively Computational Aeroacoustics in certain cases, including the case of helicopter blades without loading. This is a problem where the numerical analysis needs to allow for shock wave formation near the blade tips. Also, codes which allow for large variations in grid spacing are much to be recommended. The method has achieved good agreement with flight tests. It will now be extended to cases with loading; where (once again) it will be important, if possible, to model blade/vortex interaction satisfactorily.

The enormous importance of rotor-noise analysis for future helicopter designs as well as for the development of future advanced turbofan and propfan engines makes it in our view essential to continue to pursue the subject vigorously by means of both the main approaches to Computational Aeroacoustics.

7. Boundary-Layer and Airframe Noise

At the same time we feel strongly that noise originating in airframe boundary layers must not be neglected. Inherently, this is an aeroelastic problem, which involves interactions between boundary-layer turbulence and the flexible solid surface. These interactions are highly relevant to programs of cabin-noise minimization. In some cases, furthermore, mutual excitation between flow fluctuations and aircraft panel vibrations may significantly contribute towards community-noise radiation.

The adequate modelling of boundary-layer turbulence is not just a matter of Large Eddy Simulation. On the contrary, good models must take proper account of the repeated re-energization of the turbulence through "bursts" of intense vorticity emitted from the wall. Fortunately, some appropriate modelling for this process seems at last to be starting to emerge. Clearly, it will be essential to utilize such models when boundary-layer noise is tackled by Computational Aeroacoustics.

The acoustic interaction of boundary-layer turbulence with a flexible surface is not simple. Radiated noise is known to be almost entirely cancelled in the case of a flat surface of uniform compliance. Accordingly, it may be essential to take into account those nonuniformities of compliance (including concentrations of rigidity) which, while commonly present at the surface of an aircraft structure, can be specifically implicated as sources of boundary-layer noise.

Additionally, as engines are further quieted in the future, airframe noise will be of greater importance. The noise radiated from flow over cavities and struts and the interaction of turbulent boundary layers with trailing edges needs to be moved from the current empirical basis to a more rigorous foundation.

The possibility of "energy-level disparities" resulting from cancellations in the acoustic far field may be thought to suggest that CFD is required primarily to model the turbulence itself, with Acoustic Analogy techniques employed to infer the radiated noise. We remain convinced, on the other hand, that direct Computational Aeroacoustics also needs to be attempted; particularly, for supersonic boundary layers.

8. Model Problems

Precise validation of methodology, especially through rigorous comparisons with experiment, will form an essential foundation for all programs in the second golden age of aeroacoustics. Some essential contributions towards this objective can be made through the meticulous study of so-called "model problems."

Already, some extremely successful validations of Acoustic Analogy methods have been achieved by the study of radiation emitted when concentrations of vorticity such as vortex rings interact with each other or with solid boundaries. In such an aeroacoustic problem, one of those alternative forms of source strength which involve vorticity (but which may be shown to radiate the same sound field as does the classical quadrupole distribution T_{ij}) can give results in a valuably simple form. The method has been applied using Green's functions not only for free space but also for various internally bounded regions. In addition, a modified version of the Acoustic Analogy involving "matching" between a near field and a far field has been used successfully.

Each calculation, moreover, has been compared with data obtained in extremely careful experiments, and the confidence of aeroacousticians has been greatly increased by the ensuing demonstrations of gratifyingly close agreement. We strongly recommend continued work along these lines. In addition, recalling the successes of CFD over many years in evolving good ways of achieving fruitful interactions between computational and experimental activity, we believe that there will be similar benefits to comprehensively Computational Aeroacoustics, with "model problems" providing a substantial proportion of the needed comparisons.

9. Noise from Supersonic Jets

We have emphasized (§3) how effects of the convection of aeroacoustic sources can be described by an analysis which is valid uniformly at all Mach numbers of convection. This requires, essentially, that the finite correlation duration (as well as the finite correlation length) of aeroacoustic sources be taken into account. Then the change in character of jet noise (from quadrupole-type radiation, influenced by Doppler effect, and scaling as U^8, to monopole-type radiation, predominantly in the Mach direction, and scaling as U^3) can be recognized as a continuous development with increase in the Mach number of convection from low to supersonic values – provided that the supersonic jets are properly expanded.

On the other hand the vital goal of meeting FAR "Stage III" requirements on engine noise from future High Speed Transport aircraft will demand the most precise knowledge possible, both of the nature of disturbances to supersonic jet flows and of the magnitudes of the resulting acoustic radiation. Accordingly, these are issues to which the Workshop has given special attention.

For properly expanded supersonic jets, we are above all concerned with those *disturbances to supersonic mixing regions* whose characteristics have been intensively studied both by theory (which distinguishes between different types of disturbance at lower and at higher supersonic speeds) and by experiment. In every case the radiated sound field is strongly influenced by the Mach number of convection of the disturbances.

We recognize several promising lines of attack on these problems. For round jets, linear stability analysis indicates the forms of disturbances that can grow exponentially with distance from the orifice, and these indications are found to have real value even when disturbance magnitudes are large. In comparing the aeroacoustic importance of different forms, superiority of convection speed may be found to outweigh superiority of growth rate[3]. Some distinctly encouraging comparisons have been made between experimental data and this theoretical approach (where, admittedly, a somewhat arbitrary choice of disturbance magnitude at the orifice needs to be assumed); and these lead us to propose important new extensions of the work in which CFD will play a major role.

For round jets they will include fully nonlinear treatments of the disturbances. At the same time, it will be extremely important to investigate the potential noise-reduction advantages of supersonic jets with other (e.g. elliptical) cross-sections.

Here, CFD can massively contribute to determining the character of disturbances to such non-round supersonic jets, whether by linearized or by fully nonlinear theories. Here as in some other problems we recognize that different alternative ways of deriving the acoustic field (either in a two-stage calculation of Acoustic Analogy type or in a single-stage, comprehensively Computational Aeracoustics program) may be employed.

In addition, we recognize the continuing need for some aeroacoustic research on the noise of underexpanded supersonic jets (even though these may not be relevant in civil-aircraft applications). Here, the pattern of stationary shock waves in the jet has a dominant effect on the noise, whether it assumes the narrow-band form known as "screech" (generated by a well-established feedback mechanism) or a broad-band form of radiation associated with the passage of turbulence through shock waves.

We explicitly suggest as an important "model problem" for experimental validation of CFD approaches (compare §8) the determination of such noise generated when turbulence passes through a shock wave. We confidently expect the resulting knowledge to prove

[3]For supersonic convection speed, the acoustic radiation in the Mach direction assumes almost immediately the "saw-tooth" waveform that results from nonlinear-acoustics effects, and it is of course essential to utilize the known propagation characteristics of such closely packed assemblages of conical shock waves when extrapolating noise levels to large disturbances.

valuable in a wide range of problems. In the meantime, we also recommend further work on the avoidance of "screech," including possible investigations of "active" control aimed at de-activating the feedback loop.

10. Subsonic Jet Noise

The noise generated by subsonic airjets under laboratory conditions has been exhaustively studied for over forty years, and it may perhaps be questioned whether any intense further effort on this classical problem is required. There is, however, a continued need for further study of the harder problem of noise radiation from the exhausts of real turbofan engines; which, of course, incorporate a hot central core embedded in a much wider jet of cold "bypass" air.

Those reasons (scale separations, energy-level disparities and multipole source character) which for subsonic jets lead us to propose continued use of the Acoustic Analogy in one of its many forms were explained in §3, where some advantages of employing the classical form T_{ij} for the quadrupole strength per unit volume were also noted. For wide jets, however, it is especially important to take into account how the radiation from aeroacoustic sources is refracted by the sheared motion in the jet itself; as can be allowed for differently in different versions of the Acoustic Analogy.

Where this is applied using free-space Green's functions for "the wave equation" itself, it becomes essential to allow for modification of the radiation pattern for the higher-frequency noise (at, say, Strouhal numbers $\omega\ell/U$ greater than unity) by refraction through the sheared flow; with "ray acoustics" typically used to estimate this. Alternatively, there are several good ways of re-formulating the Acoustic Analogy by a different partition of the equations of motion into linear "propagation" terms and nonlinear "source" terms. They are designed as far as possible to take automatic account of aeroacoustic radiation through the sheared jet flow; and they require, of course, application of a different Green's function associated with the new form of the linear side of the equation.

We recommend continued use of both approaches; at the same time, we very specially emphasize the need for research devoted to characteristics of the turbulence in turbofan exhausts, including (above all) the statistical properties of quantities contributing to aeroacoustic source strengths. We look towards modern CFD methods for providing this knowledge; and, taking into account that the statistical characteristics of T_{ij} are known to be dominated by the relatively large (energy-containing) eddies in the turbulence, we are hopeful that one of the available forms of Large Eddy Simulation may help to achieve this. We strongly recommend an attempted application of LES along these lines.

11. A Brief Overview of our Recommendations

A combination of aviation's massive new noise-reduction requirements with great possibilities for fully utilizing modern CFD capabilities allow us confidently to predict a second golden age of aeroacoustics (§1), following by some four decades on the achievements (§2) of the first. Aeroacoustics must now be involved in intimate interactions with CFD (as applied not only to deterministic flows but also to the statistical characteristics of turbulence), while additionally incorporating rigorous comparisons with experiment.

The new Computational Aeroacoustics will press forward in two parallel thrusts, closely coordinated (§3). In one of them, CFD will be used along lines indicated in §§3, 6 and 10 to determine aeroacoustic source strengths, the associated radiation being derived by the Acoustic Analogy approach in one of its forms. In the other, a direct Computational Aeroacoustics will apply CFD techniques along lines indicated in §§3, 4 and 5 over a region extending beyond the flow field so as to include at least the beginnings of the acoustic far field.

There are some particularly important areas of study, including rotor noise (§6), boundary-layer noise (§7) and the noise of supersonic jets (§9), where we strongly recommend the continued use of both methods. On the other hand, important problems of the diffraction of radiation from aeroacoustic sources around complicated aircraft shapes (§5) will require the use of comprehensively Computational Aeroacoustics, while Acoustic Analogy methods seem better suited (§§3 and 10) to estimating subsonic jet noise. The study of "model" problems (§8) to allow meticulous comparisons with experiment will be valuable in both lines of attack.

For the first time, the ICASE/NASA LaRC Workshop brought together what we may call "a critical mass" of outstandingly qualified persons capable of creating and sustaining that powerful integrated attack on Computational Aeroacoustics which we envision as necessary. It will be vitally important to maintain and still further increase the strength and coherence of this group so as to ensure success in meeting the objectives of the second golden age of aeroacoustics.

MIX
Papier aus verantwortungsvollen Quellen
Paper from responsible sources
FSC® C105338

If you have any concerns about our products,
you can contact us on
ProductSafety@springernature.com

In case Publisher is established outside the EU,
the EU authorized representative is:
Springer Nature Customer Service Center GmbH
Europaplatz 3, 69115 Heidelberg, Germany

Printed by Libri Plureos GmbH
in Hamburg, Germany